# Focal-Plane Sensor-Processor Chips

Ákos Zarándy
Editor

# Focal-Plane Sensor-Processor Chips

Springer

*Editor*
Ákos Zarándy
MTA Budapest
Computer & Automation
Research Institute
PO Box 63, Budapest
Hungary
zarandy@sztaki.hu

ISBN 978-1-4419-6474-8 e-ISBN 978-1-4419-6475-5
DOI 10.1007/978-1-4419-6475-5
Springer New York Dordrecht Heidelberg London

Library of Congress Control Number: 2011921932



Printed on acid-free paper

Springer is part of Springer Science+Business Media (www.springer.com)

# Preface

Vision is our most important sensory system. It receives the largest amount of data and at the same time the most complex signal composition. Our eye and brain can put together images from photons arriving from different angles, with different densities and different wavelengths. One third of the human brain is devoted to the processing of the visual information.

The processing starts already in the retina. It is responsible for adaptation and early image processing. Along the preprocessing phases, it reaches 1:100 data reduction. It is not surprising that the efficiency of the retina stimulated many research engineers to build vision chips, which mimic this amazing performance. The common properties of these sensor-processor chips are that they can perform both data capturing and processing.

The story of the focal-plane sensor-processor (FPSP) devices started about 20 years ago, when Carver Mead built his famous silicon retina [1], which was capable of performing local adaptation. This initialized a line of different FPSP chips in the 1990s [2–6]. These chips were implemented using analog VLSI technology. The main motivation for the analog implementation was twofold. On the one hand, in the 1990s the silicon area of the basic computational elements needed for image processing (adder, multiplier, storage elements with 6- to 8-bit accuracy) was 5 times smaller in the analog domain than in the digital. On the other hand, the combination of the wide spreading CMOS sensors technology and the analog processing elements led to very efficient circuits, because no analog-to-digital conversion was needed. These chips contained 400–4,000 processing elements and could perform 10,000 FPS image capturing and processing real-time providing 10–100 times more computational power than a PC at that time. A summary of these chips and their technology can be found in [7].

In the late 1990s to early 2000s, the digital technology could profit more from Moore's law than the analog ones; hence, the digital storage and computational elements became small enough to build vision chips with digital processors [8, 9]. Reference [10] shows an analysis of the different digital processor arrangements of these chips.

Nowadays, these chips are applied to embedded devices used in industries and in some devices we use in our everyday life. The most straightforward example is the optical mouse, but these chips can also be found in 3D laser scanners, range sensors, safety equipments, and in unmanned aerial vehicles (UAVs). Besides the industry, an active scientific community focuses its effort to come out with new architectures and solution.

This book introduces a selection of the state-of-the-art FPSP array chips, design concepts, and some application examples. After a brief technology introduction (chapter *Anatomy of the Focal-Plane Sensor-Processor Arrays*), six different sensor-processor chips are introduced with design details and operation examples. The first three chips are general purpose while the last three ones are special purpose. The first in the row is the SCAMP-3 chip. It is an extremely power-efficient FPSP chip with tricky operator execution methods (chapter *SCAMP-3: A Vision Chip with SIMD Current-Mode Analogue Processor Array*). The second chip (MIPA4k, chapter *MIPA4k: Mixed-Mode Cellular Processor Array*) has both digital and analog processors and supports some special operators, such as rank-order filtering and anisotropic diffusion. The third general purpose chip (ASPA, chapter *ASPA: Asynchronous–Synchronous Focal-Plane Sensor-Processor Chip*) has a very special feature. It can implement extremely fast asynchronous binary wave propagation.

The next three FPSP array chips are special purpose chips optimized for efficient execution of different tasks. The first special purpose chip (chapter *Focal-Plane Dynamic Texture Segmentation by Programmable Binning and Scale Extraction*) is designed for dynamic texture segmentation. It is followed by a high-dynamic range high-temporal resolution chip, called ATIS (chapter *A Biomimetic Frame-Free Event-Driven Image Sensor*). Its special feature is the event based readout. The row is closed with a 1D sensor-processor (chapter *A Focal Plane Processor for Continuous-Time 1-D Optical Correlation Applications*), which is designed for continuous time optical correlation.

The next two chapters introduce future concepts. The first one describes a design, which steps through the conventional planar silicon technology, by using 3D integration. Other interesting feature of the introduced VISCUBE design (chapter *VISCUBE: A Multi-Layer Vision Chip*) is that it combines a fine-grain mixed-signal pre-processor array layer with a coarse-grain digital foveal processor array layer. Then in chapter by Jiang and Shi, the concept of a nonlinear resistive grid built from memristors is shown. The resistive grid, which can be one of the future building blocks of these FPSP chips, extracts the edges almost in the same way as the human visual system does.

The last four chapters explain different applications of the technology. The tenth chapter (*Bionic Eyeglass: Personal Navigation System for Visually Impaired People*) introduces the concept of a mobile device, called "Bionic Eyeglass," which provides visual aid for blind persons to navigate and to identify colors in their everyday life. The 11th chapter (*Implementation and Validation of a Looming Object Detector Model Derived from Mammalian Retinal Circuit*) describes the implementation of a vertebrate retina circuit, responsible for identifying looming objects, on an FPSP chip-based embedded vision system [11]. After this, an industrial application is shown in chapter by Nicolosi et al. The same embedded vision system, which was used in the previous chapter, is applied in ultra-high speed real-time visual control of a welding robot. The last chapter of this book introduces a 3D finger tracking application to control cursor in a mouseless computer.

Budapest, Hungary Ákos Zarándy

## References

1. M. A. Mahowald, C. Mead, The Silicon Retina Scientific American, Vol. 264, No. 5, pp. 76–82, May 1991
2. K. Halonen, V. Porra, T. Roska, L.O. Chua, VLSI Implementation of a Reconfigurable Cellular Neural Network Containing Local Logic (CNNL) Workshop on Cellular Neural Networks and their Application (CNNA-90), pp. 206–215, Budapest, Dec. 1990
3. H. Harrer, J.A. Nossek, T. Roska, L.O. Chua, A Current-mode DTCNN Universal Chip, Proc. of IEEE Intl. Symposium on Circuits and Systems, pp. 135–138, 1994
4. J. M. Cruz, L. O. Chua, T. Roska, A Fast, Complex and Efficient Test Implementation of the CNN Universal Machine, Proc. of the third IEEE Int. Workshop on Cellular Neural Networks and their Application (CNNA-94), pp. 61–66, Rome, Dec. 1994
5. R. Dominguez-Castro, S. Espejo, A. Rodriguez-Vazquez, R. Carmona, A CNN Universal Chip in CMOS Technology, Proc. of the third IEEE Int. Workshop on Cellular Neural Networks and their Application (CNNA-94), pp. 91–96, Rome, Dec. 1994
6. A Paasio, A Dawidziuk, K Halonen, V Porra, Minimum Size 0.5 Micron CMOS Programmable 48 by 48 CNN Test Chip Proc. ECCTD'97, 1997
7. Towards the Visual Microprocessor: VLSI Design and the Use of Cellular Neural Network Universal Machines, Edited by T Roska Á. Rodríguez-Vázquez, Wiley, NY, 2001
8. M. Ishikawa, K. Ogawa, T. Komuro, I. Ishii, A CMOS Vision Chip with SIMD Processing Element Array for 1ms Image Processing, 1999 Dig. Tech. Papers of 1999 IEEE Int. Solid-State Circuits Conf. (ISSCC'99) (San Francisco, 1999.2.16)/Abst. pp. 206–207, 1999
9. P. Földesy, Á. Zarándy, Cs. Rekeczky, T. Roska Configurable 3D integrated focal-plane sensor-processor array architecture, Int. J. Circuit Theory and Applications (CTA), pp. 573–588, 2008
10. P. Foldesy, A. Zarandy, Cs. Rekeczky, T. Roska, Digital implementation of cellular sensor-computers, Int. J. Circuit Theory and Applications (CTA), Vol. 34, No. 4, pp. 409–428, July 2006
11. A. Rodríguez-Vázquez, R. Domínguez-Castro, F. Jiménez-Garrido, S. Morillas, A. García, C. Utrera, M. Dolores Pardo, J. Listan, R. Romay, A CMOS Vision System On-Chip with Multi-Core, Cellular Sensory-Processing Front-End, in Cellular Nanoscale Sensory Wave Computing, Edited by C. Baatar, W. Porod, T. Roska, ISBN: 978-1-4419-1010-3, 2009

# Contents

# Contributors

**Felix Abt** IFSW, Stuttgart, Germany, felix.abt@ifsw.uni-stuttgart.de

**Betsaida Alexandre** Instituto de Microelectrónica de Sevilla CNM-CSIC, Universidad de Sevilla, Americo Vespucio, s/n 41092 Seville, Spain

**Norbert Bérci** Faculty of Information Technology, Pázmány University, Práter u. 50/A, Budapest, Hungary, berci.norbert@itk.ppke.hu

**Andreas Blug** Fraunhofer IPM, Freiburg, Germany, andreas.blug@ipm.fraunhofer.de

**Daniel Carl** Fraunhofer IPM, Freiburg, Germany, daniel.carl@ipm.fraunhofer.de

**Ricardo Carmona-Galán** Institute of Microelectronics of Seville (IMSE-CNM-CSIC), Consejo Superior de Investigaciones Científicas, Universidad de Sevilla, C/Americo Vespucio, s/n 41092 Seville, Spain, rcarmona@imse-cnm.csic.es

**Luis Carranza** Instituto de Microelectrónica de Sevilla CNM-CSIC, Universidad de Sevilla, Americo Vespucio, s/n 41092 Seville, Spain

**Pablo de la Fuente** Fagor Aotek, S. Coop, Paseo Torrebaso, 4 – Aptdo. Corr. 50, 20540 Eskoriatza, Guipúzcoa, Spain

**Piotr Dudek** The University of Manchester, Manchester M13 9PL, UK, p.dudek@manchester.ac.uk

**Jorge Fernández-Berni** Institute of Microelectronics of Seville (IMSE-CNM-CSIC), Consejo Superior de Investigaciones Científicas, Universidad de Sevilla, C/Americo Vespucio s/n 41092 Seville, Spain, jfberni@imse-cnm.csic.es

**Péter Földesy** Eutecus Inc, Berkeley, CA, USA
and
MTA-SZTAKI, Budapest, Hungary, foldesy@sztaki.hu

**Tamás Fülöp** Pázmány Péter Catholic University, Budapest, Hungary, fulta@digitus.itk.ppke.hu

**Soós Gergely** MTA-SZTAKI, Budapest, Hungary, soos@digitus.itk.ppke.hu

**Heinrich Höfler** Fraunhofer IPM, Freiburg, Germany, heinrich.hoefler@ipm.fraunhofer.de

**Feijun Jiang** Department of Electronic and Computer Engineering, Hong Kong University of Science and Technology, Clear Water Bay, Kowloon, Hong Kong Sar, Peoples Republic of China

**Kristóf Karacs** Faculty of Information Technology, Pázmány Péter Catholic University, Budapest, Hungary, Karacs.Kristof@itk.ppke.hu

**Mika Laiho** Microelectronics Laboratory, University of Turku, Turku, Finland, mlaiho@utu.fi

**Gustavo Liñán-Cembrano** Instituto de Microelectrónica de Sevilla CNM-CSIC, Universidad de Sevilla, Americo Vespucio s/n 41092 Seville, Spain, Gustavo.Linan@imse.cnm.es

**Alexey Lopich** The University of Manchester, Manchester M13 9PL, UK

**Tomás Morlanes** Fagor Aotek, S. Coop, Paseo Torrebaso, 4 – Aptdo. Corr. 50, 20540 Eskoriatza, Guipúzcoa, Spain, tmorlanes@fagorautomation.es

**Leonardo Nicolosi** Technische Universität Dresden, Germany, leonardo.nicolosi@tu-dresden.de

**Ari Paasio** Microelectronics Laboratory, University of Turku, Turku, Finland, arjupa@utu.fi

**Jonne Poikonen** Microelectronics Laboratory, University of Turku, Turku, Finland, jokapo@utu.fi

**Christoph Posch** AIT Austrian Institute of Technology, Vienna, Austria, Christoph.Posch@ait.ac.at

**Csaba Rekeczky** Eutecus Inc, Berkeley, CA, USA, rcsaba@eutecus.com

**Ángel Rodríguez-Vázquez** Instituto de Microelectrónica de Sevilla CNM-CSIC, Universidad de Sevilla, Americo Vespucio, s/n 41092 Seville, Spain, angel@imse-cnm.csic.es

**Tamás Roska** Pázmány Péter Catholic University, Budapest, Hungary
and
MTA-SZTAKI, Budapest, Hungary, roska@sztaki.hu

**Bertram E. Shi** Department of Electronic and Computer Engineering, Hong Kong University of Science and Technology, Clear Water Bay, Kowloon, Hong Kong Sar, Peoples Republic of China, eebert@ust.hk

**Péter Szolgay** Pázmány University, Práter u. 50/A, Budapest, Hungary, szolgay.peter@itk.ppke.hu

**Ronald Tetzlaff** Technische Universität Dresden, Dresden, Germany, ronald.tetzlaff@tu-dresden.de

**Róbert Wagner** Faculty of Information Technology, Pázmány Péter Catholic University, Budapest, Hungary

**Ákos Zarándy** Computer and Automation Research Institute of the Hungarian Academy of Sciences, (MTA-SZTAKI), Budapest, Hungary, zarandy@sztaki.hu

# Anatomy of the Focal-Plane Sensor-Processor Arrays

**Ákos Zarándy**

**Abstract** This introductory chapter describes the zoo of the basic focal-plane sensor-processor array architectures. The typical sensor-processor arrangements are shown, the typical operators are listed in separate groups, and the processor structures are analyzed. The chapter gives a compass to the reader to navigate among the different chip implementations, designs, and applications when reading the book.

## 1 Introduction

The spectrum of the focal-plane sensor-processor (FPSP) circuits is very wide. Some of them are special purpose devices, which are designed to optimally fulfill one particular task. A good industrial example for special purpose FPSP circuit is Canesta's depth sensor [1], which measures the depth information in every pixel based on the phase shift of a periodic illumination caused by the time-of-flight of the light. In our book, chapters by Fernández-Berni, Carmona-Galán, Posch, and Liñán-Cembrano et al. describe special purpose designs. The special purpose devices cannot be programmed, only their main parameters can be modified.

There are naturally general purpose FPSP devices also, which can be used in many different applications. A recently completed industrial general purpose FPSP chip is the Q-Eye, powering AnaFocus' Eye-RIS system [2]. These devices are fully programmable. This book introduces general purpose devices in chapters by Dudek, Laiho et al., Lopich, and Zarándy et al. (SCAMP-3, MIPA4k, ASPA, VISCUBE chips).

Other distinguishing feature is the domain of the processors. Some of the devices apply mixed-signal (partially analog) processors, while others use digital ones. The mixed-signal processors are smaller, consume less power, and do not require on-chip analog to digital converters. As a contrast, the digital processors are typically larger

Á. Zarándy (✉)
Computer and Automation Research Institute of the Hungarian Academy of Sciences, (MTA-SZTAKI), Budapest, Hungary
e-mail: zarandy@sztaki.hu

Á. Zarándy (ed.), *Focal-Plane Sensor-Processor Chips*,
DOI 10.1007/978-1-4419-6475-5_1, © Springer Science+Business Media, LLC 2011

and more powerful, more accurate, and more versatile. In our book, mixed-signal arrays are shown in chapters by Dudek, Fernández-Berni, Carmona-Galán, and Posch (SCAMP-3, ATIS chips), while combined mixed-signal and digital processor arrays are shown in chapters by Laiho, Lopich, Dudek, Liñán-Cembrano et al., and Zarándy et al. (MIPA4k, ASPA, VISCUBE chips).

The type of the processed image is also an important issue. Some of the FPSP circuits are optimized for handling binary images only [3, 4]. Other devices can handle both grayscale and binary images. In our book, the general purpose SCAMP-3, MIPA4k, ASPA, and VISCUBE designs (chapters by Dudek, Laiho et al., Lopich, Dudek, and Zarándy et al.) represent this approach.

We have to distinguish the processing type also according to the neighborhood involved. The simplest operation is the pixel-wise processing, while the most complex is the global processing, where all the pixels in the frame are needed for the calculation as inputs.

This chapter is devoted to discuss the different architectural variances of the FPSP circuits. In the next session, various sensor processor arrangements are listed. This is followed by the description of the typical image processor operator types. Then, the processor architectures are shown. A more specific analysis of the operators and their implementations on these architectures can be found in [5].

## 2 Sensor-Processor Arrangements

There are two major components in all FPSP devices: the photo-sensor array and the processor(s). The number, the arrangement, the density, and the interconnection of these two components define the structure of the circuits. The aggregated computational capability of the processors and the processing needs on the data flow coming from the sensors are always balanced. In some cases, the number of the processors and the sensors are the same [2–4, 6, 7], in other cases; the number of the sensors is higher than the processors [8]. The sensors are typically arranged in 1D or 2D grids. These cases are discussed in the following subsections.

### *2.1 One-Dimensional Sensor Arrangement*

One-dimensional sensor (line sensor) is used when the objects or material to be captured are moving with a constant linear speed below the camera. Typical situations are the conveyor belt, or a scanning machine. The 1D arrangement is cheaper, because it uses smaller silicon surface. Moreover, higher spatial resolution can be reached (few thousand pixel wide image), and there is no boundary problem, which would come from merging individual snapshots. The chapter (***A Focal Plane Processor for Continuous-Time 1-D Optical Correlation Applications***) introduces a linear FPSP chip in this book.

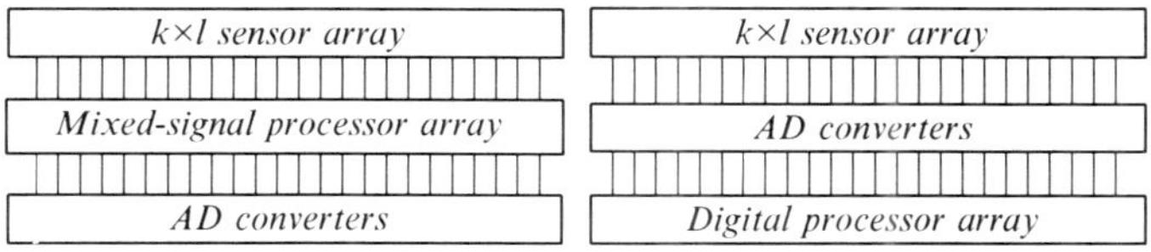


**Fig. 1** Typical 1D sensor processor arrangements with mixed-signal (*left*) or digital processors (*right*)

The one-dimensional sensor-processor arrays contain one or a few rows of sensors. In case of mixed-signal processors, the analog outputs of the sensors are directly processed. If digital processors are applied, analog-to-digital (AD) converters are needed between the sensors and the processor.

Figure 1 shows the typical 1D sensor processor arrangements. The sensor array contains one or a few long lines. The length of the lines can be from a few hundred to a few thousand pixels. Multiple lines are applied, if redundant or multi-spectral (e.g., color) information is needed.

In case of mixed-signal processor array, the number of the processors is typically the same as the number of the pixels in the sensor line(s), because the computational power and the versatility of these processors are limited.

In the digital version, the processors are more powerful and versatile. In this case, one or a few processor can serve the entire row. The number of AD converters typically matches with the number of digital processors.

## 2.2 *Two-Dimensional Sensor Arrangement*

The versatility of the 2D arrays is larger than the 1D ones. It is worth to distinguish two basic types of arrangement from this family:

1. The sensor array and the processor array are separated. In this case, typically the size or the dimension of the sensor and the processor arrays are different.
2. The sensor array and the processor array are embedded into each other. This enables close sensor-processor cooperation, since the sensors and the processors are physically very close or directly next to each other.

These cases are detailed in the next two subsections.

### 2.2.1 Separated Sensor and Processor Arrays

One of the most critical parameters of the imagers is the spatial resolution. To be able to rich high spatial resolution, one needs to use small pixel pitch. The pitch of a sensor can be as small as a few microns, while a combined sensor-processor cell starts from 25 μm pitch on inexpensive planar technologies. Therefore, to be able

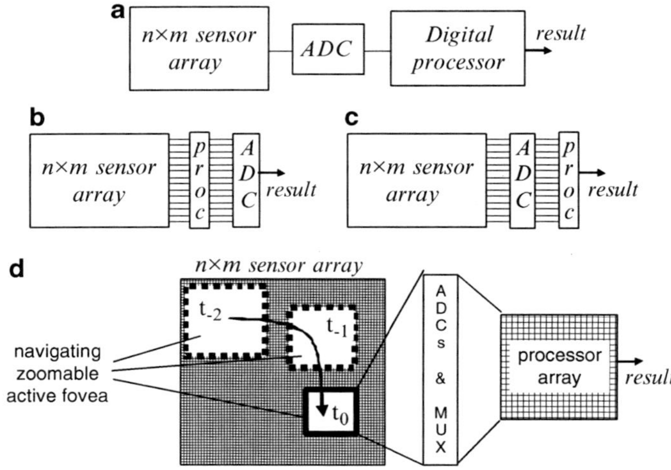


**Fig. 2** 2D sensor processor arrangements with separated sensor processor units. (**a**): one digital processor handles the entire image. Sensor arrays with mixed-signal (**b**) or digital (**c**) linear processor arrays. (**d**): foveal arrangement

apply high resolution (e.g., megapixel), one needs to separate the sensor array from the processors. The price, which is paid, is typically lower performance processing (speed or complexity), and/or reduced versatility.

We can see three different arrangements of the separated sensor-processor circuits in Fig. 2. In the first one, one digital processor serves the entire sensor array. The second variance applies linear column-wise processor arrays with mixed-signal or analog processor. The third one is the foveal arrangement. While in the previous two cases, the entire image is processed, in the foveal approach one or a few selected areas are involved into the calculations. This is an efficient approach if only some parts of the image carry relevant information to process. The description of the processor architectures and implementable operators are discussed in the next sections.

### 2.2.2 Embedded Sensor and Processor Arrays

Embedded sensor processor arrays are used when high speed is the critical parameter and not the high resolution. In this case above 10,000 visual decisions can be reached [9] in a second real-time even in complex situations on small or medium-sized images (<QVGA). We can distinguish two basic embedded processor types according to the processor density.

In the first case, there is a one-to-one correspondence between the sensors and the processors (Fig. 3). In this situation, a fine-grain processor array is used. These circuits typically apply either mixed-mode (SCAMP-3, chapter *SCAMP-3: A Vision Chip with SIMD Current-Mode Analogue Processor Array*), ACE16k [7], Q-Eye

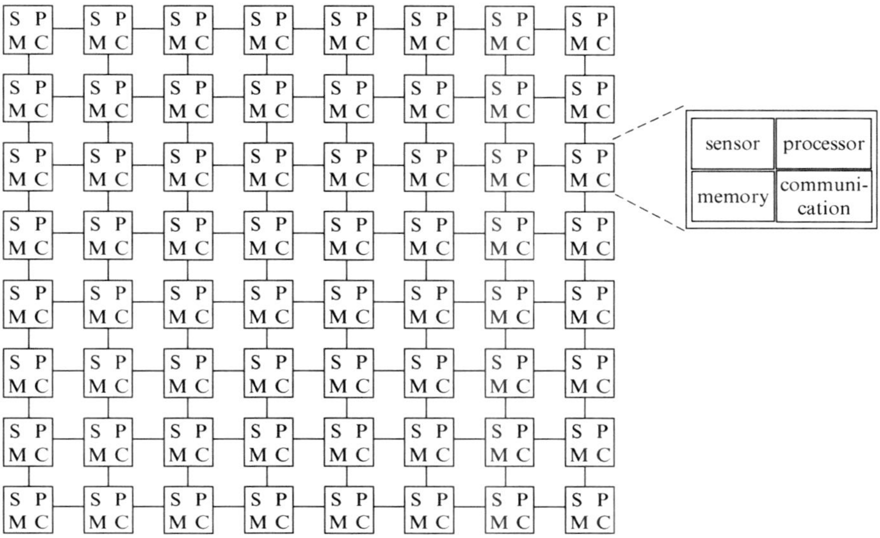


**Fig. 3** Embedded sensor-processor with fine-grain processor architecture

[2]) and/or 1 bit digital processors operating in bit-sliced mode (ASPA, chapter *ASPA: Asynchronous–Synchronous Focal-Plane Sensor-Processor Chip*; MIPA4k: chapter *MIPA4K: Mixed-Mode Cellular Processor Array*). The advantage of this structure is that it supports various efficient spatial–temporal operations such as diffusion, global OR, mean, address-event readout, etc. The description of these operations will be given in Sect. 4.3.2. Other advantage is that locally adaptive sensor control can be easily implemented such as locally changing exposure time according to illumination level or motion speeds [10, 11].

In the second case, a coarse-grain processor array is embedded to the sensor array [8]. This means that a $k \times k$ sub-array of pixels is assigned to one processor (Fig. 4). Naturally, more powerful, 8- or 16-bit digital processors are needed to process all the pixels. These processors can be more versatile than the mixed-signal or the bit sliced ones, and their communication radius is much larger, because they can reach the $k$th pixel in a single step. Besides the $k \times k$ array of pixels and the processor, each cell includes a memory and an AD converter. The size of the memory is typically enough for storing 6 or 8 pieces of the $k \times k$ image part. To squeeze the ADC to the limited area, one may use successive approximation type ADC or pixel-wise single slope one.

In both cases, the locality (mainly local data communication) plays important role. Thanks to this, these architectures are scalable, consume very low power, and suitable for further implementations with nanotechnology also, where the long communication lines are the main barriers.

Important issue of this technology is the sensor-processor tradeoff. On the one hand, both the sensor and the processors need relatively large area to be sensitive and to provide satisfactory computational power. On the other hand, the overall

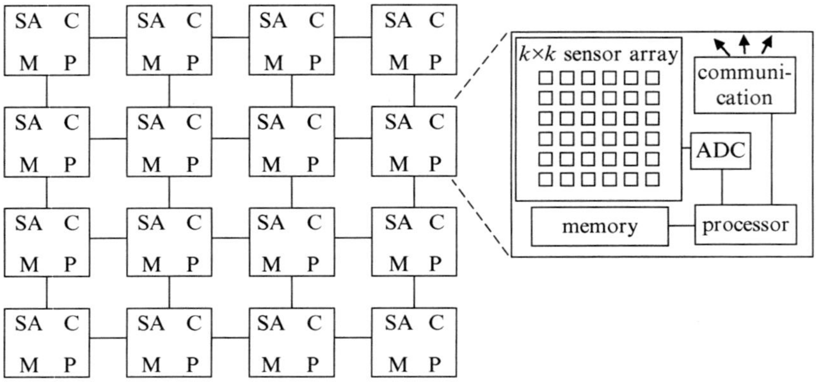


**Fig. 4** Embedded sensor-processor with coarse-grain processor architecture

area should be small; otherwise, the resolution will be too low. Moreover, the same silicon technology is not necessarily optimal for both the sensor and the processor circuits. The 3D integration can break through this bottleneck of the planar silicon technology. It introduces almost 100% fill factor, different sensor and processor materials, and extends the sensitivity gamut of the sensor. The chapter (*VISCUBE: A Multi-Layer Vision Chip*) of this book introduces a 3D sensor-processor design. Unfortunately, the 3D technology is not yet an established technology, hence it is still unreliable and expensive (year 2010). However it will change in a few years, and the 3D approach is expected to dominate the FPSP technology. A cheaper solution for increasing the fill factor nowadays without increasing the sensor area is the microlens technology [12].

## 3 Operator Types

A wide range of different operator primitives are applied in image processing and one can find many ways to classify them. Here, we apply two classification criteria (Fig. 5). The primary classification is done according to the output dimension (0D: scalar(s), 1D: row/column, 2D: or image). The third category (image → image) is further divided into three subcategories according to their relative input location and size, because for the locally interconnected processor arrays, the communication topology is one of the key properties. This section discusses these categories and lists the typical operators in the classes. The efficient implementation methods on different processor architectures will be described in the next section.

### *3.1 Image → Scalar(s) Operators*

The *image → scalar(s)* operators are typically used for feature extraction or localization. These operators can be implemented in a way that a processor scans the entire image, reading each of the pixel ones [5]. During this scan, some statistics

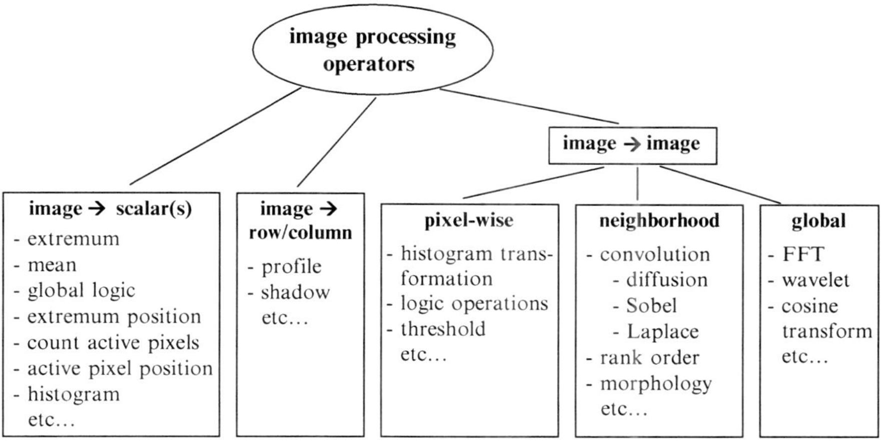


**Fig. 5** Operator classification

(min, max, mean, global OR, number of black pixels on a binary image, histogram, etc.) can be calculated. Similarly, the coordinates of extremum points or the white pixels on a binary image can be also calculated.

## 3.2 *Image → Row/Column Operators*

The *image → row/column* operators apply 1D single scans row-wise or column-wise processing. In this case, the image lines or the columns are decoupled from each other, which means that the input domain of the operator is one line or one column. Typical examples here are the profile and the shadow operators.

## 3.3 *Image → Image Operators*

The *image → image* operators are further divided into three subcategories. Some of these operators are defined for one input image (e.g., Sobel operation), others apply multiple input images (e.g., pixel-wise logic AND). In both cases, we examine here the input locality independently from the number of the input images.

### 3.3.1 Pixel-Wise

The *pixel-wise* operators include those operators, which require only the pixel itself as an input, and no information from the neighborhood is needed. These operators can be described in the following form:

$$y_{ij} = f(u_{ij}), \tag{1}$$

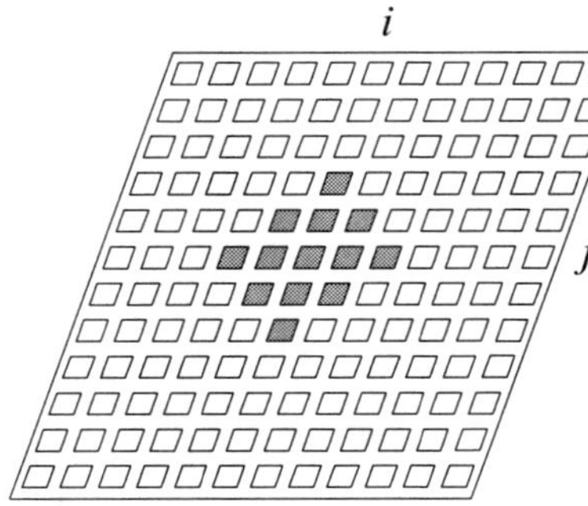


**Fig. 6** Input domain of a neighborhood operator

where

$u_{ij}$ is the input pixel;
$y_{ij}$ is the result pixel;
*f()* is a one input scalar output function (assuming one input image).

Typical operators here are the gain, offset, and contrast manipulation (histogram transformation), and the thresholding.

### 3.3.2 Neighborhood Processing

The special feature of this class of operators is that the input is coming from a relatively small neighborhood of the pixel. Figure 6 shows an example, where 13 input parameters are used to calculate the operator in the *ij* position. The neighborhood radius indicates the distance between the central pixel position and the farthest pixel in the input domain.

$$y_{ij} = g(U_{ij}), \tag{2}$$

where

$U_{ij}$ is the input domain;
$y_{ij}$ is the result pixel;
*g()* is a multiple input scalar output function (assuming one input image).

In many cases, these operators are applied multiple times. These are called iterative calculations.

These operators are also called topographic operators, because they apply local operations on topographically mapped data sets.

### 3.3.3 Global Processing

The input domain of the global *image* → *image* operators are the entire image. Typical operators here are the Fast Fourier transformation (FFT), the wavelet transformation, or a Hough transform.

# 4 Processor Arrangements

Here, we list the processor structures used on FPSP chips and examine their operator execution capabilities with different memory sizes. Here, we consider the execution of one single operator primitive only. The efficiency figures of these architectures are calculated in [5].

## 4.1 Single Processor Architectures

This processor arrangement is constructed of a processor and some memory (Fig. 7). It can be applied next to either a 1D or a 2D sensor array. The data-stream is coming from the sensor array sequentially. This means that the pixels are coming from left to right in a row, and the rows from the top to the bottom. The capability of the processor is defined by its memory size and its internal architecture. Here, we distinguish three different memory sizes, which are enough to store:

1. A few pixels.
2. A few lines.
3. A few frames.

These are discussed in the next subsections.

### 4.1.1 Single Processor with Small Memory

The simplest possible processor arrangement of an FPSP is the single processor unit with a small memory, which is enough to store a few pixels and some other data. They can execute *image → scalar(s)* operators and *pixel-wise* operators. Moreover, they can execute those *image → row* operators, which are row-wise and have left to right propagation direction, same as the pixel flow. For example, a vertical profile or a left to right shadow can be implemented, while neither vertical, nor right to left shadows can be.

In simpler case (e.g., gain or contrast modification, extremum finding), both mixed-signal or digital units can be used. More complex operators (e.g., histogram)

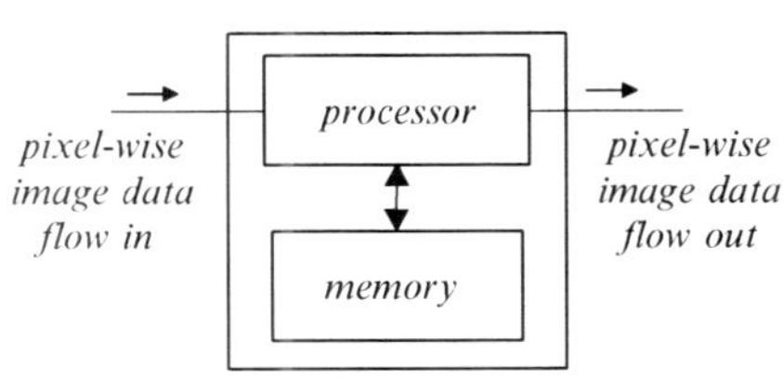


**Fig. 7** Single processor arrangement

require digital processor units. These processors are typically special purpose ones, where only the parameters and/or some arguments of the processing can be set.

### 4.1.2 Single Processor with Medium-Sized Memory

The memory size here is large enough to store a few lines from the image. This efficiently supports the execution of *image* → *scalar*, *pixel-wise*, and the *neighborhood* operators. Moreover, the *image* → *row/column* operators can be executed, where the propagation type is left to right or top to bottom. (The column wise operators can be executed also because entire row fits to the memory.) These operators are the vertical or horizontal profile and the left to right or top to bottom shadows.

Here, typical digital processor architecture is applied, because it requires complex memory management. The processor is still special purpose, with settable parameters/attributes. Since the processor receives sequential pixel stream, it cannot deal more with a pixel than the pixel clock period. This processor architecture can be efficiently used in video processing [13, 14], because most of the important operators can be implemented on them, but their memory is still small.

### 4.1.3 Single Processor with Large Memory

The simplest general purpose vision chip concept is to integrate a processor with large enough memory to store a few frames next to a sensor array. This type of processor can implement all kinds of operators, since it can access the entire image. In this case, typically fully programmable processors are applied, hence the vision chip.

The drawback of this kind of architecture is that a single processor can provide relatively small processing power, hence only simple or low frame rate applications are possible. Moreover, the sensor cannot be high (VGA or megapixel), because that would expand the required memory over a limit, which cannot fit to standard CMOS chip.

## *4.2 1D Processor Arrays*

The 1D processor arrangement (Fig. 8) is constructed of a linear processor array with local communication between the processors. The processors operate either in single instruction multiple data (SIMD) mode, or they are nonprogrammable special purpose ones. Each processor unit has a local memory. The executable operator types are defined from the aggregated memory size of the array rather than the individual memory size of the processors. The 1D processor array can be integrated with either a line sensor or a sensor array. The number of the processors can be either as many as the number of the pixels in an image line (fine-grain) or smaller (coarse-grain).

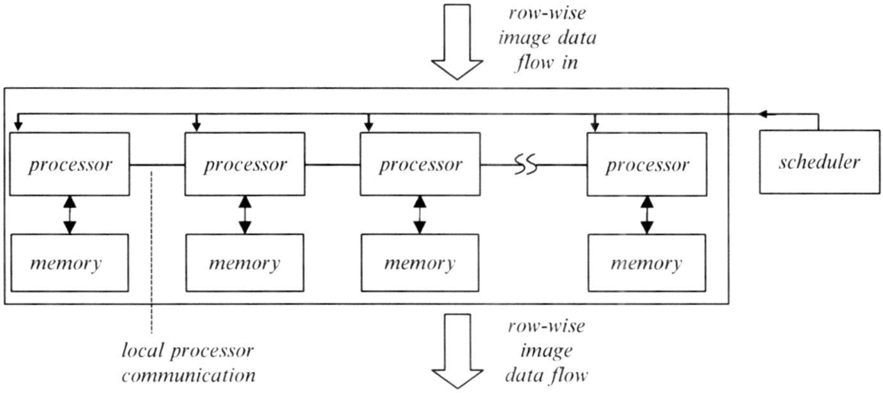


**Fig. 8** 1D processor arrangement

#### 4.2.1 1D Processor Arrays with Line Processing Capabilities

A 1D processor array with line processing capabilities has enough aggregated memory to store a few lines. This efficiently supports the execution of *pixel-wise* and the *neighborhood* operators. Moreover, those *image* → *row/column* operators can be executed, where the propagation type is left to right or top to bottom. The *image* → *scalar* operators can be executed on this architecture also; however, the execution is more difficult, because each processor generates a subresult, which should be combined. For example, in case of seeking for the maximum pixel value, each processor finds the maximum in its column(s), and after that, the absolute maximum value should be selected in a second step.

#### 4.2.2 1D Processor Arrays with Frame Processing Capabilities

The aggregated memory in this second type of 1D processor array is large enough to store entire frames. With this memory size, its capabilities become similar to a 2D coarse-grain processor architecture. It can execute practically the same operator set as the 1D with line processing capabilities, except it can calculate *image* → *row/column* operators in arbitrary direction.

### *4.3 2D Processor Arrays*

Two-dimensional processor arrays are applied either as a foveal array of a high resolution sensor, or as an embedded processor array next to a sensor array (Sect. 2). Since they can handle entire frames (or windows), we cannot distinguish them according to their aggregated memory sizes. Rather, we can separate them according to their processor density.

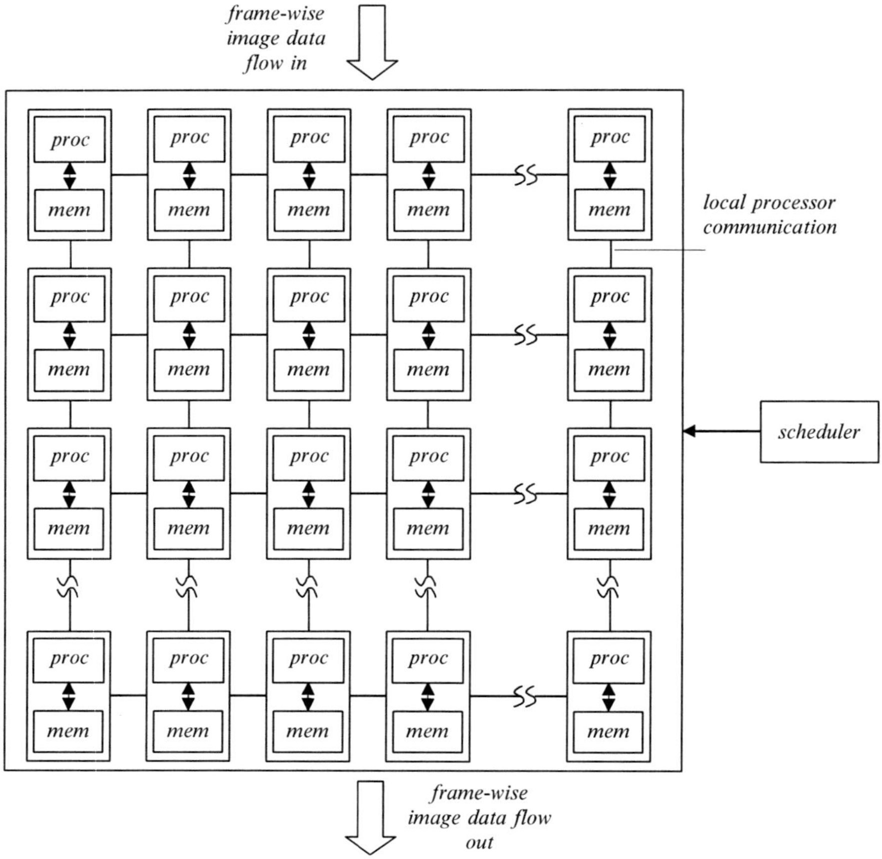


**Fig. 9** 2D processor arrangement

These processor arrays provide ultra-high processing capabilities. FPSP chips, equipped with this kind of engine, can easily reach above 10,000 FPS image capturing and evaluation (visual decision making) real time [9]. These processor arrays are SIMD architectures (Fig. 9).

### 4.3.1 Coarse-Grain 2D Processor Arrays

The coarse-grain 2D arrays can efficiently execute the *image* → *row/column* operators in all directions, the *pixel-wise* and the *neighborhood* operators. They can execute the *image* → *scalar(s)* operators also; however, the results need some post-processing as it was discussed in Sect. 4.2.1. These are typically digital processor arrays.

### 4.3.2 Fine-Grain 2D Processor Arrays

Similar to coarse-grain arrays, the fine-grain 2D arrays can efficiently execute the *image* $\rightarrow$ *row/column,* the *pixel-wise* and the *neighborhood* operators. They are not good at the *image* $\rightarrow$ *scalar(s)* operations in the conventional way; however, they can be very efficient when they use nonconventional approaches. They apply typically mixed-signal or bit-sliced digital processors.

The strength of the fine-grain processors is coming from the application of the nonconventional processing approach, because their mixed-signal and distributed asynchronous logic processor units can execute some *image* $\rightarrow$ *scalar(s)* and repetitive *neighborhood* operators with ultra-high speed. In these cases, the "let the physics do the computations" approach is used. The most important of these operators are:

- Global logic (AND/OR);
- Mean;
- Isotropic or anisotropic diffusion;
- Active pixel coordinate position readout;
- Object size estimation;
- Extremum (value and location);
- Repetitive binary morphologic operations (hole filling, recall, skeleton, centroid [15], etc).

The global logic is implemented in a way that a metal wire grid is set to weak Vcc voltage level through a resistor. In each node (processor cell), a transistor connects it to ground. The gate of the transistor is connected to the logic pixel value. Where it is high, the transistor opens. One open transistor is enough to pull the array down to zero. In this way, global OR is calculated.

By removing the pull-up transistor, and connecting a capacitor with the actual pixel value in each node, the same metal grid calculates the mean operator. After the transient decays (charge distribution is completed), the average of the pixel values will appear on each node.

Isotropic diffusion operator can be implemented on a resistive grid, by connecting a capacitance to it with the pixel value in each node. The strength of the diffusion (deviation) can be controlled by the transient time or by the resistance value. By controlling the resistance value locally, one can implement anisotropic diffusion. In case of nonlinear resistance, nonlinear diffusion can be implemented [16].

Coordinates of active pixels can be read out by scanning the pixels line-wise in an asynchronous way [2, 7]. From the introduced chips, ASPA (chapter *ASPA: Asynchronous–Synchronous Focal-Plane Sensor-Processor Chip*) is equipped with this capability.

Chips apply the operators above are described in chapters by Dudek and Laiho et al. (SCAMP-3, MIPA4k), and in [2, 6, 7] in the literature. The execution time of these operators is in the range of a few microseconds. This is 10–1000 times more power efficient than traditional digital solutions.

Extremum value and location can be identified by applying a comparator in each node. One input of the comparator receives a ramp and the other is connected to the local pixel value. The output of the comparator is connected to a global OR network. The global logic network indicates when the ramp reaches first the maximum/ minimum value in one of the node. Similar circuit for local maximum position identification is described in chpater by Zarándy et al. (VISCUBE chip).

Simple repetitive morphological operators (hole-filling, recall, etc) can be implemented either on the CNN [17] type chips [3, 4, 6, 7] in the analog domain, or by using asynchronous logic networks (chapter *ASPA: Asynchronous–Synchronous Focal-Plane Sensor-Processor Chip*, ASPA). While the execution speed of the formal one is one order of magnitude on the mentioned chips compared to a modern DSP, it is more than three orders of magnitude in the latter one [5]. This means that a grass fire type binary morphological operation can be calculated on a $128 \times 128$-sized lattice in 20 ns.

Complex morphological operators (skeleton, centroid [15]) can also be implemented on asynchronous logic networks with extremely power efficiency and ultra-high speed; however, they require larger silicon space [18].

### *4.4 Architecture Selection*

After analyzing the different architectures, a natural question arises: which one to use in certain application environment. The rule of thumb is that we need to apply:

- Embedded processor array, if high frame rate and low resolution are needed;
- Foveal processor array, if high frame rate and high resolution are needed;
- A sequence of single processors with medium size memories [5, 13, 14] in a pipeline arrangement, if high resolution and video speed are needed.

## 5 Conclusion

The anatomy of the different FPSP architectures is summarized in this chapter. This provides a help to easily navigate through the different architecture described in this book. More detailed architecture, operator, and processor structure analysis of the topographic devices can be found in [5].

## References

1. http://canesta.com
2. A. Rodríguez-Vázquez, R. Domínguez-Castro, F. Jiménez-Garrido, S. Morillas, A. García, C. Utrera, M. Dolores Pardo, J. Listan, R. Romay, A CMOS Vision System On-Chip with Multi-Core, Cellular Sensory-Processing Front-End, In Cellular Nanoscale Sensory Wave Computing, edited by C. Baatar, W. Porod, T. Roska, ISBN: 978–1–4419–1010–3, 2009

3. A. Paasio, A. Dawindzuk, K. Halonen, V. Porra, Minimum Size 0.5 Micron CMOS Programmable 48 × 48 CNN Test Chip European Conference on Circuit Theory and Design, Budapest, pp. 154–15, 1997
4. S. Espejo, R. Carmona, R. Domingúez-Castro, A. Rodrigúez-Vázquez, CNN Universal Chip in CMOS Technology, Int. J. Circ. Theor. Appl. 24, 93–111, 1996
5. A. Zarandy, Cs. Rekeczky, 2D Operators on Topographic and Non-Topographic Architectures Implementation, Efficiency Analysis, and Architecture Selection Methodology, Int. J. Circ. Theor. Appl. (CTA), Article first published online: 29 Apr 2010, DOI: 10.1002/cta.681
6. S. Espejo, R. Domínguez-Castro, G. Liñán, Á. Rodríguez-Vázquez, A 64 × 64 CNN Universal Chip with Analog and Digital I/O, In Proc. ICECS'98, pp. 203–206, Lisbon 1998
7. G. Liñan Cembrano, Á. Rodríguez-Vázquez, S. Espejo-Meana, R. Domínguez-Castro ACE16k: A 128 × 128 Focal Plane Analog Processor with Digital I/O Int. J. Neural Syst. 13(6) 427–434, 2003
8. P. Földesy, Á. Zarándy, Cs. Rekeczky, T. Roska Configurable 3D Integrated Focal-Plane Sensor-Processor Array Architecture, Int. J. Circ. Theor. Appl. (CTA), 573–588, 2008
9. Á. Zarándy, R. Domínguez-Castro, S. Espejo, Ultra-High Frame Rate Focal Plane Image Sensor and Processor, IEEE Sensor J 2(6) 559–565, 2002
10. R. Wagner, Á. Zarándy, T. Roska, Adaptive Perception with Locally-Adaptable Sensor Array, IEEE Trans Circ Syst. I, 51(5), 1014–1023, 2004
11. A. Zarandy, P. Földesy, T. Roska Per-Pixel Integration Time Controlled Image Sensor, ECCTD05 – European Conference on Circuit Theory and Design, Cork, Ireland, pp. III-149–III-152, August 2005
12. http://www.suss-microoptics.com/products/microlens.html
13. Z. Nagy, P. Szolgay Configurable Multi-Layer CNN-UM Emulator on FPGA. IEEE Trans. Circ. Syst. I: Fund. Theor. Appl. 50, 774–778, 2003
14. Cs. Rekeczky, J. Mallett, Á. Zarándy, Security Video Analitics on Xilinx Spartan – 3A DSP, Xcell J. 66, fourth quarter, 28–32, 2008
15. K. Karacs, Gy Cserey, Á. Zarándy, P. Szolgay, Cs. Rekeczky, L. Kék, V. Szabó, G. Pazienza, T. Roska, Software Library for Cellular wave Computing Engines in an era of kilo-processor chips, Version 3.1, Budapest, Cellular Sensory and Wave Computing Laboratory of the Computer and Automation Research Inst., Hungarian Academy of Sciences (MTA SZTAKI) and the Jedlik Laboratories of the Pázmány P. Catholic University, 2010
16. P. Perona, J. Malik, Scale-Space and Edge Detection Using Anisotropic Diffusion, IEEE Trans. Pattern Anal. Mach. Intell. 12(7), 629–639, 1990
17. L.O. Chua, T. Roska, Cellular Neural Networks and Visual Computing, Cambridge University Press, 2002
18. A. Lopich, P. Dudek, Architecture of Asynchronous Cellular Processor Array for Image Skeletonization, Circ. Theor. Des. 3, 81–84, 2005

# SCAMP-3: A Vision Chip with SIMD Current-Mode Analogue Processor Array

**Piotr Dudek**

**Abstract** In this chapter, the architecture, design and implementation of a vision chip with general-purpose programmable pixel-parallel cellular processor array, operating in single instruction multiple data (SIMD) mode is presented. The SIMD concurrent processor architecture is ideally suited to implementing low-level image processing algorithms. The datapath components (registers, I/O, arithmetic unit) of the processing elements of the array are built using switched-current circuits. The combination of a straightforward SIMD programming model, with digital microprocessor-like control and analogue datapath, produces an easy-to-use, flexible system, with high-degree of programmability, and efficient, low-power, small-footprint, circuit implementation. The SCAMP-3 chip integrates $128 \times 128$ pixel-processors and a flexible read-out circuitry, while the control system is fully digital, and currently implemented off-chip. The device implements low-level image processing algorithms on the focal plane, with a peak performance of more than 20 GOPS, and power consumption below 240 mW.

## 1 Introduction

The basic concept of a 'vision chip' or a 'vision sensor' device is illustrated in Fig. 1. Unlike a conventional computer vision system, which separates image acquisition and image processing, the vision chip performs processing adjacent to the sensors, on the same silicon die, producing as outputs pre-processed images, or even higher-level information, such as lists of features, indicators of presence and locations of specific objects or other information extracted from the visual scene. The advantages of this approach include the increase of sensor/processor data bandwidth, and associated reduction in power consumption and increase in the processing throughput.

P. Dudek (✉)
The University of Manchester, Manchester M13 9PL, UK
e-mail: p.dudek@manchester.ac.uk

Á. Zarándy (ed.), *Focal-Plane Sensor-Processor Chips*,
DOI 10.1007/978-1-4419-6475-5_2, 

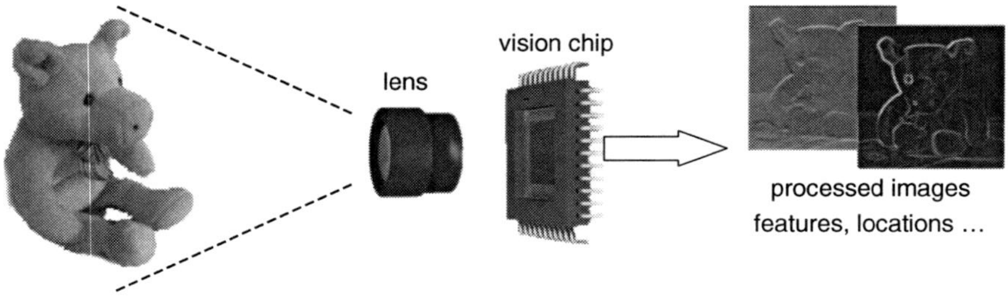


**Fig. 1** The concept of a 'vision chip'. Unlike a conventional image sensor, it outputs information extracted from the images

The idea of processing images on the focal plane, and its associated benefits, resulted over the years in many vision chip implementations [1]. From the functionality point of view, two approaches are taken: one is to produce an application-specific (or task-specific) device, performing a particular image processing operation, and the other is to construct a device based on some programmable computing hardware that can be customised to the application by software. While the application-specific devices may be highly optimised at the circuit level for performing a particular task (e.g. optical flow measurements [2], a $3 \times 3$ convolution kernel [3]) in practice, their application domain is restricted by their limited functionality. While an application-specific processing circuit may demonstrate performance or power advantages in a particular task, it is rare that only a single image processing operation (e.g. a simple filter) is required in an application. More often, a vision application requires a number of low-level image pre-processing operations, and some higher-level object-based operations, to be performed on each frame of the video stream. Hence, the advantages offered by a custom task-specific device over a more general digital processing hardware (which will usually be included in a complete vision system anyway) are rarely sufficient to merit the use of task-specific vision chips in practical applications. Such devices, despite the demonstrated high performance or efficiency figures, have thus so far largely remained confined to academic exercises and not adopted as real-world engineering solutions.

On the other hand, vision sensors combining on the focal plane, the image sensor array and general-purpose digital processors have been a subject of a number of commercial developments, and found their way into numerous applications [4, 5]. Such devices offer some of the benefits of focal-plane processing (low-power due to near-sensor processing, reduction in I/O bandwidth between the sensor/processor device and the rest of the system), but also flexibility and programmability of a digital microprocessor, and ability to perform a large number of different image processing operations on-chip. However, the current commercial digital vision processors do not make use of the fully pixel-parallel, one processor per pixel architecture, shown in Fig. 2, that conceptually offers the most optimal structure for performing pixel-parallel low-level image processing operations. Instead, they typically integrate a number of processors in a 1D array, one processor per column

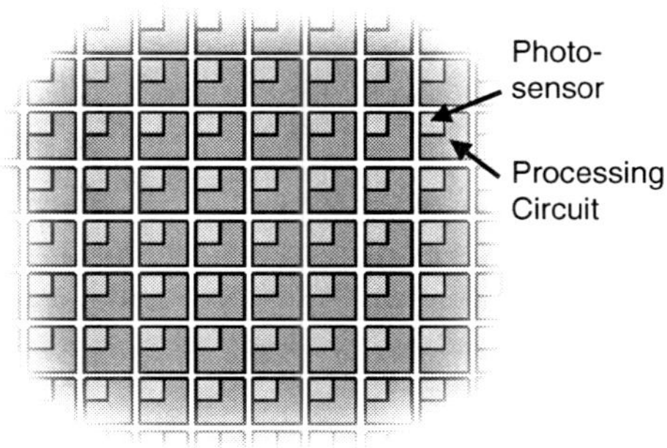


**Fig. 2** Pixel-parallel vision chip integrates image processing circuit in each pixel of the image sensor array

of pixels [5, 6]. The reason for this is the difficulty in integrating a complete programmable processor within a small pixel area in a focal-plane array. There have been a few research projects where digital in-pixel processing, having a separate processing element (PE) for each pixel in the array, has been used; from the early devices incorporating rather limited functionality, with a few memory bits per pixel and simple logic operations [7–10], to more recent devices [11], including our developments of asynchronous/synchronous vision chips [12, 13]. The technological progress of CMOS scaling and the availability of vertically stacked 3D integrated devices promise further developments in this area [14]. Nevertheless, the ability to implement reasonably complex processing cores in a small circuit area of a pixel still remains the main challenge. Another challenge is the low-power consumption requirement for a PE that is to be integrated in a massively parallel array of thousands (or millions) of pixel-processors in a single chip. A number of pioneering developments [15–17], including devices described in other chapters of this book, have investigated the issue of integrating analogue processing cores, which would offer better power/performance and area/performance ratios than digital designs, while still offering a degree of programmability usually associated with the digital processing approach.

## *1.1 SIMD Cellular Processor Arrays*

The idea of a Cellular Processor Array – a system integrating a large number of simple PEs, organised in a regular network (typically, especially in the systems considered for image processing, placed on a 2D grid), with nearest-neighbour communication – has been considered from the early days of computing. It can be traced back to von Neumann's work on Cellular Automata [18] and later work on massively parallel arrays of processors by Unger [19] and Barnes [20]. With the introduction of custom-integrated circuits, massive parallelism has become feasible, and the research resulted in the development of machines such as CLIP 4 [21],

DAP [22], MPP [23] and CM-1 [24], which used very simple processors working concurrently performing identical instructions on local data. This mode of operation, known as single instruction multiple data (SIMD), where multiple parallel processing units execute identical program operations on their individual data streams, has been popular in early parallel processor designs. It was used in 'vector' processing units of supercomputers, and has recently evolved as a 'streaming' processing mode used in commodity processors (e.g. in Intel's Streaming SIMD Extensions, IBM/Toshiba/Sony Cell Broadband Engine processor or NVIDIA GPUs). In contrast to 'fine grain' parallelism of SIMD, the 'coarse grain' parallel processing architectures generally adopt a more flexible multiple instruction multiple data (MIMD) style, in which a parallel system consists of multiple independent processing cores, executing independent programs, communicating through a shared memory or a message-passing interface.

Nevertheless, in specific application domains, where a large number of data items undergo identical operations, the SIMD mode provides the most optimal solution, using a multitude of datapath units (PEs), while sharing a single controller that issues instructions to these PEs. Low-level image processing is an example of such application, with inherent massive data parallelism. A typical operation, such as a $3 \times 3$ sharpening filter, edge detector, median filter, etc. involve executing identical instructions on every pixel in the image, and producing an output that depends on the value of the pixel, and pixels in its immediate neighbourhood. Such algorithms are easily and naturally mapped onto 2D processor arrays, providing the greatest possible speedup and the simplest control sequence if each processor is associated with one image pixel. The spatial organization of processors can be used to great advantage, minimizing the number of operations required to perform the task. The vision chip design described in this chapter builds upon the ideas of SIMD image processing, providing effective circuit and system level solutions for the design of fine-grain SIMD cellular processor arrays.

## 1.2 Chapter Overview

The remainder of this book chapter overviews the architecture and design of the SIMD current-mode analogue matrix processor (SCAMP) device that has been designed to 'emulate' the performance of a digital processor array, providing the same degree of flexibility and programming philosophy, while using compact and power-efficient analogue datapath elements. First, the architecture of pixel-parallel cellular processor array is overviewed, explaining how low-level image processing algorithms are mapped onto cellular SIMD arrays. Then the concept of a processor with analogue data path is introduced, its switched-current circuit implementation is explained and its instruction-level operation is described. Finally, the design and implementation details of the SCAMP-3 chip are presented, and application examples are shown.

# 2 SCAMP-3 Architecture

To understand the SCAMP-3 system, basic concepts of pixel-parallel SIMD approach to image processing have to be appreciated. They are introduced in the next section, followed by the description of the PE architecture implemented on SCAMP-3 and details of the I/O and control system architecture.

## 2.1 *Pixel-Parallel SIMD Array*

The three basic concepts described here are common to all SIMD cellular processor arrays, but will be described introducing the mechanisms and nomenclature used in the SCAMP-3 processor. First is the basic idea of pixel-parallel operation. Second is the idea of a nearest-neighbour communication (a 'NEWS' register). Third is the idea of conditional operation based on a local activity flag (a 'FLAG' register).

### 2.1.1 Array-Wide Operations

The basic concept of a pixel-parallel SIMD cellular processor array is that operations are performed on all array elements (e.g. image pixels) at once. The architecture of the 2D PE array is shown in Fig. 3. Each PE may contain a number of local *registers* to store data (let us assume these registers can store a real number, such as a grey-level pixel intensity). If we label registers of a PE at location $(x,y)$ as $A_{xy}, B_{xy}, C_{xy}$, etc., then we can imagine register arrays **A**, **B**, **C**, etc. being formed from corresponding registers taken from all PEs, as shown in Fig. 4.

Each PE is responsible for processing individual array elements, using the arithmetic logic unit (ALU) that represents the PEs capability to perform operations on

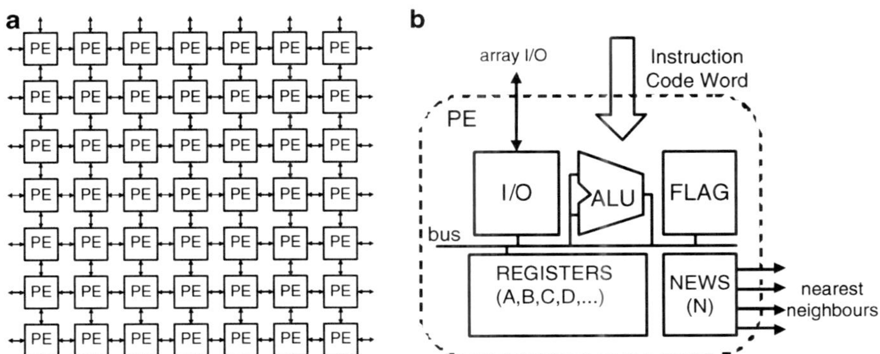


**Fig. 3** Architecture of a pixel-parallel SIMD array. (**a**) 2D arrangement of processing elements (PEs), with nearest-neighbour connectivity; (**b**) a single PE

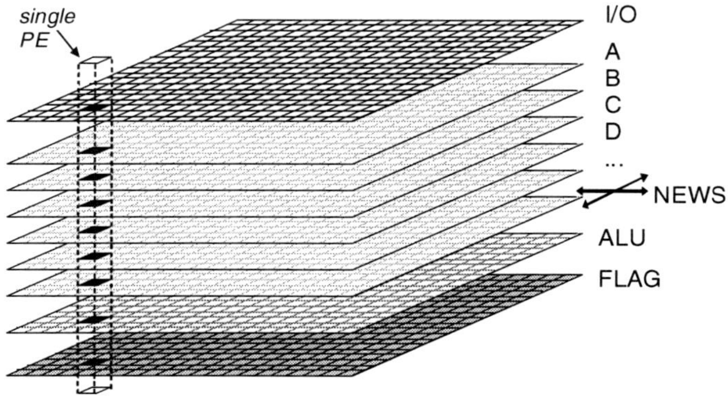


**Fig. 4** Array processor view of the SIMD architecture

**Fig. 5** Array-wide arithmetic operation

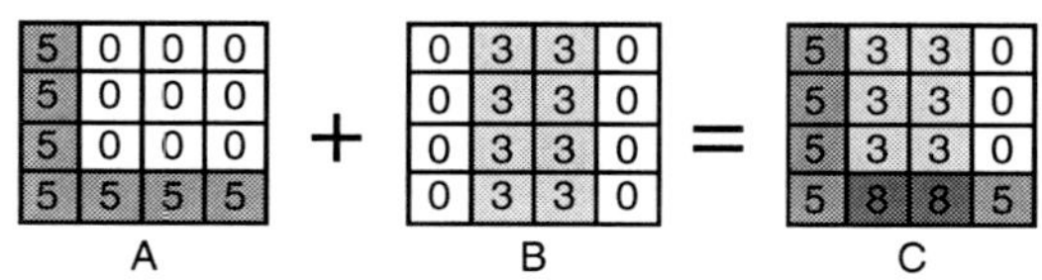


data stored in registers, according to the specific instruction set. As all PEs execute the same program, that is there is one controller in the system, issuing the same 'Instruction Code Words' to all PEs, effectively we obtain array-wide operations. For example, if we load array **A** with one image, and array **B** with another image, a single instruction '$\mathbf{C} = \mathbf{A} + \mathbf{B}$' will add the two images and put the result in register array **C** (see Fig. 5). In the following, we will simply refer to the register arrays **A**, **B**, **C**, etc. as *registers*.

The element-wise array operations are a straightforward concept that will be familiar to anyone, who has used vector/matrix-based programming languages, such as Matlab. It is important to realize though that instead of internally looping through all array elements (as would be the case on a sequential computer), the underlying pixel-parallel SIMD hardware architecture inherently supports such array operations. The individual PE at location $(x,y)$ performs a scalar operation $C_{xy} = A_{xy} + B_{xy}$. This is done concurrently in all PEs, resulting in an array-wide addition operation $\mathbf{C} = \mathbf{A} + \mathbf{B}$.

### 2.1.2 Nearest-Neighbour Communication

PEs in a cellular processor array can communicate with nearest neighbours on the grid. In the SCAMP-3 system, a 4-neighbour connectivity is used, and the communication is carried out through a special neighbour communication register **N**. The contents of this register can be loaded from any other register, for

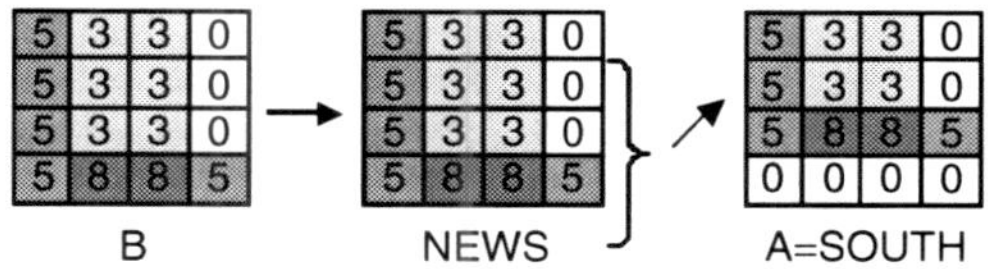


**Fig. 6** Neighbour communication operation

example in each PE $N_{xy} = A_{xy}$, but can be accessed by the neighbouring PEs when loading to another register, for example $A_{xy} = N_{x+1,y}$. This operation, when performed in all pixels at once, corresponds to shifting the entire array of data by one column (or one row), that is a one-pixel translation of the image.

By convention, the communication register **N** is called a '**NEWS** register', and when referring to the array operations geographic direction names are used to denote the contents of this register shifted by one pixel, for example assigning $A_{xy} = N_{x+1,y}$ in each array element is denoted as **A** = **EAST**; assigning $A_{xy} = N_{x,y-1}$ corresponds to **A** = **NORTH**, and so on.

As an illustration, assume an image is loaded in register **B**, and consider the following sequence of operations:

NEWS = B
A = SOUTH

The result of this sequence is illustrated in Fig. 6. In the first instruction, the **NEWS** register is loaded with corresponding pixels from image **B**. In the second instruction, register **A** is loaded with the data from the **NEWS** register so that each element takes a value of its **SOUTH** neighbour. As a result, the overall image in **A** appears shifted by one pixel up. (An obvious issue is that of a boundary condition; in the simplest case, the cells that do not have a neighbour are simply loaded with zeros, other solutions may include cyclic or zero flux boundary conditions).

As an example, consider a simple vertical edge detection that can be performed by subtracting the image from its horizontally shifted version. This can be achieved by the following program:

NEWS = B
A = B-EAST

The operation of this program (assuming zero is shifted from the boundary) is illustrated in Fig. 7.

Using the neighbour data transfer concept, filters based on convolution kernels can be easily implemented. It should be noted that many convolution kernels can be implemented in a compact way on the pixel-parallel processor arrays, in a few instructions, exploiting the kernel symmetry and decomposition possibilities. For example, consider convolution **A** = **k*****P** of the image **P** with a horizontal Sobel edge detection kernel **k**:

$$A_{xy} = \sum_{j=1}^{3}\sum_{k=1}^{3} k_{jk} P_{x+j-2,y+k-2}; \quad \mathbf{k} = \begin{bmatrix} -1 & -2 & -1 \\ 0 & 0 & 0 \\ 1 & 2 & 1 \end{bmatrix} \tag{1}$$

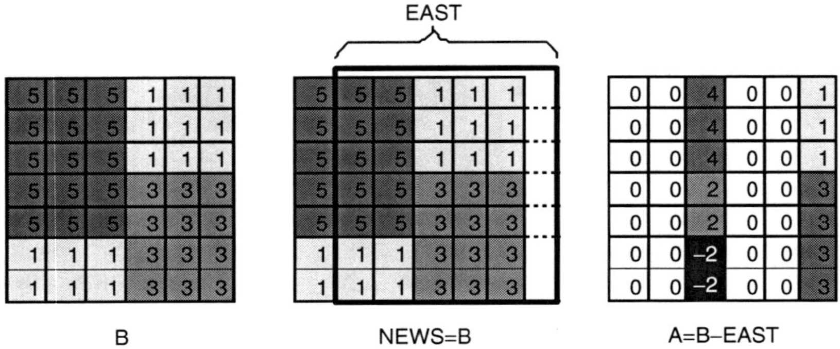


**Fig. 7** Vertical edge detection example

To implement this filter, the following operations can be executed in a pixel-parallel way:

NEWS = P
A = 2*P + EAST + WEST
NEWS = A
A = SOUTH − NORTH

Register **A** is first used as a temporary array, and ultimately it stores the result of the convolution operation. The first two instructions implement the convolution of the image with a kernel [1 2 1], the following two instructions take the result and apply the kernel $[-1\ 0\ 1]^T$, producing the final result. (Note that the actual machine-level instruction implementation of the above procedure on the SCAMP-3 system has to take into account specific instruction set restrictions, such as the fact that only one neighbour access is allowed in one instruction, every elementary instruction leads to value negation and carries an offset error that has to be cancelled out, no multiplication is available, etc. The procedure shown above has to be thus compiled into assembly code of about 30 machine-level instructions.)

### 2.1.3 Local Activity Flag

The third important concept of SIMD cellular processing is that of local activity flag. While all PEs in the array receive the same instruction stream, it is often required to provide some degree of local autonomy, so that different operations can be performed on different elements of the array, in data-dependent fashion.

For example, consider the thresholding operation. The output of this operation can be described as:

$$A_{xy} = \begin{cases} 1 & \text{if } P_{xy} > T \\ 0 & \text{if } P_{xy} \leq T \end{cases} \tag{2}$$

that is, the output register **A** should be loaded with '1' in these PEs that have their register **P** value above the threshold value $T$, and loaded with '0' in the other PEs. We could imagine that a following program executing in each PE would implement this operation:

```
if P > T then A = 1
else A = 0
end if
```

After the conditional instruction '**if**' the program branches, some PEs need to perform the assignment instruction '**A = 1**', others have to perform '**A = 0**'. However, the SIMD architecture stipulates that a single controller issues the same instructions to all PEs in the array, that is each PE receives exactly the same instruction to be executed at a given time. It is hence impossible to branch and perform different instructions in different PEs of the array at the same time.

To solve this problem, and implement conditional branching operations, a concept of a *local activity flag* is introduced. Each PE contains a one-bit FLAG register. As long as this FLAG is set, the PE is active (enabled) and executes the instructions issued by the controller. When the FLAG is reset, the PE becomes inactive (disabled), and simply ignores the instruction stream. The FLAG can be set or reset in a single conditional data-dependent instruction, and the instructions setting/reseting the FLAG are the only ones that are never ignored (so that the inactive PEs processors can be activated again).

The example thresholding operation can be thus implemented as follows:

```
A = 1
if (P > T) reset FLAG
A = 0
set FLAG
```

The execution of this program is illustrated in Fig. 8. The same instructions are delivered to every PE in the array; however, the instruction '**A = 0**' is only executed in those PEs that have the FLAG still activated after the conditional reset.

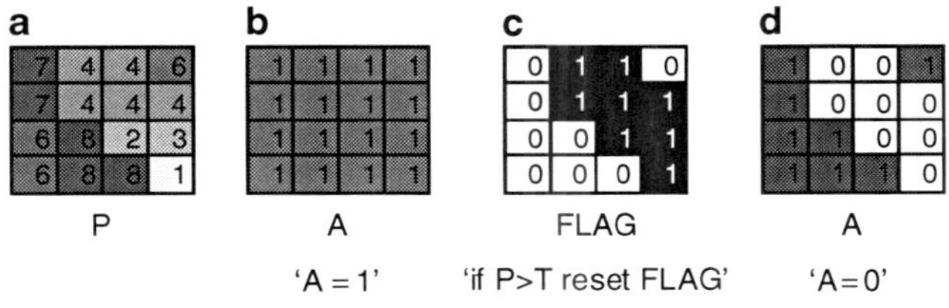


**Fig. 8** Conditional code execution (processor autonomy) using local activity flag: (**a**) register P contains the input data, (**b**) assuming all PEs are active (FLAG = 1), register A of all PEs is loaded with 1, (**c**) comparison operation, PEs that have P > T (assume T = 5 in this example) become inactive (FLAG = 0), (**d**) register A is loaded with 0 only in active PEs

It should be noted that no additional flexibility is allowed in terms of conditional branching in SIMD arrays. In particular, if the control program contains loops or jumps, these cannot be conditional on local PE data, as the same instruction stream has to be applied to all PEs. The only local autonomy is possible by masking individual PEs using the FLAG register, so that they do not execute certain instructions in the program.

## 2.2 Processing Element

The block diagram of the PE implemented on the SCAMP-3 chip is shown in Fig. 9 The classical SIMD mechanism of array-wide operation, neighbour communication, and local activity flag, described above, form the basis of the SCAMP-3 architecture. The PE contains nine registers, capable of storing analogue data values (e.g. pixel intensity values, edge gradient information, gray-level filtering results). The general-purpose registers are labeled A, B, C, D, H, K, Q and Z. The registers are connected to a local bus. The NEWS register can be also connected to local buses of the four neighbours.

The PE contains a one-bit local activity FLAG. Control signals that form the instruction code words (ICWs) are broadcast to the PEs from the external controller. The ICWs determine the type of data operation, and select individual registers to be read-from or written-to. The write control signals for all registers are gated by the contents of the FLAG, so that no register write operation is performed when FLAG is reset. This is sufficient to implement the local autonomy, as gating write control signals ensures that the state of the PE does not change as a result of broadcast instructions when the PE is in the disabled state.

The PIX register contains the pixel intensity value acquired by the photosensor intergrated within the processor. A single PE in the array is associated with a single pixel of the image sensor, that is the SCAMP-3 vision chip implements a fully pixel-parallel fine grain SIMD focal plane processor array.

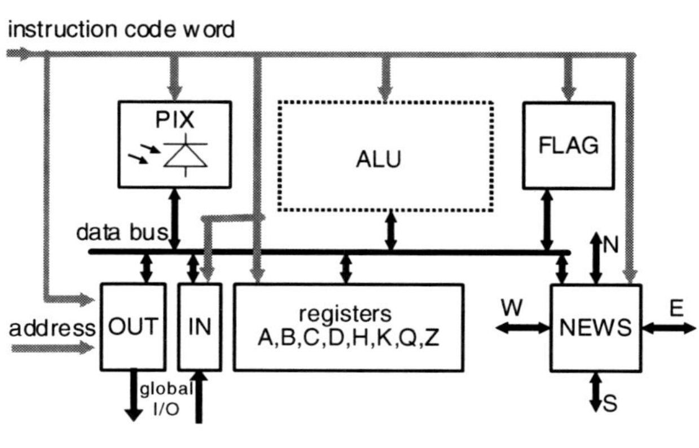


**Fig. 9** Processing element of the SCAMP-3 chip

The ALU block represents the instruction-set of the PE. The PE is capable of the following basic operations: negation, addition/summation, division by two (and three), conditional FLAG reset based on comparison with zero, and unconditional FLAG set. The results of the operations are stored in registers, and the arguments of the operations are registers, or register-like components (PIX, IN) representing the pixel-parallel input to the array. The IN register represents the global input, for example immediate argument for a register load instruction.

## 2.3 Array Structure and Global I/O

The overall system integrated on the chip contains an array of PEs, read-out circuits and control-signal distribution circuits, as shown in Fig. 10. The PE array supports random addressing. It is organized in rows and columns, and the address decoders are placed on the periphery of the array. The PE can output data via the global I/O port to the array column bus, when addressed by the row select signal, and then to the output port as selected by the column address signal. The output data is a register value (contents of any register can be read-out to the column line), or the FLAG register bit. The array can be scanned to read-out the entire register array (e.g. the output image).

A further computational capability is offered by the global read-out operations. When multiple PEs are addressed and output their data to the column lines, the PE array performs the summation operation. In particular, addressing all pixels in the array simultaneously results in the calculation of the global sum of all values in a particular array register. This can be used, for example, to perform pixel

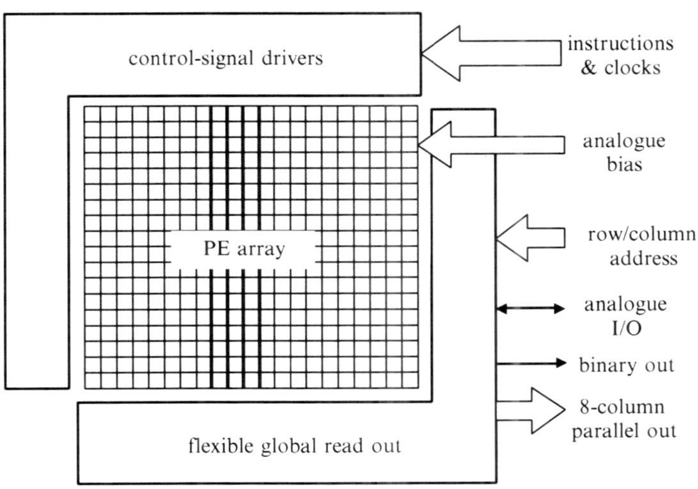


**Fig. 10** SCAMP-3 chip

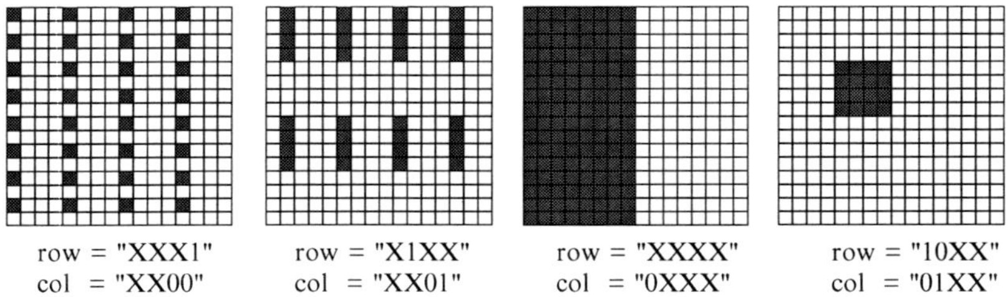


**Fig. 11** Flexible addressing scheme utilising *don't care* row/column addresses for selecting multiple PEs at the same time. The address (0, 0) points to the *bottom left* corner of the array, addressed PEs are *shaded*

counting operations (e.g. for computing histograms) or calculating some other global measures (e.g. the average brightness of the image for gain control, or overall 'amount' of motion in the image). Indeed, it is expected that when the device is used in a machine vision system, the results of global operations will often be the only data that is ever transmitted off chip, as these results will represent some relevant information extracted from the image.

Global logic operations are also supported. When reading-out the state of the FLAG register, and addressing multiple PEs at the same time, a logic OR operation is performed. Such an operation can be effectively used, for example to test the image for existence of a certain feature, or to detect a stop condition that controls the number of iterations in an algorithm.

Multiple PE selection for read-out is implemented with addressing scheme that uses *don't care* symbols as a part of the row/column address, as illustrated in Fig. 11. In practice, this is achieved by using two words to produce the address, one sets the actual address bits (e.g. '00110000') the second sets the *don't care* attribute for individual address bits (e.g. '00001111') producing together the combined word that addresses many pixels at once ('0011XXXX' in this example). The addressing of groups of pixels can be utilized in various ways. An example is the pixel-address finding routine that starts by addressing half of the array, performing a global OR operation to determine whether the pixel is in the selected region, then selecting the quarter of the array (in the half that contains the pixel, as determined in the previous step), then 1/8th, and so on, finally homing on the pixel of interest. In this way, an address of an active pixel can be determined in just a few bisection steps, without the need for reading-out (scanning) the entire image from the chip [25].

## 2.4 Control System

The PE array needs to be provided with the sequence of instructions to be executed. These are broadcast to all PEs in the array from a single controller. The control system is also responsible for read-out addressing, interfacing to external devices, and

```
C ←A              ; using C as a temporary register...
B ← C             ; ...transfer array A (original image) to array B
load s, 10        ; load control variable s with 10
: loop            ; address label
NEWS ← B          ; transfer array B to the NEWS array
B ← EAST          ; transfer EAST to B (shift NEWS by one pixel)
sub s, 1          ; subtract 1 from s
jump nz, loop     ; loop, unless s is zero (counts 10 iterations)
C ←A + B          ; add A (original image) and B (shifted by 10 pixels)
```

**Fig. 12** An example SIMD program. The program shifts an image by 10 pixels to the left, and superimposes the two images (adds original and shifted together). The operations in *italic* type are executed in the controller CPU. The operations shown in **bold** type are executed in the PE array – the controller broadcasts these instructions to the PE array. A, B and NEWS are array registers, operated in the PE datapath, while variable s is mapped to a scalar register of the controller CPU datapath

other control duties (setting up biases, etc.). A digital controller could be included on the same chip as the SIMD array; however on the SCAMP-3 prototype device, this is an external circuit, implemented on an FPGA, using a modified microcontroller core [26].

The main function of the controller is to supply the sequence of ICWs to the PEs in the array. The controller can also execute conditional and unconditional jumps, loops and arithmetic/logic operations on a set of its own scalar registers (e.g. in order to execute variants of the code based on user settings or other inputs to the system, or, for example, processing the result of a global array operation, and performing conditional program flow control based on this result). It has to be remembered that the program flow control is global. All PEs will be executing the same instructions, with the FLAG mechanism available to mask parts of the array. In practice, the program is thus a mixture of *array* operations, which are executed in the massively parallel PE datapath, and *control* operations, which are executed in the controller CPU. A simple example program, consisting of both types of operations is shown in Fig. 12.

## 3 The Analogue Processor

The pixel-parallel SIMD processor architecture, outlined above, provides the framework for designing a general-purpose 'vision chip' device. The integration of physical image sensor in each PE makes the incident image pixel data immediately accessible to the PE. A variety of low level image processing algorithms can be implemented in software. The main challenge, however, lies in the efficient implementation of the 2D PE array system in a silicon device. The solution adopted in the design of the SCAMP-3 chip, based on our idea of an analogue sampled-data processor [27], is outlined in this section.

If a reasonably high resolution device is to be considered, then the individual PE circuitry has to be designed to a very stringent area and power budget. We are interested in vision sensor devices of resolutions in the range of 320 × 240 pixels, with a pixel pitch below 50 µm, and typical power consumption in the milliwatt range. A high-performance low-power devices with lower resolutions (e.g. 128 × 128 or 64 × 64 pixels) could still be useful for many applications (e.g. consumer robotics and automation, toys, 'watchdog' devices in surveillance). Pixel sizes in the range of 50 µm have further applications in read-out integrated circuits for infrared focal plane arrays. While using deep sub-micron technologies makes it feasible to design digital processors to these specifications (typically using a bit-serial datapath), we have proposed to use analogue circuitry to implement the PE [27, 28], which can be implemented with a very good power/performance ratio in an inexpensive CMOS technology. It is well known that analogue signal processing circuits can outperform equivalent digital ones. However, while exploiting the analogue circuit techniques for superior efficiency, we still want to retain the advantages of a software-programmable general-purpose computing system.

The basic concept that allows the achievement of this goal is based on the insight that the stored-program computer does not have to operate with a digital datapath. We can equally well construct a universal machine, that is a device operating on data memory, performing data transformations dictated by the sequence of instructions fetched from the program memory, using an analogue datapath. In this scenario, the data memory is stored in analogue registers, transmitted through analogue data buses, and transformed using analogue arithmetic circuits, while the architecture remains microprocessor-like, with a fixed instruction set, and the functionality determined by a software program, and the control (program memory, instruction fetch, decode and broadcast) implemented using digital circuits. We called such a device 'the analogue microprocessor' [27].

### 3.1 Switched-Current Memory

An efficient way to design an analogue processor is to use switched-current circuit techniques [29]. A basic SI memory cell is illustrated in Fig. 13. When an MOS transistor works in the saturation region its drain current $I_{ds}$ can be, in a first-order approximation, described by the equation:

$$I_{ds} = K(V_{gs} - V_t)^2 \tag{3}$$

where $K$ is the transconductance factor, $V_t$ is the threshold voltage and $V_{gs}$ is the gate-source voltage.

The SI memory cell remembers the value of the input current by storing charge on the gate capacitance $C_{gs}$ of the MOS transistor. The operation of the memory cells is as follows. When writing to the cell, both switches, S and W, are closed (Fig. 13b). The transistor is diode-connected and the input current $i_{in}$ forces the gate-source

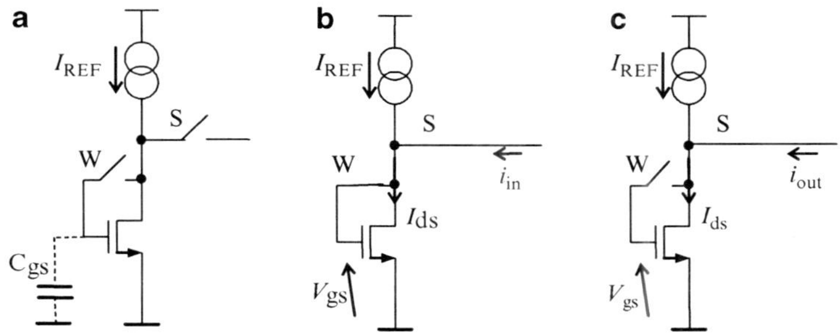


**Fig. 13** Basic SI memory cell. (**a**) during storage both switches are open, (**b**) both switches are closed when the cell is 'written to', (**c**) only switch 'S' is closed when the cell is 'read-from'

voltage $V_{gs}$ of the transistor to the value corresponding to the drain current $I_{ds} = I_{REF} + i_{in}$, according to (3). At the end of the write phase, the switch W is opened and thus the gate of the transistor is disconnected, that is put into a high impedance state. Due to charge conservation on the capacitor $C_{gs}$, the voltage at the gate $V_{gs}$ will remain constant.

When reading from the memory cell, the switch S is closed and the switch W remains open (Fig. 13c). Now, the transistor acts as a current source. As the gate-source voltage $V_{gs}$ is the same as one that was set during the write phase, the drain current $I_{ds}$ has to be the same (provided that the transistor remains in the saturation region), and hence the output current $i_{out}$ is equal to

$$i_{out} = I_{ds} - I_{REF} = i_{in} \tag{4}$$

Therefore, the SI memory cell is, in principle, capable of storing a continuous-valued (i.e. real) number, within some operational dynamic range, and subject to accuracy limitations that will be discussed later in this chapter.

## 3.2 Arithmetic Operations

In addition to data storage registers, the PE needs to provide a set of basic arithmetic operations. These can be achieved with a very low area overhead in the current-mode system. Consider a system consisting of a number of registers implemented as SI memory cells, connected to a common *analogue bus*. Each memory cell can be configured (using corresponding switches S and W) to be read from, or written to.

The basic operation is the transfer operation, as shown in Fig. 14a. Register A (configured for reading) provides the current $i_A$ to the analogue bus. This current is consumed by register B (configured for writing); register C is not selected, and hence it is disconnected from the bus. Therefore, the analogue (current) data value is transferred from A to B. This transfer is denoted as **B** ← **A**. According to the current

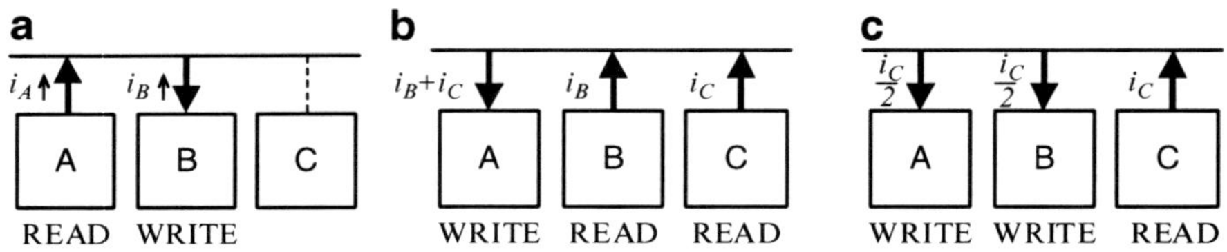


**Fig. 14** Basic operations in a current-mode processor: (**a**) register data transfer, (**b**) addition, (**c**) division by two

memory operation, register B will produce the same current when it is read-from (at some later instruction cycle). If we consider that the data value is always the one provided *into the analogue bus*, it can be easily seen that $i_\mathrm{B} = -i_\mathrm{A}$, that is the basic transfer operation includes negation of the stored data value.

Addition operation (and in general current summation of a number of register values) can be achieved configuring one register for writing, and two (or more) registers for reading. The currents are summed, according to Kirchhoff's current law, directly on the analogue bus. For example, situation shown in Fig. 14b produces operation $\mathbf{A} \leftarrow \mathbf{B} + \mathbf{C}$.

Division operation is achieved by configuring one register cell for reading and many (typically two) registers for writing. The current is split, and if the registers are identical then it is divided equally between the registers that are configured for writing, producing a division by a fixed factor. For example, in Fig. 14c both registers **A** and **B** will store current equal to half of the current provided by register **C**. We denote this instruction as $\mathbf{DIV}(\mathbf{A} + \mathbf{B}) \leftarrow \mathbf{C}$.

This completes the basic arithmetic instruction set of the analogue processor. Multiplication and division by other factors can be simply achieved by multiple application of add or divide by two operations. Subtraction is performed by negation followed by addition. Other instructions (e.g. full-quadrant multiplication) could of course be implemented in hardware using current mode circuits; however, in a vision chip application, silicon area is at premium and a simple instruction set that is sufficient for the intended application should be used. This is achieved here by the basic arithmetic instructions of negation, addition, and divide by two, which are performed entirely in the register/bus system, with no additional circuits. Consequently, more silicon area can be devoted to the register circuits, increasing the amount of local memory (or improving the accuracy of computations, which largely depends on device matching, i.e. device area).

The 'ALU' in the analogue processor, shown as a separate block in Fig. 9 is thus virtual. However, to enable comparison operations, a current comparator (detecting current provided to the analogue bus) should be provided. A combination of arithmetic operations, comparison operations (with programmable threshold) and conditional code execution can be used to perform logic operations in the analogue registers as well.

### 3.3 Accuracy of SI Circuits

The above descriptions are somewhat simplified, since they ignore the many error effects that affect the result of computation in the analogue processor. Basic switched-current memory storage operation is not entirely accurate, due to errors that are caused by the output conductance of the transistor and the current source, capacitive coupling from the bus to the memory capacitor $C_{gs}$, clock feedthrough effects due to charge injection onto the memory capacitor from the switches (implemented as MOS transistors) and capacitive coupling from the control signals, and the many noise sources in the circuit. These error effects can be to some extent mitigated using more sophisticated circuits. Extensions of the basic SI technique allow to implement more accurate circuits, with some area overhead. In particular, the $S^2I$ technique proposed by Hughes et al. [30] provides a good trade-off between the accuracy and circuit area.

Overall, the errors cannot be entirely eliminated though, and the current read from the memory cell $i_{\mathrm{out}}$ is not exactly the same as the current during the write operation $i_{\mathrm{in}}$, but can be represented as

$$i_{\mathrm{out}} = i_{\mathrm{in}} + \Delta_{\mathrm{SI}} + \varepsilon_{\mathrm{SD}}(i_{\mathrm{in}}) + \varepsilon(^{*}) \tag{5}$$

where $\Delta_{\mathrm{SI}}$ is the signal-independent (offset) error, $\varepsilon_{\mathrm{SD}}(i_{\mathrm{in}})$ is the signal-dependent component of the error, and $\varepsilon(^{*})$ is the overall random noise error. The actual situation is even more complicated, as the signal dependent error $\varepsilon_{\mathrm{SD}}$ depends not only on the current value written to the cell, but also on the analogue bus state during the read operation. Nevertheless, through the combination of circuit techniques and careful design, the errors can be made small enough to yield useful circuits in a reasonable circuit area, with accuracies corresponding to perhaps about 8-bit digital numbers.

It is important to point out that, on the one hand, the signal-independent offset errors are very easy to compensate for algorithmically, subtracting offsets during the operation of the program. The signal-dependent errors, are also systematic, and can be to some extent also accounted for by the system, although they are more cumbersome to handle. The random errors, on the other hand, put the limit on the achievable accuracy. The random noise has a temporal component, that is the value of $\varepsilon(^{*})$ changes on each memory operation (thermal noise, shot noise, etc.) and a spatial component which represents the variation in the offset and signal-dependent errors in each memory cell in the system due to the component variability. It has to be mentioned that ultimately these random errors limit the accuracy of the analogue processor techniques and scalability of the designs to finer scale CMOS technology nodes.

As shown by (5), each storage operation (and consequently each transfer and arithmetic operation) is performed with some errors. In general, these errors depend on the signal value and type of operation both during register write and subsequent read. In general, most of these errors are small, and simply have to be accepted, as long as they do not degrade the performance of the processor beyond the useable

**Table 1** Compensation of signal-independent offset errors

| Operation | Instructions | Current transfers showing signal-independent error |
|---|---|---|
| Transfer | $\mathrm{C} \leftarrow \mathrm{A}$ | $i_{\mathrm{C}} = -i_{\mathrm{A}} + \Delta_{\mathrm{SI}}$ |
| $\mathrm{B} := \mathrm{A}$ | $\mathrm{B} \leftarrow \mathrm{C}$ | $i_{\mathrm{B}} = -i_{\mathrm{C}} + \Delta_{\mathrm{SI}} = i_{\mathrm{A}}$ |
| Addition | $\mathrm{D} \leftarrow \mathrm{A} + \mathrm{B}$ | $i_{\mathrm{D}} = -(i_{\mathrm{A}} + i_{\mathrm{B}}) + \Delta_{\mathrm{SI}}$ |
| $\mathrm{C} := \mathrm{A} + \mathrm{B}$ | $\mathrm{C} \leftarrow \mathrm{D}$ | $i_{\mathrm{C}} = -i_{\mathrm{D}} + \Delta_{\mathrm{SI}} = i_{\mathrm{A}} + i_{\mathrm{B}}$ |
| Subtraction | $\mathrm{D} \leftarrow \mathrm{A}$ | $i_{\mathrm{D}} = -i_{\mathrm{A}} + \Delta_{\mathrm{SI}}$ |
| $\mathrm{C} := \mathrm{A}–\mathrm{B}$ | $\mathrm{C} \leftarrow \mathrm{B} + \mathrm{D}$ | $i_{\mathrm{C}} = -(i_{\mathrm{B}} + i_{\mathrm{D}}) + \Delta_{\mathrm{SI}} = i_{\mathrm{A}} - i_{\mathrm{B}}$ |
| Negation | $\mathrm{C} \leftarrow$ | $i_{\mathrm{C}} = \Delta_{\mathrm{SI}}$ |
| $\mathrm{B} := -\mathrm{A}$ | $\mathrm{B} \leftarrow \mathrm{A} + \mathrm{C}$ | $i_{\mathrm{B}} = -(i_{\mathrm{A}} + i_{\mathrm{C}}) + \Delta_{\mathrm{SI}} = -i_{\mathrm{A}}$ |

limit. There are exceptions, however, where simple compensation techniques may significantly improve the accuracy at a small cost in the size of the program.

First, consider the offset error, denoted as $\Delta_{\mathrm{SI}}$ in (5). This is typically a relatively large systematic error (e.g. several percent of the maximum signal value). However, it can be very easily cancelled out. Since every instruction introduces current negation, the offset error is eliminated using double-instruction macros for transfer, addition, subtraction and negation operations, as illustrated in Table 1.

Second, consider the mismatch error associated with the division instruction. It should be noted that mismatch errors associated with register transfer/storage operations are very small since in the SI memory cell the same transistor is used to copy the current from input cycle to output cycle, and hence it does not matter much if the individual registers are not exactly matched. However, the accuracy of the current splitting in two (as shown in Fig. 14c) relies on matching of the transistors in different cells. In practical designs, the current mismatch of these transistors, given the same terminal voltages, can be as large as several percent. Assuming the mismatch error $\varepsilon$, and ignoring the $\Delta_{\mathrm{SI}}$ error (that can always be compensated in the way described in Table 1), the division such as **DIV A + B ← C** results in

$$i_{\mathrm{A}} = -i_{\mathrm{C}}(1+\varepsilon)/2 \tag{6}$$

$$i_{\mathrm{B}} = -i_{\mathrm{C}}(1-\varepsilon)/2. \tag{7}$$

If an accurate division is required, then the following five-step compensation algorithm should be used. This is based on the scheme proposed in [31], and works by finding the actual error resulting from the division result (note that after **DIV A + B ← C** we get $i_{\mathrm{B}} - i_{\mathrm{A}} = \varepsilon i_{\mathrm{C}}$), adding that to the original dividend, and performing mismatched division again. For example, to perform an accurate $\mathbf{A} = \mathbf{C}/\mathbf{2}$ operation the program shown below should used:

$$\begin{aligned}
\mathbf{DIV\,A+B \leftarrow C;}\quad & i_{\mathrm{A}} = -i_{\mathrm{C}}(1+\varepsilon)/2 + \Delta, i_{\mathrm{B}} = -i_{\mathrm{C}}(1-\varepsilon)/2 + \Delta \\
\mathbf{H \leftarrow B+C;}\quad & i_{\mathrm{H}} = -(i_{\mathrm{B}} + i_{\mathrm{C}}) + \Delta = -i_{\mathrm{C}}(1+\varepsilon)/2 \\
\mathbf{D \leftarrow H+A;}\quad & i_{\mathrm{D}} = -(i_{\mathrm{H}} + i_{\mathrm{A}}) + \Delta = i_{\mathrm{C}}(1+\varepsilon) \\
\mathbf{DIV\,A+B \leftarrow D;}\quad & i_{\mathrm{B}} = -i_{\mathrm{D}}(1-\varepsilon)/2 + \Delta = -i_{\mathrm{C}}(1+\varepsilon)(1-\varepsilon)/2 + \Delta \\
\mathbf{A \leftarrow B;}\quad & i_{\mathrm{A}} = -i_{\mathrm{B}} + \Delta = i_{\mathrm{C}}(1-\varepsilon^2)/2.
\end{aligned}$$

At a cost of a few extra instructions and additional local memory usage (in the above, the temporary register D can be replaced by C if its exact value is no longer needed), a significant division error improvement is achieved, theoretically from $\varepsilon$ mismatch down to $\varepsilon^2$ error. In practice, secondary error effects influence the result and, for example, on the SCAMP-3 processor the improvement in division accuracy from mismatch-limited value of 2.3% to below 0.2% is achieved.

# 4 SCAMP-3 Circuit Implementation

The analogue microprocessor idea, combined with the SIMD pixel-parallel architecture, provides foundations for the implementation of an efficient vision chip device. The PE design on the SCAMP-3 chip follows the switched-current processor design concept outlined in the previous section. A simplified schematic diagram of the complete PE is shown in Fig. 15.

## 4.1 Registers

The register cells have been implemented using the S$^2$I technique [30], with additional transistors to manage power consumption (switching the DC currents off when a register is not used) and provide conditional 'write' operation (to implement the activity flag operation). The detailed schematic diagram of the register cell is

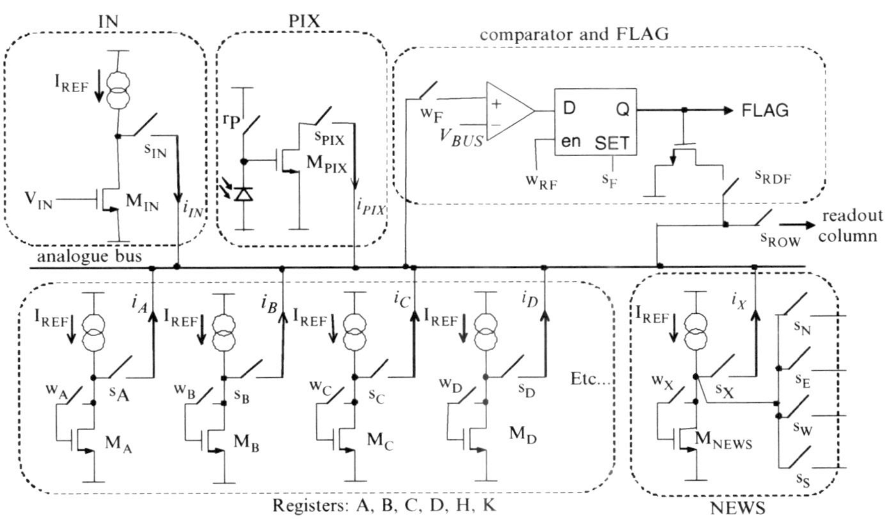


**Fig. 15** Simplified schematic diagram of the PE

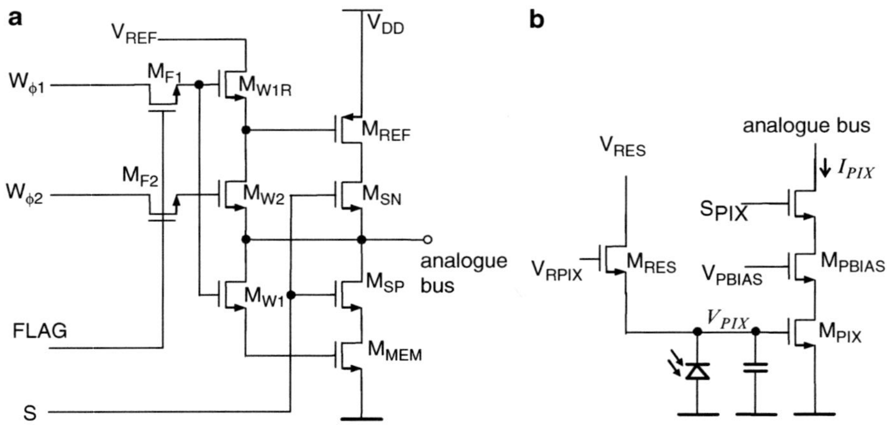


**Fig. 16** Schematic diagram of the cells implemented on the SCAMP-3 chip (**a**) register cell, (**b**) pixel circuit

shown in Fig. 16a. $M_{MEM}$ is the storage transistor, $M_{REF}$ is the current source/error storage transistor following the $S^2I$ technique, and $M_{W1}$, $M_{W2}$ and $M_{W1R}$ are the switches controlling writing/error correction. $M_{SN}$ and $M_{SP}$ connect the register to the analogue bus and shut-down the current path if register is not selected. $M_{F1}$ and $M_{F2}$ are used to gate write-control signals by the FLAG signal.

The NEWS register is build in a similar way to a regular register, but providing output current to the analogue buses of the neighbouring PEs. Furthermore, the four neighbour communication network provides means for executing a fast array-wide diffusion operation (e.g. to low-pass filter or 'blur' the image).

## *4.2 Photodetector*

The photodetector circuit is shown in Fig. 16b. In integration mode, $M_{RES}$ provides a reset signal that charges the capacitance so that $V_{PIX} = V_{RES}$. During photocurrent integration, the capacitance is discharged. The voltage $V_{PIX}$ is transformed to current $I_{PIX}$ (using $M_{PIX}$ biased in linear region with cascode transistor $M_{PBIAS}$), which is provided to the analogue bus at a required time selecting the PIX register for read-out, that is switching $S_{PIX}$. Therefore, instruction such as **A** $\leftarrow$ **PIX** can be executed to sample the pixel current $I_{PIX}$ at some time after the reset. This can be done multiple times during the integration time, and also just after reset (e.g. to perform differential double sampling in order to reduce fixed pattern noise of the imager). A typical frame loop, overlapping integration and program execution is shown in Fig. 17. However, it has to be noted that other possibilities exist (e.g. sampling at multiple times for enhanced dynamic range [32], dynamically adjusting integration time for achieving constant overall brightness, locally adjusting integration time for adaptive sensing).

```
: start
B←PIX            ; sample PIX value to B
RESPIX           ; reset PIX
A←B + PIX        ; subtract the acquired value from the reset value
...              ; user program here
OUT Z            ; output results
: loop start
```

**Fig. 17** A typical frame loop. The pixel value is sampled first (assuming it was reset in the previous frame), then the photodetector is reset and the reset level subtracted from the sampled value. The user program is executed, while the photocurrent is integrated for the next frame. The processing results are output (this forces the controller to halt the program and execute a read-out sequence, it can also wait at this point for the next frame-synchronisation pulse, if operation at a fixed frame rate is required)

The pixel circuit also enables a 'logarithmic mode' sensing, which is instantaneous (not integrating). In this mode, voltage $V_{RPIX}$ is constantly applied, and the sub-threshold current of $M_{RES}$ produces a voltage $V_{PIX}$ (note that $V_{RPIX}-V_{PIX}$ is a gate-source voltage that varies exponentially with current for a transistor in weak inversion).

## 4.3 Comparator and FLAG

A comparator, connected to the analogue bus, has the capability to determine the overall sign of the current provided to the analogue bus by registers connected to the bus. The result of the comparison operation is latched in the **FLAG** register, build as a static D-type latch. The value stored in the **FLAG** register is then used to gate the write-control signals of all registers (this is done using a single-transistor gates $M_{F1}$ and $M_{F2}$, as shown in Fig. 16a), thus ensuring that the state of the processor does not change in response to broadcast instructions when **FLAG** signal is low. The D-latch can be set by a global signal to activate all PEs.

## 4.4 Input Circuit

The **IN** circuit (see Fig. 15) provides input current to the analogue bus of the PE. The value of this current is set by transistor $M_{IN}$ driven by a global signal $V_{IN}$ that has to be set accordingly. Currently, an off-chip digital-to-analogue converter and a lookup table are used in the system, to set this value to achieve the desired input current. The **IN** register is used in instructions such as $\mathbf{A} \leftarrow \mathbf{IN(35)}$, which load all registers in the array with a constant value (by convention, numerical value of 35 used in this example corresponds to 35% of the reference current of approximately 1.7 μA).

As all PEs receive the same voltage $V_{IN}$, the mismatch of transistors $M_{IN}$ and any systematic voltage drops across the array will affect the accuracy of the input current. This can be to some extent reduced using a differential scheme, subtracting the IN(0) value from the desired value, over two cycles, as shown below:

$$\mathbf{A} \leftarrow \mathbf{IN}(\mathbf{0})$$

$$\ldots$$

$$\mathbf{B} \leftarrow \mathbf{A} + \mathbf{IN}(\mathbf{35})$$

The first line loads register A with the input circuit offset error (and the $\Delta_{SI}$ offset). The second line subtracts the offsets from the desired input value. Register B is thus loaded with 35% of the bias current.

Although loading individual PEs with different values is possible (through setting FLAG registers of individual PEs in turn, via read-out selection circuit, and then using global input), this mode is not practical and the optical input via the **PIX** circuit remains the primary data source for the chip.

## *4.5 Output Circuits*

The analogue bus of the selected PE (selected through a row select signal) can be connected to a read-out column, and then (depending on the column select signal) to the output of the chip. The current from any of the registers can be thus read-out. If multiple PEs are selected, output currents are summed.

The output of the FLAG register can also be used to drive the read-out column line. An nMOS only drive, with precharged read-out bus, enables OR operation if multiple PEs are selected. In addition, an 8-bit parallel output (reading out eight PEs in one row concurrently) is also provided.

## *4.6 Silicon Implementation*

We have implemented the 128 × 128 pixels SCAMP-3 chip [33], shown in Fig. 18a. The layout plot showing floorplan of a single PE is shown in Fig. 18b. The device has been fabricated in an inexpensive 0.35 μm 3-metal layers CMOS technology, with PE pitch below 50 μm. The chip size is 54 mm$^2$, and it comprises 1.9 M transistors. The basic parameters of the device are included in Table 2.

When clocked at 1.25 MHz, the processing array provides over 20 GOPS (Giga operations per second), with individual ‘operation’ corresponding to one analogue instruction, such as summation or division by two. The chip dissipates a maximum of 240 mW (when continuously executing instructions). When processing video streams at typical frame rates of 30 fps, the power dissipation depends on the length of the algorithm, but can be as low as a few mW for simple filtering or target tracking tasks.

**Fig. 18** SCAMP-3 device implementation: (**a**) fabricated integrated circuit, (**b**) PE layout

**Table 2** SCAMP-3 chip specifications

| | |
|---|---|
| Number of parallel processors: | 16,384 |
| Performance: | 20 GOPS |
| Power consumption: | 240 mW max |
| Supply voltage: | 3.3 V and 2.5 V (analogue) |
| Image resolution: | 128 × 128 |
| Sensor technology: | Photodiode, active pixel, 0.35 μm CMOS |
| Pixel size: | 49.35 μm × 49.35 μm (includes processor) |
| Pixel fill-factor: | 5.6% |
| Imager fixed pattern noise: | 1% |
| Accuracy of analogue processors (instruction error): | |
| Storage linearity: | 0.52% |
| Storage error fixed pattern noise: | 0.05% rms |
| Division-by-2 fixed pattern noise: | 0.12% rms |
| Random noise: | 0.52% rms |
| Image processing performance benchmarks (execution time @ 1 MHz clock): | |
| Sharpening filer 3 × 3 convolution: | 17 μs |
| Sobel edge detection: | 30 μs |
| 3 × 3 median filter: | 157 μs |

## 5 Applications

The SCAMP-3 applications are developed using a PC-based simulator (Fig. 19a) that models the instruction set of the processor array and the control processor, including behavioural models of analogue errors. A development system, allowing operation and debugging of SCAMP-3 hardware in a PC-based environment (including user front-end Graphical User Interface, and a hardware system with a USB interface, as shown in Fig. 19b), has been developed. Ultimately, the target applications for the device are in embedded systems.

The general-purpose nature of the SIMD array allows implementation of a wide range of image processing algorithms. Some examples of filtering operations are

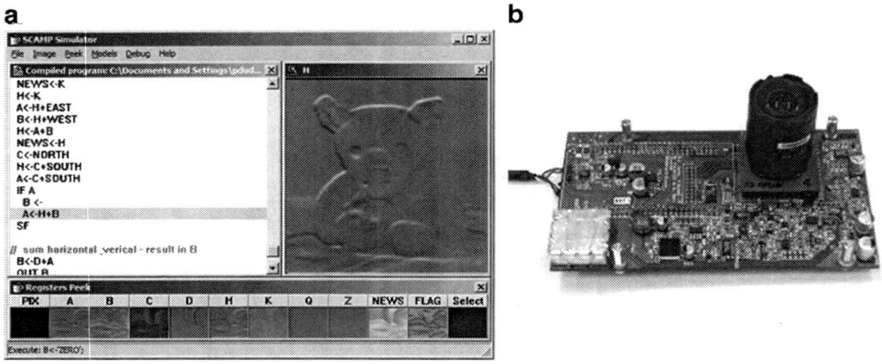


**Fig. 19** SCAMP-3 development environment: (**a**) simulator software, (**b**) hardware development kit

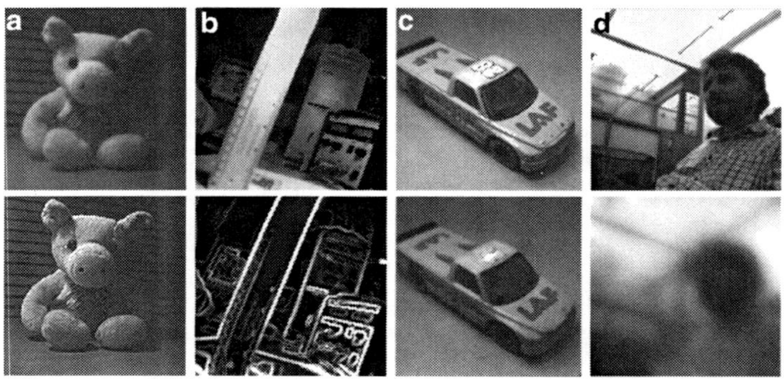


**Fig. 20** Image processing algorithms executed on SCAMP-3. *Top*: original image, *Bottom*: processing result; (**a**) Sharpening filter, (**b**) Sobel edge detection, (**c**) Median filter, (**d**) Blur (diffusion)

shown in Fig. 20. Linear and non-linear filters presented in that figure require between 17 and 157 program instructions (with the exception of blur, which uses a resistive grid scheme and only requires two instructions), with maximum execution speed of 0.8 μs per instruction (higher instruction rates are possible, but with increased errors [33]).

More complex algorithms, for example adaptive thresholding, in-pixel A/D conversion and wide dynamic-range sensing [32], various cellular automata models [34], skeletonisation, object detection and counting, can also be executed at video frame rates. An example of image segmentation via active contours (using a Pixel Level Snakes algorithm [35]) is shown in Fig. 21. In this example, all processing is done on the vision chip. For illustration purposes, the frames shown are the gray-level images that are read-out from the device, where the results are superimposed on the input images on chip, although in practical applications, only the extracted information (i.e. binary contours in this case) would be read-out. We have considered

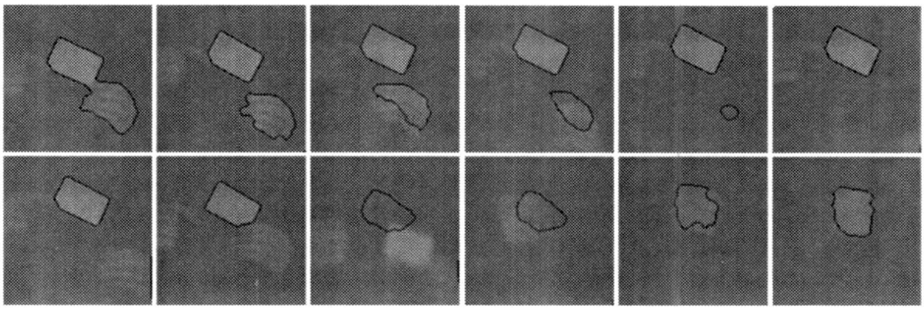

**Fig. 21** Results of executing the Pixel level snakes algorithm on SCAMP-3. The contours evolve shifting at most 1 pixel at a time. Every 30th frame of the video sequence is shown

the application of the chip in several more complex tasks such as video compression [26], comparisons using wave distance metric [36], angiography image segmentation [37], binocular disparity [38] and neural network models [39].

## 6 Conclusions

This chapter presented the design of a SCAMP-3 pixel-parallel vision chip. The chip operates in SIMD mode, and the PEs are implemented using analogue current-mode scheme. This combination of digital architecture and analogue PE implementation provides flexibility, versatility and ease of programming, coupled with sufficient accuracy, high performance, low-power consumption and low cost of the implementation. The main applications of this technology are in power-sensitive applications such as surveillance and security, autonomous robots, toys, as well as other applications requiring relatively high computing performance in machine vision tasks, at low power consumption and low cost. The $128 \times 128$ prototype has been fabricated in 0.35 μm CMOS technology, and reliably executes a range of image processing algorithms. The designed processor-per-pixel core can be easily integrated with read-out/interfacing and control circuitry, and a microcontroller IP core, for a complete system on a chip solution for embedded applications.

**Acknowledgement** This work has been supported by the EPSRC; grant numbers: EP/D503213 and EP/D029759. The author thanks Dr Stephen Carey and Dr David Barr for their contributions to testing and system development for the SCAMP-3 device.

## References

1. A. Moini, Vision chips, Kluwer, Boston, 2000
2. A.A. Stocker, Analog integrated 2-D optical flow sensor, Analog Integrated Circuits and Signal Processing, vol 46(2), pp 121–138, Springer, Heidelberg, February 2006

3. V. Gruev and R. Etienne-Cummings, Implementation of steerable spatiotemporal image filters on the focal plane, IEEE Transactions on Circuits and Systems II: Analog and Digital Signal Processing, 49(4), 233–244, April 2002
4. R.P. Kleihorst et al., X et al: A low-power high-performance smart camera processor, IEEE International Symposium on Circuits and Systems, ISCAS 2001, May 2001
5. L. Lindgren et al., A multiresolution 100-GOPS 4-Gpixels/s programmable smart vision sensor for multisense imaging. IEEE Journal of Solid-State Circuits, 40(6), 1350–1359, 2005
6. S. Kyo and S. Okazaki, IMAPCAR: A 100 GOPS in-vehicle vision processor based on 128 Ring connected four-way VLIW processing element, Journal of Signal Processing Systems, doi:10.1007/s11265-008-0297-0, Springer, Heidelberg, November 2008
7. J.E. Eklund, C. Svensson, and A. Åström, VLSI implementation of a focal plane image processor – A realisation of the near-sensor image processing concept, IEEE Transactions on Very Large Scale Integration (VLSI) Systems, vol 4(3), pp 322–335, September 1996
8. F. Paillet, D. Mercier, and T.M. Bernard, Making the most of $15k\lambda^2$ silicon area for a digital retina, Proc. SPIE, vol 3410, Advanced Focal Plane Arrays and Electronic Cameras, AFPAEC'98, 1998
9. M. Ishikawa, K. Ogawa, T. Komuro, and I. Ishii, A CMOS vision chip with SIMD processing element array for 1ms image processing, Proc. International Solid State Circuits Conference, ISSCC'99, TP 12.2, 1999
10. J.C. Gealow and C.G. Sodini, A pixel-parallel image processor using logic pitch-matched to dynamic memory, IEEE Journal of Solid-State Circuits, 34(6), 831–839, June 1999
11. M. Wei et al., A programmable SIMD vision chip for real-time vision applications. IEEE Journal of Solid-State Circuits, 43(6), 1470–1479, 2008
12. A. Lopich and P. Dudek, ASPA: Focal plane digital processor array with asynchronous processing capabilities, IEEE International Symposium on Circuits and Systems, ISCAS 2008, pp 1592–1596, May 2008
13. A. Lopich and P. Dudek, An $80 \times 80$ general-purpose digital vision chip in $0.18\,\mu m$ CMOS technology, IEEE International Symposium on Circuits and Systems, ISCAS 2010, pp 4257–4260, May 2010
14. P. Dudek, A. Lopich, and V. Gruev, A pixel-parallel cellular processor array in a stacked three-layer 3D silicon-on-insulator technology, European Conference on Circuit Theory and Design, ECCTD 2009, pp 193–197, August 2009
15. A. Dupret, J.O. Klein, and A. Nshare, A DSP-like analogue processing unit for smart image sensors, International Journal of Circuit Theory and Applications, 30, 595–609, 2002
16. G. Liñán, S. Espejo, R. Domínguez-Castro, and A. Rodríguez-Vázquez, Architectural and basic circuit considerations for a flexible $128 \times 128$ mixed-signal SIMD vision chip, Analog Integrated Circuit and Signal Processing, vol 33, pp 179–190, 2002
17. M. Laiho, J. Poikonen, P. Virta, and A. Paasio, A $64 \times 64$ cell mixed-mode array processor prototyping system, Cellular Neural Networks and Their Applications, 2008. CNNA 2008, July 2008
18. J. von Neumann, A system of 29 states with a general transition rule, A.W. Burks (Ed.), Theory of Self-reproducing Automata, University of Illinois, IL, 1966
19. S.H. Unger, A computer oriented to spatial problems, Proc. IRE, vol 46, pp 1744–1750, 1958
20. G.H. Barnes, R.M. Brown, M. Kato, D.J. Kuck, D.L. Slotnick, R.A. Stokes, The ILLIAC IV computer, IEEE Transactions on Computers, 17(8), 746–757, August 1968
21. M.J.B. Duff, Review of the CLIP image processing system, Proc. National Computer Conference, pp 1055–1060, 1978
22. S.F. Reddaway, The DAP approach, Infotech State of the Art Report on Supercomputers, 2, 309–329, 1979
23. K.E. Batcher, Design of a massively parallel processor, IEEE Transactions on Computers, 29(9), 837–840, September 1980
24. D. Hillis, The connection machine, MIT, Cambridge, MA, 1985
25. P. Dudek, A flexible global readout architecture for an analogue SIMD vision chip, IEEE International Symposium on Circuits and Systems, ISCAS 2003, Bangkok, Thailand, vol III, pp 782–785, May 2003

26. D.R.W. Barr, S.J. Carey, A. Lopich, and P. Dudek, A control system for a cellular processor array, IEEE International Workshop on Cellular Neural Networks and their Applications, CNNA 2006, Istanbul, pp 176–181, August 2006
27. P. Dudek and P.J. Hicks, A CMOS general-purpose sampled-data analogue processing element, IEEE Transactions on Circuits and Systems – II: Analog and Digital Signal Processing, 47(5), 467–473, May 2000
28. P. Dudek and P.J. Hicks, A general-purpose processor-per-pixel analog SIMD vision chip, IEEE Transactions on Circuits and Systems – I, 52(1), 13–20, January 2005
29. C. Toumazou, J.B. Hughes, and N.C. Battersby (Eds.), Switched-currents: An analogue technique for digital technology, Peter Peregrinus, London, 1993
30. J.B. Hughes and K.W. Moulding, $S^2I$: A switched-current technique for high performance, Electronics Letters, 29(16), 1400–1401, August 1993
31. J.-S. Wang and C.-L. Wey, Accurate CMOS switched-current divider circuits, Proc. ISCAS'98, vol I, pp 53–56, May 1998
32. P. Dudek, Adaptive sensing and image processing with a general-purpose pixel-parallel sensor/processor array integrated circuit, International Workshop on Computer Architectures for Machine Perception and Sensing, CAMPS 2006, pp 18–23, September 2006
33. P. Dudek and S.J. Carey, A general-purpose 128 × 128 SIMD processor array with integrated image sensor, Electronics Letters, 42(12), 678–679, June 2006
34. M. Huelse, D.R.W. Barr, and P. Dudek, Cellular automata and non-static image processing for embodied robot systems on a massively parallel processor array, Automata-2008, Theory and Applications of Cellular Automata, pp 504–513, Luniver Press, 2008
35. P. Dudek and D.L. Vilarino, A cellular active contours algorithm based on region evolution, IEEE International Workshop on Cellular Neural Networks and their Applications, CNNA 2006, pp 269–274, Istanbul, August 2006
36. D. Hillier and P. Dudek, Implementing the grayscale wave metric on a cellular array processor chip, IEEE Workshop on Cellular Neural Networks and their Applications, CNNA 2008, pp 120–124, July 2008
37. C. Alonso-Montes, D.L. Vilariño, P. Dudek, and M.G. Penedo, Fast retinal vessel tree extraction: A pixel parallel approach, International Journal of Circuit Theory and Applications, 36(5–6), 641–651, July–September 2008
38. S. Mandal, B. Shi, and P. Dudek, Binocular disparity calculation on a massively-parallel analog vision processor, IEEE Workshop on Cellular Nanoscale Networks and Applications, CNNA 2010, Berkeley, pp 285–289, February 2010
39. D.R.W. Barr, P. Dudek, J. Chambers, and K. Gurney, Implementation of multi-layer leaky integrator networks on a cellular processor array, International Joint Conference on Neural Networks, IJCNN 2007, Orlando, FL, August 2007

# MIPA4k: Mixed-Mode Cellular Processor Array

**Mika Laiho, Jonne Poikonen, and Ari Paasio**

**Abstract** This chapter describes MIPA4k, a $64 \times 64$ cell mixed-mode image processor array chip. Each cell includes an image sensor, A/D/A conversion, embedded digital and analog memories, and hardware-optimized grey-scale and binary processing cores. We describe the architecture of the processor cell, go through the different functional blocks and explore its processing capabilities. The processing capabilities of the cells include programmable space-dependent neighbourhood connections, ranked-order filtering, rank identification and anisotropic resistive filtering. For example, asynchronous analog morphological reconstruction operation can be performed with MIPA4k. The image sensor has an option for locally adaptive exposure time. Also, the peripheral circuitry can highlight windows of activation, and pattern matching can be performed on these regions of interest (ROI) with the aid of parallel write operation to the active window. As the processing capabilities are complemented with global OR and global sum operations, MIPA4k is an effective tool for high-speed image analysis.

## 1 Introduction

The Cellular Non-linear Network Universal Machine (CNN-UM) structure proposed in [1] offers a computing paradigm that can be used to describe spatial–temporal interaction of locally connected processing elements (cells). The local nature of the interconnections makes the CNN architecture inherently suitable for integrated hardware realization. The key idea in the CNN-UM model is performing "analogic" computation: sequences of analog and logic operations can be performed locally and the results can be combined using local intermediate result storage. This makes the CNN-UM hardware potentially very effective in terms of real-time computation of complex algorithms.

M. Laiho (✉)
Microelectronics laboratory, University of Turku, Turku, Finland
e-mail: mlaiho@utu.fi

Á. Zarándy (ed.), *Focal-Plane Sensor-Processor Chips*,
DOI 10.1007/978-1-4419-6475-5_3, © Springer Science+Business Media, LLC 2011

Even though the CNN model can compactly describe many locally active processing tasks, hardware realization has turned out to be difficult. Fully digital hardware realizations suffer from large area and high power consumption, whereas analog implementations struggle with accuracy issues that also lead to large cell areas. The ACE16k CNN chip [2] is an example of an analog implementation that remains rather faithful to the original CNN theory. Analog multipliers are used for programmable neighbourhood couplings, and the circuit operates in continuous time. The main challenge of this "orthodox" CNN approach is the difficulty in effectively performing robust, fully programmable analog neighbourhood operations and binary operations with the same analog hardware. Using this scheme, the binary neighbourhood operations tend to be slow and inefficient in terms of energy usage.

Because of the difficulties in implementation, different processors that are inspired by the CNN theory (or just do CNN-like operations in a different fashion) have been proposed. An example are the SCAMP chips [3, 4], in which, instead of having analog multipliers for couplings, neighbourhood contributions are gathered by time-multiplexing different directions and combining the results. This way, a compact cell structure can be obtained. The downside with the time-multiplexed approach is that continuous time operation is not possible, which affects the speed.

The approach taken in the MIPA4k chip discussed in this chapter is to separate the processing tasks into different categories and to optimize the circuitry for each of these categories. Using task-specific processing cores with limited programmability, in contrast to full analog programmability of the couplings in the ACE16k, has turned out to be effective. For example, in a typical binary image processing operation, the corresponding multiplications can be made very inaccurately, whereas in grey-scale image processing tasks, a much higher accuracy (larger multiplier area) is required. Note that a binary operation denotes here an operation with black/white (BW) images. Furthermore, a larger multiplier yields also slower processing speed. Now, consider that both binary and grey-scale operations be performed with high accuracy analog multipliers, and further consider that most of the processing tasks in the algorithm are of BW nature. In that case, the total speed of performing the algorithm with a general purpose analogue computational core is much slower compared to executing the algorithm with processing cores dedicated for each particular task.

The difficulty is in selecting an effective combination of processing cores and choosing a proper level of programmability, while optimizing area usage. The Mixed-Mode Processor Array (MIPA) architecture has been introduced to test these ideas. The concept is based on digital programming, combined with highly optimized analog and digital (mixed-mode) cell level processing hardware for implementing specific but widely applicable operations. The approach mitigates robustness problems compared to traditional universal analog CNN circuit models. On the other hand, the circuit-level implementation of the MIPA cell realizes a higher degree of functional parallelism than, e.g., the very versatile and robust SCAMP or ASPA [5, 6] architectures. MIPA is capable of asynchronous analog (grey-scale) information propagation through the cell array, which is one of the great potential advantages of the CNN model. Such operations are effective in tasks such as segmentation. The set of functionalities that has been targeted in the MIPA4k cell and array design has been inspired by the examination of [7,8], where projected and desirable capabilities for different generations of CNN-type processor arrays were examined and specified. According to the classification in [8], the MIPA4k processor can be functionally seen to be roughly in the CNN class of 5b. This means, e.g., that non-linear templates and space-variant programming for template plasticity are possible. Although the MIPA architecture does not offer the same level of theoretical universality as a fully programmable CNN-UM, the approach with dedicated cores and carefully considered programmability makes it highly effective in a broad range of processing tasks.

This chapter presents the MIPA mixed-mode array processor architecture and overviews the MIPA4k prototype chip. The concept is based on the architecture proposed in [9, 10] and the chip has been introduced in [11]. As many parts of the circuit have been described in detail elsewhere, the aim here is to familiarize the reader with the general concept and provide understanding to design approach. More information on the binary processing part can be found in [12] (programming scheme), [13] (space-dependent templates, associative search), [14] (wave processing) and [15] (peripheral circuits). Detailed information on the grey-scale circuits can be found in [16] (locally adaptive sensing), [17–20] (rank filter/identification), and [20–23] (fuzzy/anisotropic filtering).

## 2 MIPA4k Cell Architecture

The MIPA4k processor cell consists of different processing *cores*, as proposed in [10]. The idea is to implement a selected set of useful low-level image processing operations with very efficient, functionally dedicated hardware resources, i.e., *cores*. The cores are optimized, yet programmable hardware components for different fundamental operations, such as non-linear grey-scale filtering or binary (BW) CNN-type operations. In practice, especially in the case of grey-scale processing circuitry, some of the large analog transistors have been shared among the different cores to save silicon area. This is made rather straightforward by adopting a current-mode computation approach; the reconfiguration of current-mirror structures can be done simply with digitally controlled switches. The configuration of the cell for different functionalities is actually a large part of the programming of the processor cell.

The cell architecture of the MIPA4k is illustrated in Fig. 1. In each cell of the MIPA4k array, the processing cores have been combined with a photodiode, in-cell A/D/A converter circuitry, a global digital I/O bus, multiple in-cell analog and digital memories and the global sum/OR operation. The lines in Fig. 1 illustrate the analog and digital (binary) connectivity between different hardware blocks within the cell. For example, the contents of the digital memories are accessible by both grey-scale and binary cores.

The MIPA4k cells are simultaneously connected to 4 local neighbours (N,E,S,W) in grey-scale operations and 8-connected to the local neighbourhood in BW

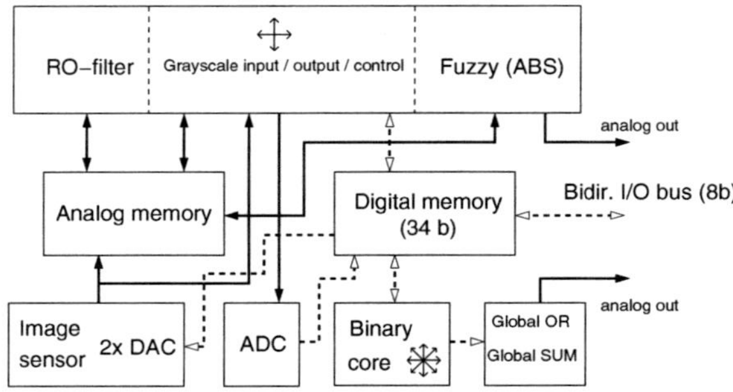


**Fig. 1** Cell architecture with the main functional blocks and communication with the blocks illustrated. The *continuous lines* are analog current mode signals and the *dashed lines* represent binary signals

processing. The 4-connected local neighbourhood was chosen for the grey-scale operations, since it allows information propagation in all directions, but saves greatly in analog device area compared to an 8-connected neighbourhood. Increasing the number of cell inputs would not only linearly increase the number of analog transistors in the grey-scale cores, but would also require better accuracy (larger area) from the circuitry to guarantee sufficient robustness. In the BW core, due to the small size of the circuitry per input, the inclusion of the full first neighbourhood is not a problem.

## 2.1 *Photosensors and Digital I/O*

Each cell of the MIPA4k processor array contains a simple integrating photodiode sensor. The sensor circuitry is illustrated in Fig. 2. The sensors are reset globally with PMOS transistors, after which the diode voltage is integrated and converted with the transistor $M_{01}$ into a current input signal for the processing circuitry within the cell. Alternatively, the same sensor output can be taken directly into an in-cell ADC (switches $S_1$ and $S_2$). The second parallel current output through $M_{02}$ in the sensor readout circuitry enables the processing of the input image during integration, while at the same time, retaining an unaffected version of the image.

To facilitate easy and robust testing of the MIPA4k prototype, two separate binary weighted DACs have been included in each cell for providing input currents to the grey-scale circuitry. The nominal resolution of the DACs is 7 bits. The cell also includes a 7-bit successive approximation ADC, based on a third similar DAC. The global control signals for the in-cell ADCs are generated asynchronously on-chip, i.e., no high-speed external clock is required for controlling the ADCs. In a more practical cell implementation, it could be possible to replace the three DACs with a single D/A converter, which would also be used as a part of the SAR ADC. In the MIPA4k prototype, multiple separate converters were seen as the safest choice, to make testing easier and to take into account the possibility that the analog current memories would not work properly.

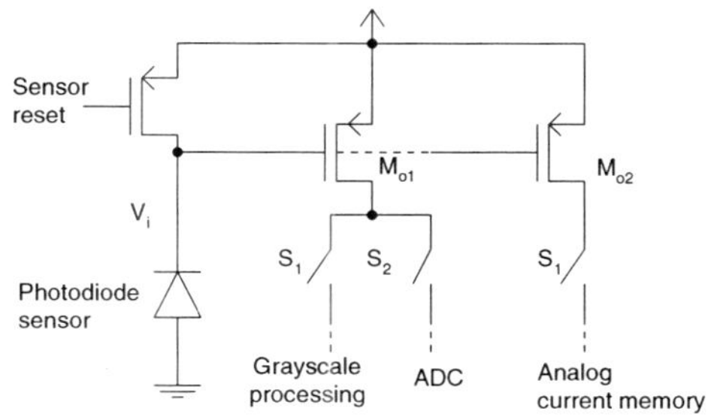


**Fig. 2** Sensor setup

Digital input/output to the array cells is implemented through a global bidirectional 8-bit bus; 7-bit grey-scale data and one binary image can be transferred through the bus to/from the cell. The contents of the ADC output register can be written into the input DAC registers and the contents of any of the additional 13 static digital memories for storing binary data can be interchanged with the help of in-cell read/write memory decoding. These transfers take place simultaneously for all cells.

## *2.2 Local Adaptivity*

Local adaptivity or plasticity is possible for both grey-scale and binary cores in the MIPA4k. The active grey-scale input directions, i.e., the neighbourhood connectivity to both the RO-filter and the fuzzy block can be selected individually. The control bits for the input directions can be set globally for the whole array or read from local binary memories. This means that the active inputs can be selected separately for each cell, e.g., through binary results from a previous operation or by writing different input mappings to the cells externally through the global I/O bus. Grey-scale bias currents used for threshold operations or for ranked-order filter programming can also be set locally in each cell by using analog current memories.

The binary input coefficients for the BW core can be controlled globally or stored in the in-cell memories, and therefore they can be set individually for each cell of the array. Similarly, the local bias value in the BW processing can be set individually for each cell. This can also be done based on previous processing results, since the contents of the digital registers can be interchanged.

## *2.3 Chip Implementation*

Figure 3 shows a chip microphotograph. The $64 \times 64$ processor array has been implemented on a $0.13\,\mu$m standard digital CMOS technology with six metal layers. The size of the chip is approximately $5.1 \times 4.5\,\text{mm}^2$ and the size of a single

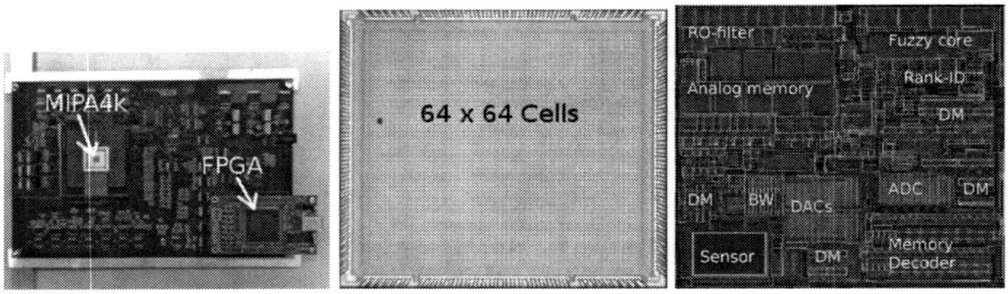


**Fig. 3** Prototype board, chip microphotograph and cell layout

**Table 1** Areas of the main building blocks of a MIPA4k cell (% of cell total area)

| Block name | Cell area (%) |
|---|---|
| Analog memory | 17 |
| Decoder | 10 |
| Fuzzy core | 9 |
| ADC | 9 |
| RO-filter | 9 |
| Digital memory | 9 |
| Sensor | 7 |
| Rank ID | 6 |
| 2 × DAC | 5 |
| BW | 3 |
| Others | 16 |

processor cell is $72 \times 61\,\mu m^2$, containing approximately 1,500 transistors. A large part of these are small digital devices, in essence switches and simple logic, which are used for configuring (programming) the cell circuitry for different grey-scale and BW operations. The major cell sections corresponding to different functionalities have been indicated in the cell layout of Fig. 3. Digital memory (DM) was distributed into different parts of the cell layout after placing the other functional blocks, for the sake of efficient area usage.

The chip is bonded on a PGA256 package; approximately 170 of the package pins are used for digital control and I/O signals. The prototyping board for the chip contains voltage regulators for creating the power supply voltages for the chip (the chip uses two separate power supply voltages: 1.3 V and 0.95 V) and FPGA, a set of registers for storing control signals off-chip, as well as some data converters for creating nine bias currents for the chip with a fully digital board-level control scheme. A XILINX Spartan FPGA is used to control the chip and to relay data to/from a computer via USB2 interface.

Table 1 shows the approximate areas of the main building blocks of a cell (in percents of total cell area). It can be observed that analog memories are the biggest individual area consumer. The others in the table refers to the rest of the building blocks, wiring overhead and spacing between blocks.

## 3 Grey-scale Processing Cores

The optimized circuitry for grey-scale processing within the MIPA4k cell is designed to implement a selected set of efficient non-linear filtering operations for low-level image data reduction, e.g., noise reduction and grey-scale image segmentation, in a fully parallel (analog) manner. The grey-scale processing cores in the cell are (1) a fully parallel, programmable 5-input ranked-order (RO) filter, and (2) a "fuzzy" processing block based on absolute value extraction. The input transistors providing neighbourhood connectivity for grey-scale operations, as well as the grey-scale output section, are shared between the two cores. Therefore, RO-filter and absolute value operations cannot be performed simultaneously. The local input to the grey-scale cores can be brought directly from the in-cell photodiode, from one of the in-cell DACs or from an analog current memory, as shown in Fig. 1.

Grey-scale processing, which requires more than one basic operation, can be performed iteratively by storing intermediate results into the analog current memories. Also, some simple arithmetic operations, such as sum, difference and limited multiplication/division, can be performed on the processed grey-scale values by using a combination of multiple current memories (e.g., write result into two parallel memories and read out only one). The analog current memories are simple sampled-current circuits, implemented as either P- or N-type devices. A common current-mode memory bus enables access to/from different N- or P-type analog current mirror inputs/outputs in the grey-scale circuitry, as well as sum/difference extraction or transfers between separate memories. There are altogether ten analog current memories in the cell, five of which are P-type and five N-type. The current memory circuitry and the related capacitors use 17% of the MIPA4k cell area.

The grey-scale output current resulting from the cell operations can be compared to a (locally or globally defined) threshold value with a simple current comparator, resulting in a binary output signal, or stored into a current memory. Alternatively, the analog current value can be converted into digital form with the SAR ADC and read out of the cell through the global binary bus. For testing purposes, the current can be directed to a global current output node for external measurement.

### 3.1 Ranked-Order Filtering

The first of the MIPA4k's grey-scale processing cores consists of a programmable ranked-order filter circuit and associated circuitry. A ranked-order filter (also known as *rank* order filter) is a circuit, that selects from a set of inputs the one with a predetermined rank. i.e., the largest, second largest,..., smallest. This *order statistic* operation can be used as a basis for many efficient non-linear image processing operations, such as all basic operations of grey-scale mathematical morphology or median filtering. Complex combined morphology operations can be implemented with in-cell storage of intermediate values.

Order statistic operations, ranked-order filtering in the most general case, can be implemented according to the CNN model by using difference controlled non-linear templates [7,24,25]. For example, a multiple-step approach for implementing minimum and maximum extraction for grey-scale morphology, based on the use of a non-linear $A$ template with a transient mask, was proposed in [25]. However, this method cannot be directly extended to the extraction of any programmable rank.

Order statistic filtering can be defined without any sorting operations, in a manner which can be directly implemented with analog circuitry. The signal with the desired rank, i.e., the $k$th largest element $a_k$ among $N$ inputs, is found by solving the equation

$$\sum_{i=1}^{N} H(a_i - y) = b, \tag{1}$$

where the function $H(x)$ is a step function given by

$$H(x) = \begin{cases} 1 & (x > 0) \\ 0 \sim 1 & (x = 0) \\ 0 & (x < 0). \end{cases} \tag{2}$$

The bias term $b$ value for extracting rank $k$ is $k-1 < b < k$, which is not dependent on the number of filter inputs.

Equation (1) can be directly solved by using a parallel combination of analog circuits, which carry out the non-linear operations of (2). The analog ranked-order filter circuitry will simply settle into an equilibrium condition, fulfilling the rank equations above, without any iteration or external control in addition to correct biasing. Compared to massively parallel iterative sorting, this is a very efficient approach, in terms of performance and/or hardware complexity. The very simple 5-input current-mode ranked-order filter circuit employed in the MIPA4k, which has been described and analysed in detail in [17,20], together with the input/output PMOS current mirrors for grey-scale connectivity, takes up less than 10% of the cell area.

The required condition defined above for the bias term, for extracting rank $k$, leads to robust programming of the ranked-order filter circuitry, as long as $k$ is not very large. The circuit allows a comfortable margin for variation (due to device mismatch) in the bias term [20]. Figure 4 shows an example of different ranked-order filtering operations performed on the MIPA4k.

### 3.1.1 Rank Identification

In addition to "copying" the correct-ranked input current to the cell output, the RO-filter also includes rank identification circuitry, which can be used to point out the input direction corresponding to the correct rank. Being able to also identify the input with the desired rank can considerably extend the range of possible functionalities, e.g., by enabling asynchronous grey-scale maximum value propagation for morphological reconstruction. The capability of identifying the input with the

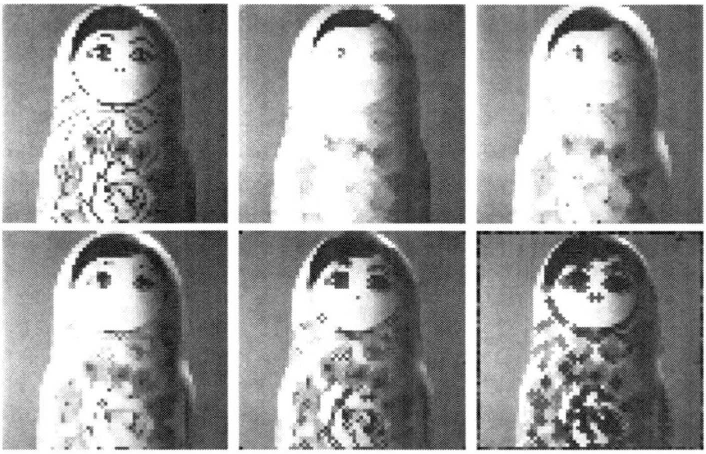

**Fig. 4** Ranked-order filtering with the MIPA4k. From *left* to *right*: Original input image from the photosensors, maximum, second largest, median, second smallest, minimum

correct rank can also be used in the normal ranked-order extraction operation. By identifying the correct input source, the input current or voltage can be redirected through a switch, thus reducing the effects of device mismatch compared to copying the values [26].

The ranked-order filter in MIPA4k has been extended to realize the full identification of any programmed rank. The identification circuit takes up approximately 5% of the MIPA4k cell area. The operating principle of the rank identification circuitry is described in detail, together with its application in asynchronous grey-scale morphology, in [18–20].

## 3.2 Asynchronous Grey-scale Morphology

MIPA4k can perform operations that require asynchronous propagation of grey-scale (analog) values. Analog implementation is preferred here since a digital system would need very high processing speed for matching performance. Even analog iterative processing, using the intermediate result of each processing step as the input of the next iteration, is much less efficient compared to an asynchronous system. An asynchronous implementation works so that once the propagation has been initiated, no further controls are needed. The downside with asynchronously propagating analog operations is the inherent inaccuracy of an analog implementation. In the MIPA4k, the grey-scale propagation is implemented in a way that makes the realization very robust against circuit imperfections.

Asynchronously propagating morphological grey-scale reconstruction with two input images (*marker* and *mask*) is an example of a globally propagating grey-scale operation. The first known analog implementation of this operation was presented in detail in [19], and the same type of circuitry, although with some improvements, is also included in the MIPA4k processor cell.

Morphological reconstruction is a constrained propagating dilation or erosion operation, which can be used, e.g., for image segmentation [27]. Reconstruction requires two input images, namely a *Marker* and a *Mask*. In a dilation-based reconstruction (R), the Marker image ($MR$) is consecutively dilated, and the result of each dilation step is compared to the corresponding pixel value in the Mask image ($MS$). The minimum between the two values for each pixel location is selected as the output, that is

$$R(MR, MS) = \min\{D(MR, S), MS\} \text{ until stability,} \tag{3}$$

where $S$ is the structuring element defined by the neighbourhood connectivity, and $D$ refers to the dilation operation. The Marker image for the reconstruction can be created from an initial input image e.g., by eroding it a number of times, or a Marker obtained by some other means can be used. The Mask is typically the original uneroded input image from pixel sensors, local current memory or in-pixel DACs. An ideal reconstruction operation converges into a stable state, where further steps do not change the output. Pixel values in the resulting image cannot reach intensities above a regional maximum in the marker image, or the intensity of the same pixel in the mask image.

The implementation of asynchronous reconstruction on the MIPA4k, illustrated in a simplified manner in Fig. 5, is based on controlled gate voltage diffusion in a ranked-order filter network. The RO-filter in each cell simultaneously both extracts the maximum value within the local cell neighbourhood and identifies its input direction. If the neighbourhood maximum is smaller than the local mask value, the corresponding input will be connected to the output of the cell. This output is conveyed to the inputs of the neighbouring cells, thus propagating the maximum value within the cell network. When the Mask value is smaller than the extracted local maximum, it will be connected to the cell output. Because the method is implemented via gate voltage diffusion into a capacitive load, it is not subject to detrimental positive feedback effects, which could saturate the grey-scale output. A detailed description on the asynchronous reconstruction method and circuitry can be found in [19].

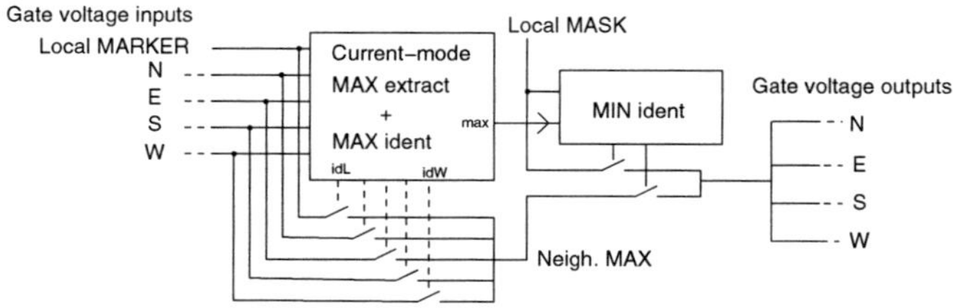


**Fig. 5** Implementation principle of asynchronous grey-scale reconstruction on the MIPA4k

**Fig. 6** From *left* to *right*: Original input image. Threshold result on the original input. Eroded and reconstructed input image. Threshold result of the reconstructed image

The asynchronously propagating reconstruction can be applied in the grey-scale pre-processing phase of an object tracking algorithm to help in the extraction of binary objects. By reconstructing flat grey-scale areas within the image, based on a suitable Marker image, the image contents can be more easily segmented into prominent features and objects. This can help to robustly extract image regions and, e.g., reduce noise caused by unimportant details. Figure 6 shows an example of a grey-scale morphological reconstruction for image segmentation, performed with the MIPA4k chip. The original image, captured with the in-cell photodiodes, was first successively eroded five times, after which the result of the erosion operations was applied to the cell input as the Marker image. The original image, kept in a current memory during the erosions, was used as the Mask for the reconstruction. The figure also shows BW images resulting from a threshold operation on the original and reconstructed image. The threshold value was manually selected in both cases in such a way that the head area in the foreground was completely extracted. It can be clearly seen that the grey-scale segmentation helps in better extracting the large-scale foreground objects. With simple thresholding of the sensor output, significant spurious detail remains. Raising the threshold value would also remove most of the head area in the image.

## *3.3 Absolute Value-Based Operations*

The "fuzzy" processing block gets its name from the fundamental circuit component in the block, which is an absolute value (ABS) circuit, in this case a simple full-wave current rectifier [28]. The process of fuzzification, i.e., generating a *fuzzy membership function* based on the pixel's neighbourhood values, relies heavily on absolute value extraction. The MIPA4k cell includes five absolute value circuits. Four of these can be used to extract the absolute values of the difference between the local cell value and the neighbouring cells. The fifth ABS circuit is connected to the analog current memory bus. The fuzzy block also includes some additional digitally reconfigurable current mirror circuitry for processing the extracted current-mode absolute value signals.

The inclusion of fully parallel in-cell fuzzy processing circuitry within an array processor cell was examined in [22]. This approach was not directly implemented in the MIPA4k array cell. However, as the fundamental absolute value circuitry is included, fuzzy membership operations can be approximated in a serial fashion with the help of current memories. The absolute values of the neighbour differences can be used as grey-scale signals or compared to a globally or locally set threshold to create a binary comparison result. One of the key operations based on absolute value extraction in the MIPA4k is *non-linear anisotropic diffusion filtering*, which is described here in more detail. The fuzzy core can be used to effectively implement, e.g., edge-preserving non-linear filtering, direct edge detection, area-based segmentation or locally adaptive grey-scale operations.

### 3.3.1 Anisotropic Diffusion

One of the most important low-level image processing operations is the filtering of noise from a grey-scale image. This is useful in image segmentation, enhancing the most prominent features and removing less important local small-scale details. A coarser resolution image can be created by applying detail-reducing filtering, which removes high-resolution artifacts, such as texture and noise, while preserving larger-scale image objects. In terms of image segmentation, some of the most important features are usually the edges defining the boundaries of different large-scale image regions.

Although noisy high-resolution features are suppressed with the popular linear isotropic diffusion, e.g., a Gaussian kernel, the downside is that it does not differentiate a large intensity difference, i.e., an edge, from an essentially flat image region containing only noise or texture components. Smoothing blurs the regional edges, which should be preserved for effective segmentation. The edges can also be dislocated by filtering, because regions with very different intensity levels influence each other through diffusion, without respect to an edge separating the regions [29].

To overcome these difficulties, *non-linear anisotropic diffusion* [29], and other functionally equivalent approaches have been proposed under different names [30]. The basic idea of non-linear anisotropic diffusion is to only filter small differences in an image while preserving large changes in intensity. Because large neighbour pixel differences are usually associated with edges, this approach leads to better segmentation, by emphasizing intra-region smoothing over inter-region smoothing. Because the diffusion is prevented at edges in the image, the locations of region boundaries are also preserved. The filtering can also be effectively used for edge detection, since the edge information is available from the filter. A processor array capable of edge detection by resistive fuse operation was proposed in [31]; however, the chip was limited to this particular functionality.

An efficient way of implementing anisotropic diffusion directly within an array processor cell is the *resistive fuse* [21, 32]. Anisotropic diffusion can be characterized (1-D network examples are used for clarity) by defining the interaction between two locally connected neighbouring pixels with intensities $I_n$ and $I_{n+1}$. The intensity of pixel $n$ after the filtering operation can be defined as

$$I'_n = I_n + [D(I_n, I_{n+1}) \cdot g(\Delta I)], \tag{4}$$

where *D(x)* denotes any general diffusion operation between the two pixel values (in this case, a resistive network), $\Delta I = I_{n+1} - I_n$ is the difference in pixel intensities and *g(x)* is an *edge-stopping function* dependent on the local gradient value so that

$$g(x) = \begin{cases} 1, & |\Delta I| \leq I_{\mathrm{TH}} \\ 0, & |\Delta I| > I_{\mathrm{TH}} \end{cases} \tag{5}$$

Therefore, a diffusive connection is present at the boundary between two pixels if the absolute value of the intensity difference is smaller than the threshold value $I_{\mathrm{TH}}$. This response can be implemented with a resistive fuse. It has a finite resistance value while the input difference is less than a certain threshold, and very large resistance when the difference is larger than the threshold [32]. The binary control signals can also be used to directly determine edge locations (unfiltered pixel connections) in the image. In the MIPA4k processor cell, a common threshold value can be applied globally to all cells of the array, or the filtering thresholds can be set individually for each pixel cell.

### 3.3.2 Resistive and Capacitive Anisotropic Filtering

The anisotropic filtering can be configured to be either "resistive" of "capacitive" with somewhat different capabilities. Figure 7 illustrates the different network configurations and their transistor-level implementations.

In the resistive case of Fig. 7a, the resistive network is composed of vertical (diode-connected NMOS transistor) and horizontal resistors (resistive fuse devices). The fuse device is composed of an absolute value circuit and a threshold circuit that controls the horizontal resistor as $R_{\mathrm{F}} = R_{\mathrm{H}}/g(x)$, i.e., sets it into a zero conductance mode when neighbour difference is larger than the set threshold value. The response of the network is determined by the ratio of the vertical and horizontal resistor values. The network settles into an energy minimum, leading to a smoothing operation with a decay rate proportional to $e^{-nR_{\mathrm{H}}/R_{\mathrm{V}}}$, where $n$ is the distance in cells from the input location. The behavior of both vertical and horizontal resistors (that are realized with transistors) is very non-linear, as is discussed in more detail in [20]. However, the effective filtering operation can be made sufficiently effective despite the non-linearity.

Another way to use the same non-linear resistor network is to store the local input into a current memory created with the NMOS current mirror transistor and an additional gate capacitor (MOS) device, as shown in Fig. 7b. Now, during filtering, the horizontal resistors are connected between capacitive nodes. This means that when a diffusive connection is present, the voltages of connected cell nodes

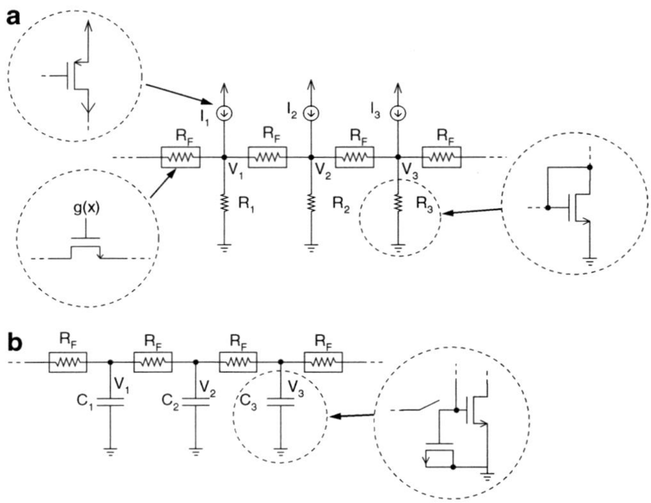


**Fig. 7** Anisotropic filtering with a (**a**) "resistive" cell network, i.e., with both horizontal and vertical resistors. (**b**) A "capacitive" network

will be equalized, leading to a flattening filtering operation. Although very effective segmentation operation can be achieved, this form of filtering is naturally more vulnerable to e.g., noise and leakage effects, since the input values are not actively maintained.

Both ("resistive" or "capacitive") anisotropic filtering operations are possible within the MIPA4k cell, by a simple digitally controlled configuration of the NMOS current mirror device either as a diode connected transistor or a capacitive memory device. In both modes of operation, the two (N/E directions) binary control bits from the resistive fuse circuitry can be written into static in-cell binary memories for determining edge pixel locations.

Figure 8 shows the results of a resistive anisotropic filtering operation on an input image captured with the in-cell photodiodes. Each top/bottom image pair in Fig. 8 is the result of a different resistive fuse threshold bias current, which was controlled globally with an off-chip DAC. Both the resulting filtered image and the edge detection result, extracted from the resistive fuse control signals, are shown. As can be seen, the edge magnitude for the anisotropic filtering can be effectively controlled.

The amount of smoothing realized by the filter when the pixel differences are diffused depends on the common-mode input level of the pixels in a non-linear manner [20]. Control over the horizontal resistor values would allow selecting different smoothing magnitudes, which could be very useful, e.g., for scale-space examination of an image. Unfortunately, this was not implemented in the current MIPA4k chip, but the physical resistor values within the cells are fixed. However, adding this feature would be very simple in a revised version of the cell circuitry.

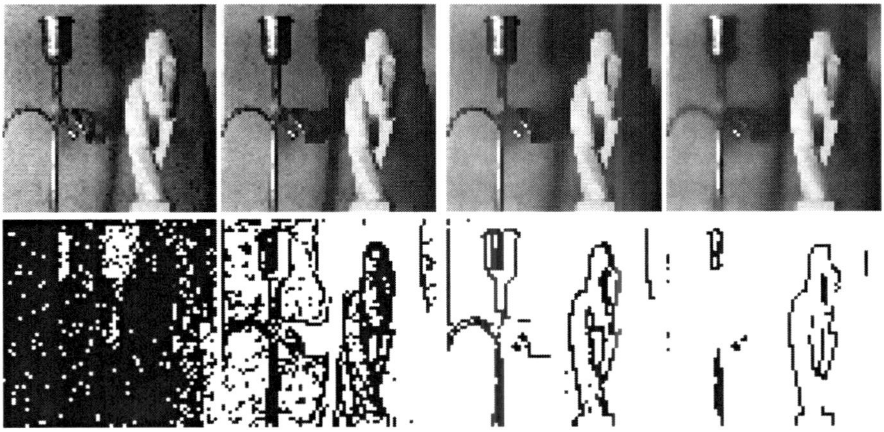

**Fig. 8** Resistive anisotropic filtering with different threshold values (*top row*), and the corresponding edge detection result from the filter circuitry for each threshold (*bottom row*)

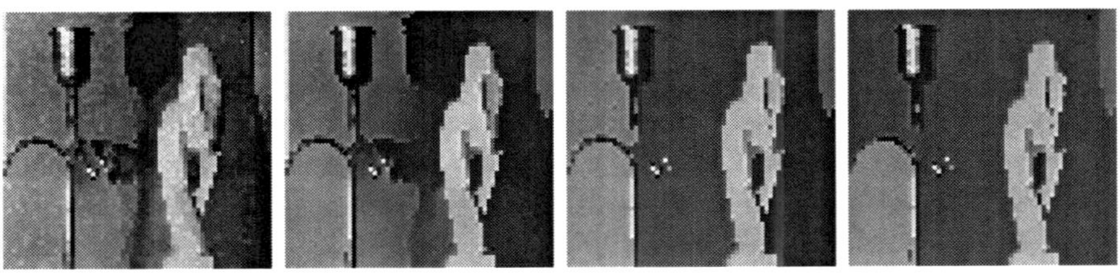

**Fig. 9** Capacitive anisotropic filtering with a single threshold and different diffusion transient lengths of ≈80 ns, ≈1 μs, ≈10 μs, ≈160 μs

Figure 9 shows an example of capacitive anisotropic diffusion. As can be seen, the capacitive operation leads to a much more effective segmentation of the image by completely removing the low-intensity (relative) details and areas, whereas the resistive filtering merely smooths the image. On the other hand, the capacitive operation is much more sensitive to errors in the threshold value. Thus, poorly defined regions (e.g., with a "hole" in the surrounding edge) can be completely flattened with sufficient diffusion time. However, the powerful segmentation provided by the capacitive filtering, realized as a single asynchronous operation, could be very useful when applied correctly (with careful control over the diffusion time). The diffusion transient can be controlled with an accuracy determined by the maximum control signal rate for the chip. In the current MIPA4k prototype board, the clock-rate of the FPGA, which is digitally controlling chip operations, is 25 MHz, i.e., the minimum time step for the delay control is 40 ns.

Although, in all the previous examples, a global threshold value was used for the anisotropic filtering, locally controlled threshold values can also be applied within the MIPA4k array.

## *3.4 Locally Adaptive Image Sensing*

Natural environments are challenging for image capture since the illumination levels can greatly vary within a scene. Because of this, conventional image sensors tend to lose important details from the extreme ends of the dynamic range. Imagers that produce high dynamic range (HDR) images overcome this problem to some extent. A HDR image can be captured by imaging the same scene with different exposure times and combining the results, e.g., by applying predetermined integration time slots of different lengths for the pixels by predicting pixel saturation during integration [33]. A very high dynamic response can also be achieved by using a time-domain (time to threshold) imaging method [34].

A true HDR image may not be optimal for image analysis since the aim is to understand contents of the visual scene, not to archive a photorealistic impression of the scene. Compression of the HDR scene to low dynamic range (LDR), in such a manner that important visual characteristics are maintained, is preferred for image analysis. An effective way to capture a HDR image and compress it into a lower dynamic range version for further use is to set the integration times of individual sensors according to local average illuminance. The integration time can be adapted by using an external processor [35] or the adaptive mechanism can be built into the image sensor [36]. In [36, 37], the local average of the image, obtained from a resistive network, is stored in the pixel and used to control the exposure times of pixels individually. The weakness of the method is that it uses a global time-evolving reference for controlling the local integration and requires separate computation of the local averages from a previous frame. These steps take time and may cause motion lag due to the multi-frame dependency.

### 3.4.1 Implementation of Adaptive Sensing on the MIPA4k

A locally adaptive image sensing approach can be demonstrated on the MIPA4k, where the adaptation is performed automatically during integration, without the need for any global ramps or information from a previous frame. The real-time operation during integration means that the adaptation is free of lag. Therefore, artifacts associated with motion are avoided. The presented *compressive* sensing method is not intended for capturing an actual HDR image, but rather an LDR input image for additional analysis, comprising the important (edge) information from both extremes of an HDR input scene.

The basic idea of the adaptive sensing in MIPA4k is to control the integration time locally for each pixel based on the average intensities in different parts of the scene. When a pixel in the filtered image reaches a preset threshold, the integration is stopped, i.e., the corresponding pixel value will be stored into memory. At this point, the original unfiltered sensor outputs are centered around the threshold value, so that their relative differences, i.e., edge information, will be maintained. Image regions with lower overall intensity will continue integration until their average reaches the threshold or until some maximum integration time. This process results in a captured

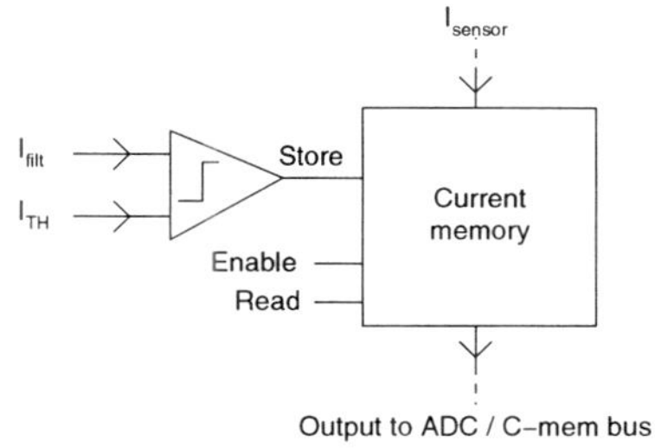


**Fig. 10** Current memory write and readout scheme for the adaptive sensing

image where the inter-scene large-scale dynamics are compressed, but local details are preserved. The basis for the presented adaptive sensor implementation was the examination in [38], which adapted the principles proposed in [39] into a form more suitable for circuit implementation. Thus, parts of the visual scene with lower light intensity will be integrated for a longer time than those with very high light intensity, compressing the scene into LDR. This will help to extract detail from low intensity areas of the scene while preventing information in high intensity areas from being lost due to sensor saturation.

When the integration starts, after globally resetting the photodiodes, the output of the sensor is applied via $M_{01}$ (Fig. 2) into the grey-scale processing circuitry within the cell, which performs a spatially variant non-linear anisotropic filtering operation asynchronously during the integration. Simultaneously, the original sensor values are available through the parallel sensor output $M_{02}$ (Fig. 2) to an in-cell current memory as shown in Fig. 10. The current memory is a simple NMOS transistor with an additional gate capacitor device; the memory transistor is operating as a diode-connected device as long as both *Store* and *Enable* are "1." When the filtered local pixel value exceeds the threshold $I_{\mathrm{TH}}$, the corresponding unprocessed pixel value is stored into the current memory. The integration then continues as long as the global maximum integration time is reached. After this, the remaining pixel values which have not reached the threshold will be stored into the memories by setting *Enable* to "0." Finally, the stored pixel values can be read out into the in-cell ADC by setting *Read* to "1" (and $S_1$ in Fig. 2 to "0").

Figure 11 shows an example of normal and adaptive sensing performed on the MIPA4k focal plane processor array. The scene contained a very bright LED lamp pointed towards the sensor. The upper left-hand image in Fig. 11 shows a regular image capture with a fixed integration time of $t_1 \approx 2.6$ ms. The output pixel values were converted directly from the sensors outputs with the in-cell current-input ADC. The lamp area is effectively a uniform white area, due to sensor saturation in the bright region.

The top middle image in Fig. 11 shows the result of the adaptive image sensing, where detail within the lamp is clearly visible, while also the hand in the foreground can still be perceived. It can be observed that the adaptive sensing tends to reduce the signal range and increase noise. If the limited signal range creates problems for further processing, some of the compression can be compensated for within the MIPA4k cell by using a reduced DAC input range in the in-cell SAR ADC,

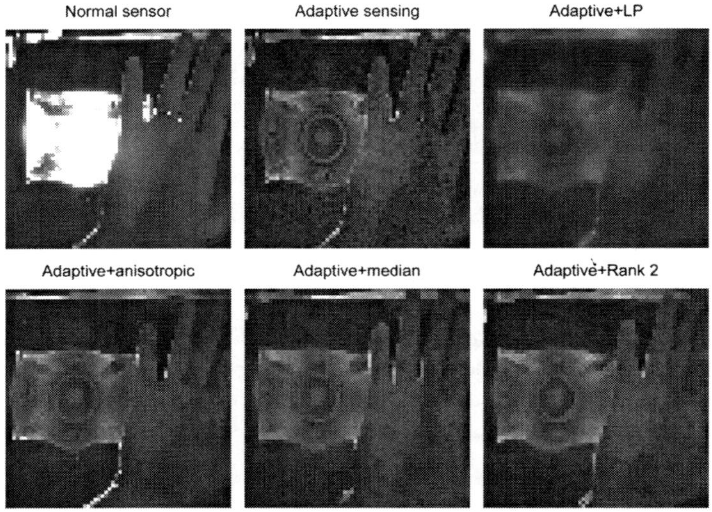


**Fig. 11** Demonstration of adaptive sensor operation. *Top left* subfigure shows an image captured with a fixed integration time $t_1$ of 2.6 ms. The *top middle* subfigure shows an image captured using the adaptive sensing procedure. The other subfigures show filtered versions of the result of the adaptive sensing (lp filter, anisotropic filter, median filter, 2nd largest filter)

expanding the compressed image dynamics closer to the full 7-bit (nominal) output range of the ADC. The outputs of the ADC can be locally written into the inputs of an in-cell DAC and the original analog signal range can be restored when applying the input to the cell for processing via the DAC. Image noise can be reduced with e.g., anisotropic-, mean- or median filtering performed by the processor array in a fully parallel manner. The rest of the images in Fig. 11 show some examples of filtering performed directly on the adaptive sensor output, prior to A/D conversion. The low-pass (LP) filtering was performed with the anisotropic diffusion network, by setting the non-linear threshold to a very large value. Slight differences in the images are due to the fact that the different filtering operations were performed separately, i.e., the hand moved between image captures. The adaptive sensor operation on the MIPA4k is far from optimal; both the photodiode sensors and the locally adaptive control scheme could be easily improved; however, it clearly demonstrates the feasibility of the approach.

## 4 Binary Processing Core

The binary processing core is vitally important for implementing efficient early image analysis operations. Useful image content data (e.g., object size, shape or position) can be effectively determined in the binary domain after necessary greyscale filtering and threshold operations for reducing unimportant image content or for enhancing particular image properties have been carried out. Because binary (BW) domain operations are inherently robust, the circuit implementation can be made significantly more compact than for grey-scale operations. This also yields good power efficiency.

The BW core in MIPA4k is very small compared to the rest of the cell circuitry, occupying approximately only 3% of the total cell area, when the static digital memories are not taken into account. However, the BW core can implement all binary CNN operations with very robust and simple current-mode threshold logic operations using 1-bit programmable weights [12, 40]. Although some CNN templates have to be divided into subtasks, a single iteration can be performed quickly (≈40 ns) and the 1-bit programming is fast (determined ultimately by the global control clock frequency). The binary core can also perform local logic operations during template evaluation and is capable of performing asynchronously and synchronously propagating neighbourhood logic operations, such as binary reconstruction. The binary core has access to 13 static memories for storing intermediate images and can also access binary outputs resulting from the grey-scale cores. In addition, the binary core can be used without neighbourhood inputs to implement local logic OR operations with 9-bit data. Space-dependent weights are stored in the local digital memories within the MIPA4k cell.

## *4.1 BW Processing Core*

Figure 12 shows the structure of the BW processing core within MIPA4k. A basic description is provided here, whereas the interested reader is referred to [12] for a broader description of the operation.

The state and output of the BW core are denoted by $x$ and $y$, respectively. When A-templates are processed, $yu$ shown in the figure will be the same as the state, whereas with B-templates, $yu$ will act as the CNN input. Nodes $\overline{y_s}$ and $e$ are dynamic

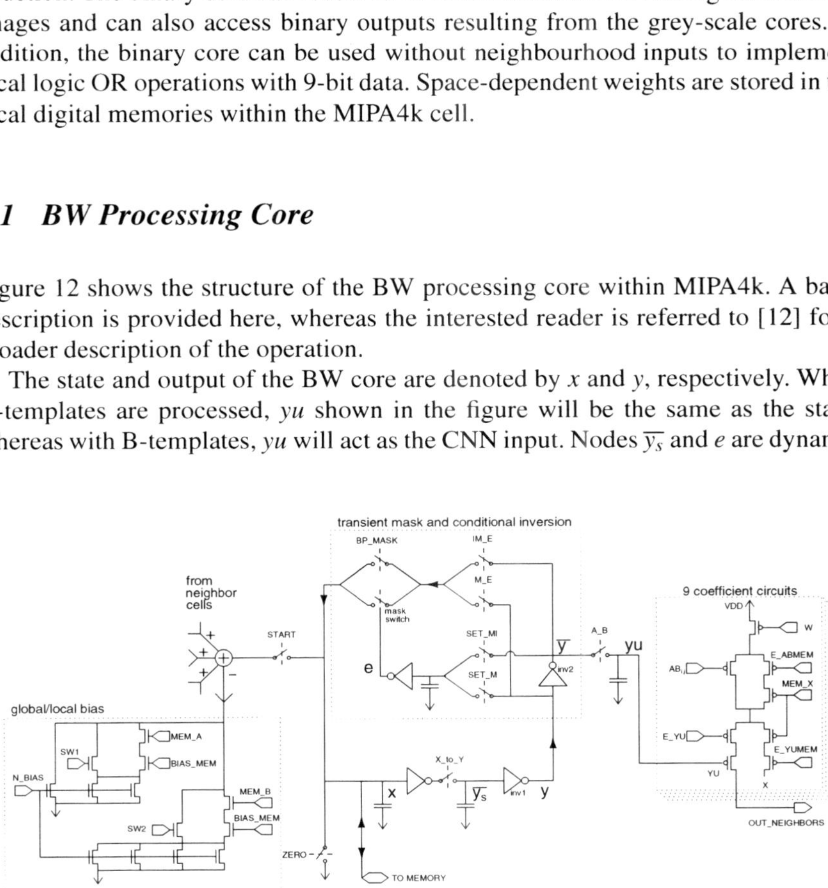


**Fig. 12** The BW core of MIPA4k

memory nodes. These are used to store the output and transient mask control bits, respectively. When the transient mask is off, the sign of the sum of bias and coefficients becomes new state $x$. Control signals $START$, $X\_to\_Y$ and $A\_B$ initiate the transient, control the writing of state to output, and control writing the output to the coefficient circuits, respectively.

The writing of dynamic memory $e$ is controlled by control signals $SET\_M$ and $SET\_MI$ (either $y$ or $\overline{y}$ is written to $e$). Control signals $M\_E$ and $IM\_E$ are used to determine whether $y$ or $\overline{y}$ will be driven to $x$ in case either $BP\_MASK$ (bypass mask) or $e$ is at an active state. Therefore, an active transient mask leaves the state of the cell unaffected by the coefficient circuits and bias.

The left side of Fig. 12 shows the bias circuit. When control signal $BIAS\_MEM$ is LO, the bias value is programmed globally (space independently) with control signals $SW1$ and $SW2$. Bias is programmed to 0.5, 1.5, 2.5 or 3.5 unit currents. Bias voltage $N\_BIAS$ is used to properly set the unit current of the bias to a desired value. When control signal $BIAS\_MEM$ is HI, the two bits that control the magnitude of bias come from a local memory, allowing space-dependent biasing schemes.

The right side of Fig. 12 shows the coefficient circuit of the BW cell. Bias voltage $W$ determines the unit current of the coefficient circuits. When control signals $E\_ABMEM$ and $E\_YUMEM$ are inactive, and control signal $E\_YU$ is active, the coefficient circuits operate conventionally: the template coefficients $AB_{i,j}$, $i, j \in [1,3]$ are globally determined allowing processing with space-independent templates. On the other hand, when all template coefficients $AB_{i,j}$ are turned off, and $E\_ABMEM$ is turned on, the template coefficients are controlled by local memory, resulting in a space-dependent template operation. Furthermore, consider that $E\_ABMEM$ is off, and $E\_YUMEM$ is on. In this case, the template coefficients are determined globally by $AB_{i,j}$, but the coefficient circuits are controlled by the local memory instead of $yu$. In addition, the output currents of the coefficient circuits are directed back into the state of the same cell. Therefore, 9-input OR function with globally programmable local inputs can be performed for the memory contents of the cells.

The hardware costs of including the space-dependent templates and local multi-input OR operation are 5 transistors per coefficient circuit and 4 transistors for the bias. Noting that all these transistors are switches, the hardware costs, including wiring are rather low provided that a modern CMOS process is used.

The nine-input NOR operations for contents of local memory can be used for efficient content-addressable searches in MIPA4k [13]. It is well known that the general pattern matching template [41] can be used to find locations with a matching neighbourhood condition. Here, the coefficient circuits operate on memory data in contrast to neighbourhood pixels and perform the general pattern match. This makes possible content-addressable 9-bit search by carrying out two tests in MIPA4k. The first task is to test whether any of the supposedly HI memory contents are LO. The second test is to find out whether any of the supposedly LO memory contents are HI. The results of the tests are ANDed to get the final result of the content-addressable search. These tests are carried out using OR operations for inverted and non-inverted memory contents.

A CNN with conventional multipliers could also carry out an associative search by performing the required logic operations bit-serially. As this would be done at the cost of processing time, it may be worth while to use coefficient circuits for CAM searches for search-intensive algorithms.

## 4.2 *Peripheral Circuits*

The MIPA4k is capable of identifying regions/objects of interest (ROIs), determining their size and location, and classifying objects of interest in a multi-object scene. At the core of this is that attention can be pointed to a part of the image plane by a ROI processing window. The window can be used for purposes of pattern matching and can be spread to different scales, to provide size invariance to some extent. The peripheral circuits of MIPA4k are tailored to facilitate these tasks: the decoder is controllable also by boundary cell outputs, whereas the peripheral circuits are equipped with the capability of writing the ROI window in parallel.

### 4.2.1 Row/Column Access Circuitry and Parallel Write

Figure 13 shows the row access circuitry of row *i* excluding the dedocer and encoder circuits [15].

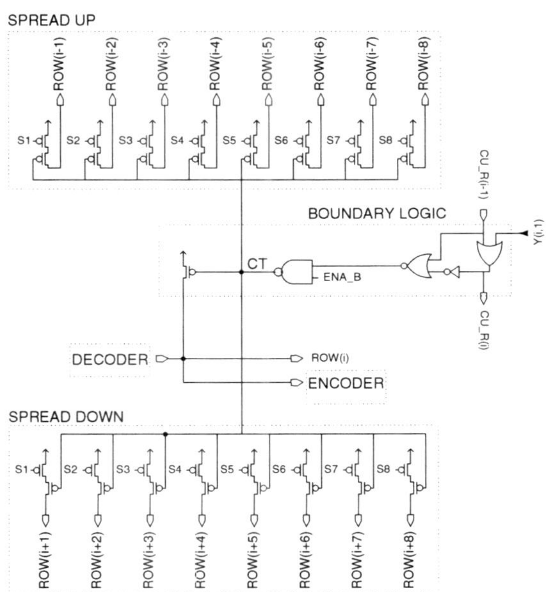


**Fig. 13** Row access circuitry

The boundary logic conveys the binary output $Y(i,1)$ of the boundary cell to the row access circuitry. $Y(i,1)$ can be used to control the row select signal $ROW(i)$ the decoder. Outputs of cascaded logic OR gates in the boundary logic circuitry, namely the $CU\_R(i)$ signals, indicate whether there is an active decoder output in an upper row. The $CT(i)$ bit indicates an active boundary, and that bit can be active on only one row at a time. The $CT(i)$ bit is obtained from

$$CT(i) = \overline{ENA\_B \wedge \overline{CU\_R(i-1) \vee \overline{CU\_R(i)}}},$$

where

$$CU\_R(i) = Y(i,1) \vee CU\_R(i-1)$$

and $ENA\_B$ is a signal that indicates that boundary cell controls the row select signal.

The spreading circuits can change the states of the decoders above and below the row of an active boundary bit $CT(i)$. This spreading is controlled by signals $S1-S8$. For example, if S8, S4 and $CT(i)$ are low, the decoder outputs $ROW(i)$, $ROW(i-8)$ (output of decoder eight rows above), $ROW(i-4)$, $ROW(i+8)$ and $ROW(i+4)$ are pulled high. Only two of signals $S1-S8$ can be active at the same time so that a maximum of five decoder outputs are high at the same time. Since the column access circuit is similar in structure, a group of 25 cells can be simultaneously selected. By properly choosing the active control signals, the row and column spreading circuits can activate different patterns of cells, where the location of the pattern is determined by the row/column boundary bits. The activity patterns can be scaled for window sizes ranging from $17 \times 17$ to $5 \times 5$.

The periphery also has parallel write control circuitry so that object prototypes $F[1:5,1:5]$ can be written in parallel to the 25 cells selected by the row/column access circuitry. The parallel write circuitry assures that the columns and rows of the object prototype maintain their order. For example, the left column of the object prototype $F[1:5,1]$ is always written to the leftmost selected cells at all scales of the window.

### 4.2.2 Processing Example Using Parallel Write

The following processing example demonstrates how the peripheral circuits work in practice. Processing steps that are carried out using regular templates are not described in detail here. More information on these can be found from [12] and [41]. Figure 14a shows the input scene, obtained with the optical sensor of MIPA4k. Threshold results of this scene are shown in Fig. 14b in a grey on black background. The threshold result is then processed with *hole filler* and *concave location filler* templates. The latter fills concave locations of objects to obtain a better centroid extraction result. Centroid extraction is carried out by successively applying conditional erosion operations that are sensitive to different directions. The erosion is conditional in order not to remore the last pixel of the object [14].

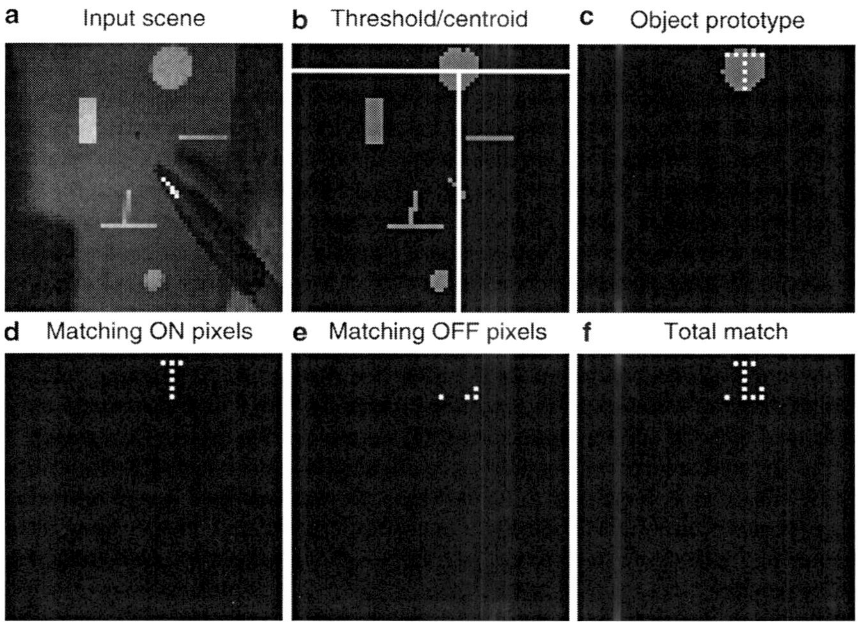


**Fig. 14** Measured processing example

Figure 14b also shows lines crossing through the object that is located highest and leftmost in the image. These lines represent the centroid coordinates that were extracted with a sequence of wave, and logic operations [14]. The row address is obtained with a shadow template operation (propagates to S, W, E directions simultaneously) that is initiated by the centroid image, followed by edge detection template. The one row of active pixels that result from this operation is ANDed with the centroid image so that only pixel(s) on the highest row remain. Consecutively, a northern shadow is taken, the result of which is ORed with the row address. After this processing sequence, the row and column addresses of the selected centroid are available at the boundary cells.

Figure 14c shows in grey the selected object that has been obtained with binary reconstruction template (thresholded image as mask and selected centroid as marker). Next, the boundary logic and spreading circuits are activated and an object prototype (a letter T in this case) is written over the selected object using the parallel write functionality. This is shown in white in Fig. 14c. Note that in this case the $5 \times 5$ object prototype is extended to a window of $9 \times 9$. Next, object pixels are matched using local logic operations and the result is shown in Fig. 14d. Similarly, pixels that are OFF in the $5 \times 5$ object prototype are matched to the inverted object image as shown in Fig. 14e. The results of Fig. 14d, e are ORed, and the result is shown in Fig. 14f. The number of active pixels represents how well the object matches the object prototype. This measure can be quantified using the global sum functionality of MIPA4k.

## 5 Speed and Power Performance

The choice of using current-mode processing leads inevitably to some static power consumption. However, by limiting the maximal in-cell signal currents to approximately 5 μA, even the worst case static current consumption per cell is relatively low, approximately in the order of 100 μA. It has to be noted that because the static power consumption is input-dependent, the average values are much lower.

Because the dedicated hardware makes it possible to perform efficient non-linear grey-scale filtering operations in a single step, a low frequency global clock rate for controlling the cell operations can be used, which helps to keep dynamic power consumption low. Since the processing time for a single grey-scale operation can be as low as 100–200 ns, and even the duration of a worst-case propagating operation is in the order of microseconds, real-time processing with a high frame rate can be performed with the current inputs turned off for most of the time to save power.

The targeted accuracy of the grey-scale circuitry is in the range of 5–6 equivalent digital bits. This is presumed to be sufficient for the low-level image operations. The presented chip offers an excellent platform for practically investigating the efficiency and robustness of different parallel image algorithms on limited-accuracy analog hardware.

The binary template operations are processed as described in [12]. Processing time of the binary feedforward templates is currently limited to 40 ns, which is the control cycle of the external FPGA controller. The processing times of asynchronously propagating templates depend on the type of operation. For example, a typical hole filler transient time is 2.6 μs, whereas in the processing example of Fig. 14, the transient time of the asynchronous hollow template was 800 ns. For very large objects, the transient time of the hollow may need to be longer.

## 6 Conclusions

This chapter gives an overview of a 64 × 64 mixed-mode array processor (MIPA4k). It is a pixel-parallel processor in which each pixel (cell) has separate processing cores for different grey-scale and binary processing tasks. MIPA4k excells at high speed, low lag analysis of visual scenes. It can also perform low power pixel-parallel image processing by switching the processing cores into power saving mode after a frame has been processed. Since lots of processing capabilities are included in each cell (yielding a large area), in this approach the array size is smaller compared to conventional imagers. Even if the MIPA4k cell density has a lot to optimize, and future 3D fabrication (in which different processing cores are put in different layers) may help reduce the resolution gap, prospective applications are likely to be tailored for below-megapixel resolutions.

**Acknowledgements** This work was partly funded by the Academy of Finland grants 106451, 117633 and 131295. The authors also thank Turku Science Park and The Turku University Foundation for their help in funding the MIPA4k chip manufacturing.

## References


1. L.O. Chua, L. Yang, *Cellular Neural Networks: Theory*, IEEE Transactions on Circuits and Systems-I, vol. 35, no. 10, pp. 1257–1272, October 1988
2. A. Rodriguez-Vazquez, G. Linan-Cembrano, L. Carranza, E. Roca-Moreno, R. Carmona-Galan, F. Jimenez-Garrido, R. Dominguez-Castro, S.E. Meana, *ACE16k: The Third Generation of Mixed-Signal SIMD-CNN ACE Chips Toward VSoCs*, IEEE Transactions on Circuits and Systems-I: Regular Papers, vol. 51, no. 5, pp. 851–863, 2004
3. P. Dudek, P.J. Hicks, *A General-Purpose Processor-per-Pixel Analog SIMD Vision Chip*, IEEE Transactions on Circuits and Systems-I: Regular Papers, vol. 52, no. 1, pp. 13–20, 2005
4. P. Dudek, *Implementation of SIMD Vision Chip with* $128 \times 128$ *Array of Analogue Processing Elements*, Proceedings of the IEEE International Symposium on Circuits and Systems, ISCAS 2005, vol. 6, pp. 5806–5809, 2005
5. A. Lopich, P. Dudek, *Architecture of a VLSI Cellular Processor Array for Synchronous/Asynchronous Image Processing*, Proceedings of the 2006 IEEE International Symposium on Circuits and Systems, pp. 3618–3621, 2006
6. A. Lopich, P. Dudek, *ASPA: Focal Plane Digital Processor Array with Asynchronous Processing Capabilities*, Proceedings of the 2008 IEEE International Symposium on Circuits and Systems, pp. 1592–1595, 2008
7. C. Rekeczky, T. Roska, A. Ushida, *CNN-Based Difference-Controlled Adaptive Nonlinear Image Filters*, International Journal on Circuit Theory and Applications, 26, pp. 375–423, 1998
8. C. Rekeczky, A. Tahy, Z. Vegh, T. Roska, *CNN-based Spatio-Temporal Nonlinear Filtering and Endocardial Boundary Detection in Echocardiography*, International Journal of Circuit Theory and Applications, vol. 27, pp. 171–207, 1999
9. A. Paasio, A. Kananen, M. Laiho, K. Halonen, *A Compact Computational Core for Image Processing*, Proceedings of the European Conference on Circuit Theory and Design, ECCTD'01, pp. I-337–I-339, 2001
10. A. Paasio, A. Kananen, M. Laiho, K. Halonen, *An Analog Array Processor Hardware Realization with Multiple New Features*, Proceedings of the International Joint Conference on Neural Networks, IJCNN, pp. 1952–1955, 2002
11. J. Poikonen, M. Laiho, A. Paasio, *MIPA4k: A* $64 \times 64$ *Cell Mixed-Mode Image Processor Array*, Proceedings of the IEEE International Symposium on Circuits and Systems, ISCAS2009, pp. 1927–1930, Taipei 2009
12. M. Laiho, A. Paasio, J. Flak, K. Halonen, *Template Design for Cellular Nonlinear Networks With 1-Bit Weights*, IEEE Transactions on Circuits and Systems I, vol. 55, no. 3, pp. 904–913, 2008
13. M. Laiho, J. Poikonen, A. Paasio, *Space-Dependent Image Processing Within a* $64 \times 64$ *Mixed-Mode Array Processor*, IEEE International Workshop on Cellular Nanoscale Networks and Applications, Berkeley, 2010
14. M. Laiho, J. Poikonen, A. Paasio, *Object Segmentation and Tracking with Asynchronous Grayscale and Binary Wave Operations on the MIPA4k*, European Conference on Circuit Theory and Design, Antalya, 2009
15. M. Laiho, J. Poikonen, A. Paasio, K. Halonen, *Centroiding and Classification of Objects Using a Processor Array with a Scalable Region of Interest*, IEEE International Symposium on Circuits and System, Seattle, pp. 1604–1607, 2008
16. J. Poikonen, M. Laiho, A. Paasio, *Locally Adaptive Image Sensing with the* $64 \times 64$ *Cell MIPA4k Mixed-Mode Image Processor Array*, European Conference on Circuit Theory and Design, Antalya, 2009
17. J. Poikonen, A. Paasio, *A Ranked Order Filter Implementation for Parallel Analog Processing*, IEEE Transactions on Circuits and Systems I: Regular Papers, vol. 51, no. 5, pp. 974–987, 2004
18. J. Poikonen, A. Paasio, *Rank Identification for an Analog Ranked Order Filter*, Proceedings of the 2005 IEEE International Symposium on Circuits and Systems, ISCAS 2005, vol. 3, pp. 2819–2822, 2005

19. J. Poikonen, A. Paasio, *An 8 × 8 Cell Analog Order-Statistic-Filter Array With Asynchronous Grayscale Morphology in 0.13 μm CMOS*, IEEE Transactions on Circuits and Systems I: Regular Papers, vol. 56, no. 8, pp. 1541–1553, 2009
20. J. Poikonen, *Absolute Value Extraction and Order Statistic Filtering for a Mixed-Mode Array Image Processor*, Doctoral thesis, Turku Centre for Computer Science (TUCS) Dissertations, no. 80, November 2006
21. J. Poikonen, A. Paasio, *Current Controlled Resistive Fuse Implementation in a Mixed-Mode Array Processor Core*, Proceedings Of the 8th IEEE International Workshop on Cellular Neural Networks and their Applications, pp. 76–81, 2004
22. L. Vesalainen, J. Poikonen, A. Paasio, *A Fuzzy Unit for a Mixed-Mode Array Processor*, Proceedings of the 8th IEEE International Workshop on Cellular Neural Networks and their Applications, CNNA 2004, pp. 273–278, 2004
23. J. Poikonen, M. Laiho, A. Paasio, *Anisotropic Filtering with a Resistive Fuse Network on the MIPA4k Processor Array*, IEEE International Workshop on Cellular Nanoscale Networks and Applications, Berkeley, 2010
24. B.E. Shi, *Order Statistic Filtering with Cellular Neural Networks*, Proceedings of the 3rd IEEE International Workshop on Neural Networks and their Applications, CNNA-94, pp. 441–443, 1994
25. A. Zarandy, A. Stoffels, T. Roska, L.O. Chua, *Implementation of Binary and Gray-Scale Mathematical Morphology on the CNN Universal Machine*, IEEE Transactions on Circuits and Systems-I: Fundamental Theory and Applications, vol. 45, no. 2, pp. 163–168, 1998
26. I.E. Opris, *Analog Rank Extractors and Sorting Networks*, Ph.D. Thesis, Stanford University, 1996
27. L. Vincent, *Morphological Grayscale Reconstruction in Image Analysis: Applications and Efficient Algorithms*, IEEE Transactions on Image Processing, vol. 2, no. 2, pp. 176–200, 1993
28. J. Poikonen, A. Paasio, *An Area-Efficient Full-Wave Current Rectifier For Analog Array Processing*, Proceedings of the 2003 IEEE International Symposium on Circuits and Systems, ISCAS 2003, vol. 5, pp. 757–760, 2003
29. P. Perona, J. Malik, *Scale-Space and Edge Detection Using Anisotropic Diffusion*, IEEE Transactions on Pattern Analysis and Machine Intelligence, vol. 12, no. 7, pp. 629–639, 1990
30. A. Blake, A. Zisserman, *Visual Reconstruction*, MIT, Cambridge, MA, 1987
31. J. Schlemmel, K. Meier, M. Loose, *A Scalable Switched Capacitor Realization of the Resistive Fuse Network*, Analog Integrated Circuits and Signal Processing, 32, pp. 135–148, 2002
32. P. Yu, S. Decker, H. Lee, C. Sodini, J. Wyatt, *Resistive Fuses for Image Smoothing and Segmentation*, IEEE Journal of Solid-State Circuits, vol. 2, pp. 894–898, 1992
33. P. Acosta-Serafini et al., *A 1/3 VGA Linear Wide Dynamic Range CMOS Image Sensor Implementing Predictive Multiple Sampling Algorithm With Overlapping Integration Intervals*, IEEE Journal of Solid State Circuits, pp. 1487–1496, 2004
34. C. Posch et al., *An Asynchronous Time-Based Image Sensor*, Proceedings of the IEEE International Symposium on Circuits and Systems, ISCAS 2008, pp. 2130–2133, 2008
35. A. Zarandy et al., *Per Pixel Integration Time Controlled Image Sensor*, European Conference on Circuit Theory and Design, pp. III/149–III/152, 2005
36. C.M. Dominguez-Matas et al., *3-Layer CNN Chip for Focal-Plane Complex Dynamics with Adaptive Image Capture*, IEEE International Workshop on Cellular Neural Networks and their Applications, pp. 340–345, 2006
37. R. Carmona et al., *A CNN-Driven Locally Adaptive CMOS Image Sensor*, IEEE International Symposium on Circuits and Systems, ISCAS'04, vol. V, pp. 457–460, Vancouver, 2004
38. M. Laiho, J. Poikonen, K. Virtanen, A. Paasio, *Self-adapting Compressive Image Sensing Scheme*, IEEE International Workshop on Cellular Neural Networks and their Applications, pp. 125–128, 2008
39. V. Brajovic, *Brightness Perception, Dynamic Range and Noise: A Unifying Model for Adaptive Image Sensors*, IEEE Computer Society Conference on Computer Vision and Pattern Recognition, 2004

40. M. Laiho, V. Brea, A. Paasio, *Effect of Mismatch on the Reliability of ON/OFF Programmable CNNs*, IEEE Transactions on Circuits and Systems-I, vol. 56, no. 10, pp. 2259–2269, October 2009
41. T. Roska et al., *CNN Software Library, Version 1.1*, Analogical and Neural Computing Laboratory, Hungarian Academy of Sciences, 2000, http://lab.analogic.sztaki.hu/Candy/csl.html

# ASPA: Asynchronous–Synchronous Focal-Plane Sensor-Processor Chip

**Alexey Lopich and Piotr Dudek**

**Abstract** This chapter describes the architecture and implementation of a digital vision chip, with asynchronous processing capabilities (asynchronous/synchronous processor array or ASPA). The discussion focuses on design aspects of cellular processor array with compact digital processing cell suitable for parallel image processing. The presented vision chip is based on an array of processing cells, each incorporating photo-sensor with a one-bit ADC and simple digital processor, which consists of 64-bit memory, arithmetic and logic unit (ALU), flag register and communication unit. The chip has two modes of operation: synchronous mode for local and nearest-neighbour operations and continuous-time mode for global operations. The speed of global image processing operations is significantly increased by using asynchronous processing techniques. In addition, the periphery circuitry enables asynchronous address extraction, fixed pattern addressing and flexible random access data I/O.

**Keywords** Vision chip · Smart sensor · Asynchronous image processing · Cellular processor array

## 1 Introduction

Image processing has always been one of the main applications for parallel computing. Because of data-parallelism and computational locality, the implementation of early image pre-processing algorithms on parallel architectures can gain significant speed-up, which is expressed by the Amdahl's law [1]. To varying degrees, data-parallel computations are facilitated by such groups of devices as multiple instruction multiple data (MIMD) supercomputers, digital signal processors (DSPs), application-specific solutions (e.g. systolic arrays) and single instruction multiple data (SIMD) array processors.

P. Dudek (✉)
University of Manchester, Manchester M13 9PL, UK
e-mail: p.dudek@manchester.ac.uk

Á. Zarándy (ed.), *Focal-Plane Sensor-Processor Chips*,
DOI 10.1007/978-1-4419-6475-5_4, 

However, from the perspective of early vision system design, all these devices have certain disadvantages. The cost, size and power consumption make supercomputers unsuitable for the majority of computer vision applications. The level of concurrency provided by modern DSPs still does not fully exploit the inherent parallelism of low-level image processing. The lack of versatility in application-specific integrated circuits (ASICs) makes them unsuitable for general-purpose use. SIMD arrays, which exhibit high degree of parallelism, allow a very large number of processors to be integrated on a single chip and provide enough programmability to accomplish a wide range of early vision tasks. Since the early developments [2, 3], such massively parallel processor arrays were demonstrated to provide a very high performance solution for image pre-processing applications. However, they suffer from a bottleneck caused by the data transfer between the sensor unit and the array. To solve this problem, an integration of image sensing and processing into a single silicon device has been considered [4, 5]. The architecture of software programmable 'vision chips' is typically based on SIMD processor arrays, with either linear processor-per-column architecture for line-by-line processing [6, 7] or fine-grain processor-per-pixel arrays, where each cell combines a photo sensor and a processor [8–10] (Fig. 1). The latter approach benefits from eliminating the I/O bottleneck as the sensory data is fed directly into the corresponding processing elements (PEs). Smart sensors also benefit from small silicon area and reduced power consumption as compared to conventional vision systems with separate sensing and processing units.

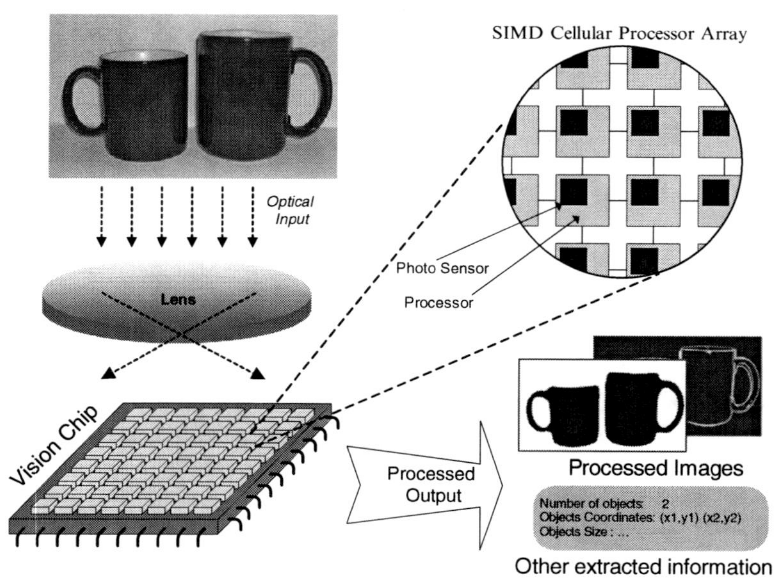


**Fig. 1** A vision chip placed in the focal plane of the optical system uses a pixel-parallel array to sense and process incoming images

This chapter presents an example of the architecture design and implementation of a digital vision chip, based on asynchronous/synchronous processor array (ASPA). Each processing cell of the ASPA chip is based on universal synchronous digital architecture, but also enables generation and processing of asynchronous locally controlled trigger-waves [11], continuous-time execution of a distance transformation and global data transfers. A mixed asynchronous/synchronous approach facilitates efficient execution of both local and global operations in a compact massively parallel architecture. Experimental results indicate that the chip delivers real-time performance not only in low-level image pre-processing, but also in more complex medium-level image analysis tasks.

## 1.1 Background

A number of research projects have been dedicated to the development of processor array architectures. The first SIMD processor (LAPP1100) to be integrated with a sensor array was designed in Linköping University [12]. LAPP comprises 128 PEs, each incorporating 14-bits of internal memory and binary processing unit. Further development of this array resulted in PASIC processor [13] with 256 PEs, each endowed with an arithmetic and logic unit (ALU), 8-bit shift registers and 128 bits of memory. Adding more internal memory and improving functionality has brought further successors [14, 15]. The latest chip presented in [7], providing up to 100 GOPS, is capable of processing images with resolution of up to 1,535 pixels wide, which is sufficient for the majority of industrial applications. Many SIMD processor arrays have been initially designed as image processing units separate form the image sensor, due to difficulty in implementing large resolution arrays on a single device. Examples include the SPAR processor described in [16], implemented as a single chip, based on a $16 \times 16$ array of bit-serial PEs with 512 bits of memory per cell. Another architecture with improved performance and explicit orientation towards low-level image processing was described in [17]. One of the first commercial realizations of a SIMD image processor was presented by Philips [18] with latest version of its Xetal device descried in [6]. The Xetal linear processor array was reported to achieve peak performance of 107 GOPS.

A number of integrated sensor/processor vision chips that are based on processor-per-pixel SIMD array have also been presented in literature [8–10, 19]. It should be noted that application-specific vision chips, e.g. [20] can provide better performance to area/power ratio; however, in this chapter we focus only on programmable general-purpose chips that can fulfil a wide range of image processing tasks specified by software code. All vision chips can be categorized according to the domain in which they process image data: analogue and digital. Many analogue vision chips such as the SCAMP [8], ACE16k [10] and MIPA4k [20] demonstrate high utilization of silicon area and low power consumption with respect to achieved performance. In contrast to analogue vision chips, their digital counterparts operate with image data represented in digital format, which provides an absolute accuracy, limited only by the word length. However, digital hardware realization of arithmetic operations usually occupies large area as compared to analogue solutions and due to strict pixel area requirements digital design is often limited to a bit-serial approach. For example, the NSIP chip [19] contains a simple ADC, logical unit and 8-bit memory in every pixel cell. Although cells operate with binary data, simple grey-scale operations are possible, thanks to A/D conversions in the temporal domain. A somewhat similar processor array architecture called PVLSAR, where in addition to in-pixel binary operation grey-scale processing is combined with sensing was reported in [21]. The latest version of the PVLSAR device incorporates $200 \times 200$ array of processing cells with increased internal memory of 48 bits. Another work described in [22] reports a $16 \times 16$ array binary vision chip, SRVC, with $30 \times 40\,\mu\text{m}^2$ pixel pitch, where each cell can store 4 bits of data and perform basic logic operations. A $64 \times 64$ digital vision chip presented in [9] contains PEs with a $3 \times$ 8-bit memory, an adder, latches and a photo-sensor with an ADC, and the cell pitch of $67.4\,\mu\text{m}$.

## 1.2 Asynchronous Processor Arrays

In early image pre-processing, many algorithms have a common feature of the computational locality, i.e. all pixels in the output image are calculated based on local neighbourhood. Such algorithms have a straightforward implementation on parallel processor-per-pixel arrays, where the pixel function is implemented in every processing cell. The performance improvements in these systems can be as high as $O(N^2)$ for $N \times N$ arrays when compared to serial architectures.

Yet, there are a number of algorithms that are based on a global data-flow across the pixel network. In these algorithms, the result in each pixel, while iteratively relying on local data, implicitly depends on the entire image. The most characteristic feature in these operations is that the actual processing is carried out in the limited number of active pixels, placed at the front of the wave-propagating activity. Figure 2 shows a simple example of closed curve detection algorithm. In this example useful processing is occurring only in the wave front (marked by dark grey in

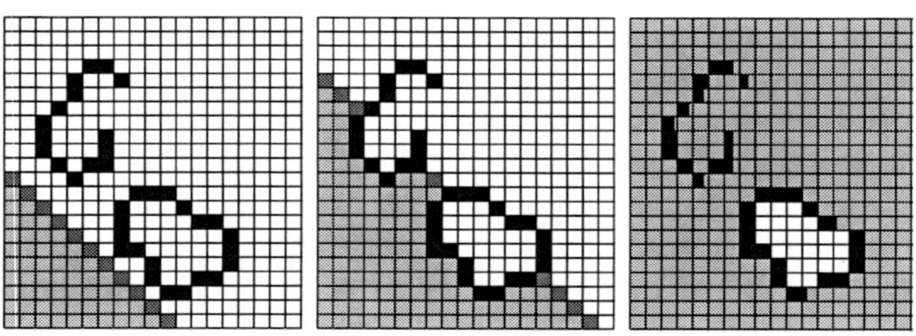

**Fig. 2** Asynchronous trigger-wave propagation algorithm (see text for details)

Fig. 2), whereas the rest of the array is idle. The propagation starts from the initial corner marker and spreads across the propagation space (white pixels). Black pixels (representing objects' borders) prohibit the propagation, thus leaving the internal space of closed contours 'un-triggered'. This propagation splits the entire array into two subsets, which consist of pixels that have been involved in the propagating activity (light grey) and pixels that remain in the original state (white). Although feasible on SIMD arrays, due to iterative data-dependent data-flow, the wave-propagating algorithms result in inefficient power and time consumption when executed in a sequential manner on such hardware, as only small subset of wave-front cells perform useful processing at any time, while other cells still receive broadcast instructions and must conditionally de-activate on each iteration cycle. Moreover, synchronous iterations are relatively slow. To increase the speed of global operations, it is necessary to either introduce a global reconfigurable network of interconnections or optimize the data-flow through the pixel network. The first approach becomes very difficult for a physical implementation due to a limited number of routing layers for these interconnections, whereas the complexity of interconnections, which is proportional to the processed image size, is continuously increasing. A potential solution could be based on FPGA-like architectures, but would result in significant silicon resources being utilized for flexible routing. The second approach is more suited for VLSI implementation. A promising architecture could be based on the array of processing cells triggered only by their local neighbours, so that they perform computations only when the neighbours' states are updated. For this purpose, asynchronous methodology provides a suitable solution, due to its inherent speed and low latency. In addition, locally triggered asynchronous circuits simplify distribution of control signals and allow cells to stay in low power mode until they are triggered. The aforementioned merits make asynchronous approach advantageous for realization of global image processing operations.

Consequently, to execute low- and medium-level operations in the most efficient way, one should accommodate reconfigurable asynchronous circuitry and versatile processing power into a very small area, ideally comparable with the optical pixel pitch. A number of works explored the various approaches towards mixed synchronous/asynchronous architecture. The NSIP chip, described in [19], contains a $32 \times 32$ array of simple Boolean processors, each incorporating global logic unit (GLU), which enables execution of global operations such as threshold with hysteresis and object reconstruction in a parallel asynchronous manner. We have fabricated an evaluation chip with somewhat similar architecture of asynchronous processing unit, but with optimized circuitry [11]. With each cell occupying $25 \times 25\,\mu\text{m}^2$ of silicon area, the extrapolated performance of this chip on $128 \times 128$ images is $16.3 \times 10^6$ images s$^{-1}$ with an assumption that the worst propagation length is 128.

Another example of a synchronous vision chip with global asynchronous capabilities has been reported in [23]. In addition to a conventional SIMD operation mode, the $64 \times 64$ array enables reconfiguration of its hardware by chaining PEs into a single asynchronous network. This makes it feasible to calculate the sum of $m$-bit numbers during $m$ clock cycles (provided the signal propagation delay along the chain of cells is shorter than the clock cycle). Thus, by endowing each cell with a full-adder the chip is also capable of performing global summation operations. Another example of global regional computations on cellular arrays with local interconnection was presented in [24]. A cellular processor array, based on so-called 'convergent micro pipelines', allows computing the sum over a region without a combinatorial adder in every cell. The number of iterations in this approach depends on the shape of the object.

The idea to match the architecture with functional concept of asynchronism has been further explored in [25–29]. A graph-based computational model, which reflects fine-grain data parallelism, typical for image analysis, has been developed and described in [26]. The model was designed for hardware or software implementation and it was shown that some synchronous operations, such as distance transformation, computation of Voronoi diagram, watershed segmentation and opening or closing by reconstruction can be replaced by their asynchronous analogues, thus benefiting from operation speed-up and reduced power consumption. Some application-specific hardware realizations were also presented in a number of research works [27–30]. Reported in [28] $16 \times 16$ processor array not only allows asynchronous execution of global morphological operations, but also supports limited programmability so that either dilation or erosion can be performed. Another study for application-specific solution for asynchronous watershed segmentation has been demonstrated in [30]. Alternative investigation on mapping fully parallel watershed segmentation algorithm onto fine-grain cellular automata can be found in [31]. That work describes an attempt to employ a mixed synchronous/asynchronous approach to simplify asynchronous operations, thus reducing hardware overhead for their realization. An application-specific cellular-processor array for binary skeletonization has been presented in [29]. By utilizing 'grassfire' transformation, the skeletons of binary images are extracted asynchronously during single iteration.

While asynchronous trigger-wave execution does provide the best power/ performance result, it is important to realize that the physical implementation is unique for every executed operation. In other words, it is problematic to provide 'programmability' for asynchronous cells and every required global operation will need to be realized in hardware separately, thus leading to the increased cell size. Such limited functionality and restriction on programmability make asynchronous cellular logic networks unsuitable for general-purpose image processing on their own. Therefore, a suitable solution for a vision chip should adopt a mixed synchronous/asynchronous architecture, thus benefiting in terms of programmability and versatility while in the SIMD mode (discrete-time, synchronous operation) and achieving the maximum performance in global operations while configured as a single asynchronous network.

## *1.3 Digital vs. Analogue Vision Chips*

When designing any hardware system, it is important to define a suitable signal processing domain, e.g. analogue, digital or mixed signal. In the case of building massively parallel sensor-processor arrays, this issue is especially important, as it affects all aspects of hardware implementation.

Up until recent progress in fabrication technologies (sub-100 nm feature size, and 3D IC integration), the analogue designs had distinct advantages for building focal-plane processor arrays. First, analogue implementation led to higher area utilization. The space needed for implementation of various arithmetic functions could be two orders of magnitude less than that in digital logic. For example, addition and subtraction could be realised based on Kirchhoff's current law by appropriate wiring without any additional gates [8]. Second, quite complex nonlinear functions can be performed with the help of a just few devices. The result is that imaging properties (such as fill-factor and resolution) of analogue vision chips are superior to their digital counterparts with similar functionality. However, the circuit area, their operational speed and power consumption are directly related to the required signal-to-noise ratio (SNR). While for analogue designs, dynamic range is restricted by noise, for digital counterparts it is limited only by local memory. Noise also affects the accuracy of analogue calculations. Device variability (mismatch), thermal and shot noise, electromagnetic coupling and substrate noise – all these factors affect the result of circuit operation. While certain types of distortion (e.g. mismatch) can be to some extent compensated by applying differential circuit techniques or performing additional operations, the noise imposed during analogue data manipulation has an accumulative effect and eventually can significantly reduce the SNR. In digital circuits, the noise issue is not relevant. The operation of such circuits is robust and the only potential source of noise is during AD conversion of the photo signal, i.e. quantisation error.

In summary, although current analogue vision chips [8, 10] are more area- and power-efficient than the digital ones, the progress in CMOS fabrication process and poor scalability of analogue circuits reduce their advantage making the digital design option more attractive for future massively parallel architectures.

## 2 Architecture

The architecture of the ASPA vision chip, shown in Fig. 3, is based on a 2D processor-per-pixel array of locally interconnected processing cells. The cells are placed on the 4-connected rectangular grid, so that every cell is connected to its four nearest neighbours (North, East, South and West). All cells in the array operate in the SIMD mode, by executing instruction words (IW) broadcast by the central controller. Both controller and vision chip program memory are stored off-chip. Although all PEs receive identical instructions, not all of them may execute the instructions, thanks to local flag indicators. In addition to providing local autonomy feature, local flags are used to constrain continuous-time operation of the array. Data I/O and instruction distribution is provided by peripheral instruction drivers.

Since outputting entire frames out of the sensors contribute significant part to the overall processing time, it is important to provide the facility to extract more abstract

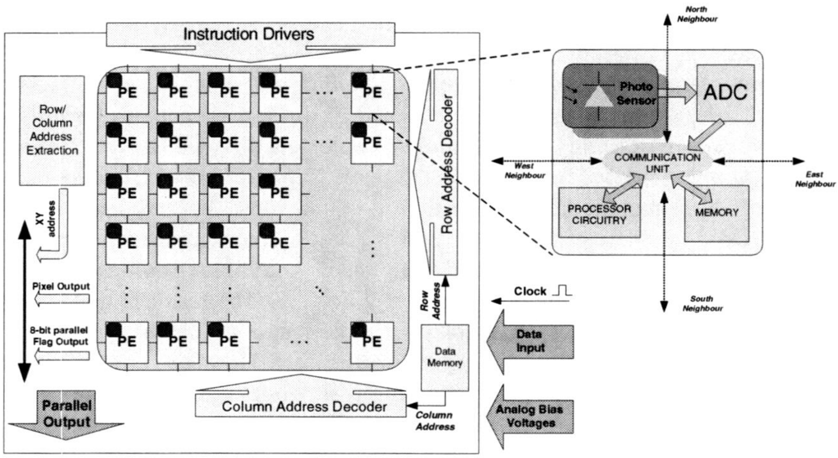


**Fig. 3** The ASPA chip architecture

descriptors from the sensor, thus optimizing the output throughput. This data can include binary representations of processed image, sub-frames, coordinates of the object, area of the segmented part of the image, presence of the object in the frame, etc. To enable such features, the dedicated circuitry multiplexes various data into output buffers. Random pixel access as well as flexible pattern addressing [32] is facilitated by dedicated addressing circuitry. To enable Address Event Representation read-out mode [33], an additional functional unit, which performs asynchronous pixel address extraction, is introduced.

When defining the architecture of the processing cell, it is important to look into functional requirements, imposed by target applications. Early image processing algorithms can be split into two categories, according to data-dependencies between input and output results. The first group is represented by local convolutions, morphological operators and other routines, where the pixel output is represented by a function of defined local neighbourhood. In the second group, the output result (which may not necessarily be a 2D array of values) depends on the entire image. Although it is possible to represent such algorithms as an iterative process, where during each iteration the pixel output is determined as a function of local neighbourhood data, the number of iterations is data dependent and can significantly vary for different inputs.

To facilitate the execution of both local and global operations, the organization of a basic cell is similar to a simple microprocessor with added functionality to support global asynchronous data-processing (although it is necessary to identify only fundamental operations with a compact hardware implementation, which can be used as a basis for the majority of global algorithms). It benefits from programmability and versatility when in synchronous SIMD mode, and from low-power consumption and ultra fast execution when configured as a combinatorial circuit in continuous-time

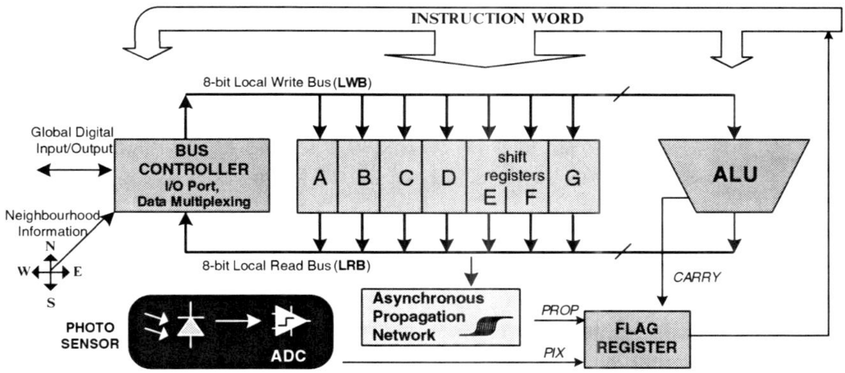


**Fig. 4** The ASPA processing cell

mode. The architecture of the cell is presented in Fig. 4. It comprises an I/O port, digital ALU, register file, flag register (FR), photo-sensing unit and auxiliary circuitry.

When in SIMD mode, the cell operates with digital values based on the register-transfer principle. The instruction specifies source and destination for the data: for arithmetic operations – transfer between the memory and the ALU or a shift register; for I/O operations – transfer between memory and global I/O port; for neighbourhood communication – transfer between memory registers in neighbouring cells (*North, East, South, West*). Additionally, the local FR supports conditional branching so that a set of cells can skip certain instructions based on local flag values.

The amount of data memory is of vital concern when designing a pixel-level PE. On the one hand, the amount of local memory influences overall functionality and performance of the cell. Storage capacity determines how many intermediate results can be stored within a pixel. On the other hand, memory utilizes a significant part of the pixel area. Therefore, it is important to find a trade-off between the memory capacity and cell size. To set memory constraints, factors such as silicon usage area required precision and algorithm requirements have to be considered. The first two factors are usually constrained by external requirements. The size of the chip is determined by parameters such as fabrication costs required image resolution and yield. Bit-precision is set by the types of image processing algorithms executed on a vision chip. For the majority of image processing algorithms, an 8-bit grey-level precision is generally thought to be sufficient. Therefore, this word length is taken as a general constraint. However, adoption of bit-serial approach removes this limitation, i.e. it is possible to process data of arbitrary data width.

The local memory of the ASPA PE consists of five 8-bit digital general-purpose registers (A, B, C, D, G) and two 8-bit shift registers (E, F). Analysis of in-pixel memory usage in low- and medium-level processing shows that this amount is sufficient for the majority of early vision applications, including basic inter-frame analysis, where in addition to storing intermediate results during pixel data processing, it is necessary to store information about the previous frame (or several frames).

Digital registers can also be used to store binary values so that complex algorithms based on binary processing, for example Pixel-Level Snakes [34], can be implemented. Additionally, the ALU contains an 8-bit accumulator for storing temporary data, thus every PE can store up to eight bytes of data. The processing is not restricted to this width however, as it is possible to store and process data of arbitrary width. This feature provides the flexibility of the 64-bit memory usage for bit-serial and bit-parallel processing.

In order to optimize the processor cell area, all arithmetic operations are executed in bit-serial. However, local and global transfers are performed with bit parallel data. Such feature enables certain data manipulation during transfers, increase global communication throughput and improve the speed of communication between local neighbours.

## 3 Circuit Implementation

### *3.1 Local Memory and Datapath*

Internal memory in a pixel significantly contributes to the PE size. A logical choice in attempting to minimize this impact is to base the local memory on dynamic logic, i.e. one bit latch is based on a three-transistor memory cell.

An *i*th bit slice of the 8-bit wide pixel data-path, including registers A, B, C, D, shift register E and bus controller (BC), is shown in Fig. 5. When the memory write switch, controlled by the load signal (i.e. LA, LB, LC, ...) and gated by the Flag signal is closed, the storage capacitance is either charged or discharged depending on input data from a Local Write Bus ($LWB_i$). Presented in Fig. 6 is an exam-

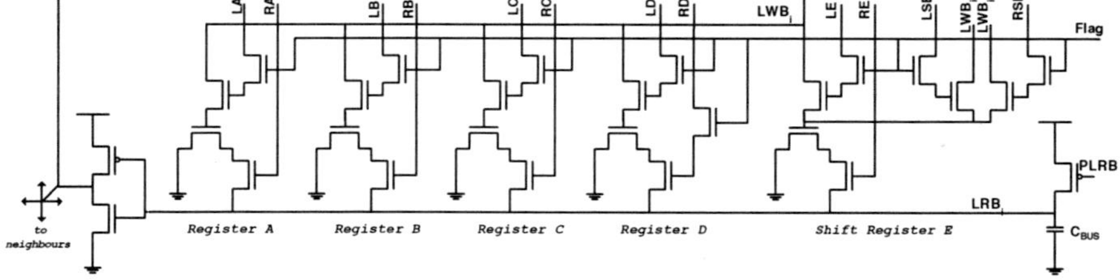


**Fig. 5** One bit slice of the pixel data path, including local memory and bus controller

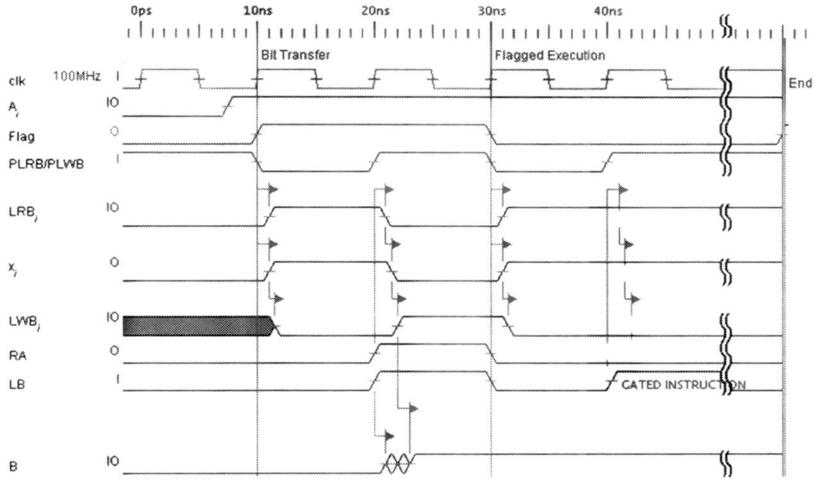


**Fig. 6** Timing diagram for data-path during basic bit transition at 100 MHz

ple of data-path operation during bit transition. During the first two clock cycles (10–30 ns), the logic '1' is transferred from memory $A_i$ to $B_i$. The following clock cycles (30–50 ns) demonstrate gating of the load instruction LB. The output of each memory cell is connected to the corresponding bit of the precharged Local Read Bus ($LRB_i$). If both the stored value and read control signal (RA, RB, RC, ...) are logic '1', then the transmission path from $LRB_i$ to ground is formed so that capacitance $C_{BUS}$ is discharged, otherwise the value at the output node $LRB_i$ is preserved. The leakage of the stored charge due to the reversed *pn* junction bias and sub-threshold leakage currents will eventually corrupt the value stored in a register; however, memory retention time is comparable with real-time frame rates and, if required, simple data refreshment can be performed. Logic level on the $LRB_i$ is sensed by inverter that drives input $INT_i$ of the BC. Depending on $INT_i$, the internal node $x_i$ is preserved to the precharged value, or discharged to logic '0'. The inverted value of $x_i$ is then transferred to $LWB_i$ and to the corresponding memory element.

Apart from multiplexing internal and external data into the LWB, BC can also perform bit-wise logic OR on input data. The inputs to the BC are a global input ($GLOB_i$), inverted LRB signal ($INT_i$), and inverted LRB signals from the four neighbours ($S_i, N_i, W_i, E_i$). Control signals ($sel_{GLOB}, sel_{INT}, sel_S, sel_N, sel_W, sel_E$) specify the value to be transferred to the LWB, thus enabling the following set of operations: GPR → GPR (ALU), Global I/O Bus → GPR (ALU), Neighbor → GPR (ALU).

In addition to standard memory, there are two bi-directional shift registers. Based on a modified memory cell, these registers allow data shift in both directions with a shift-in signal. Memory cells assigned to store the most and least significant bits (MSB and LSB) can be reconfigured to allow the two registers to act as a single 16-bit register, i.e. it is possible to shift data in two registers separately, as well

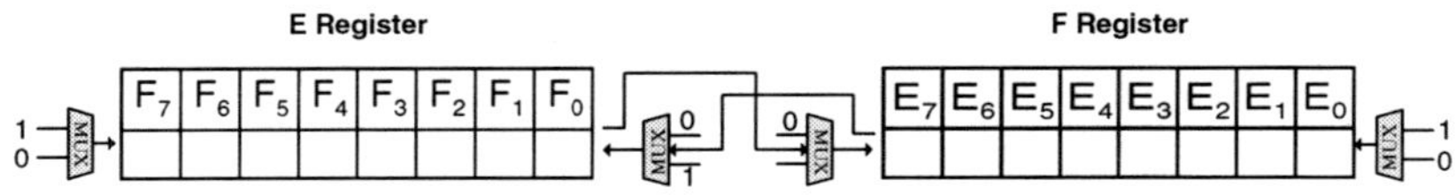


**Fig. 7** Shift register interconnection

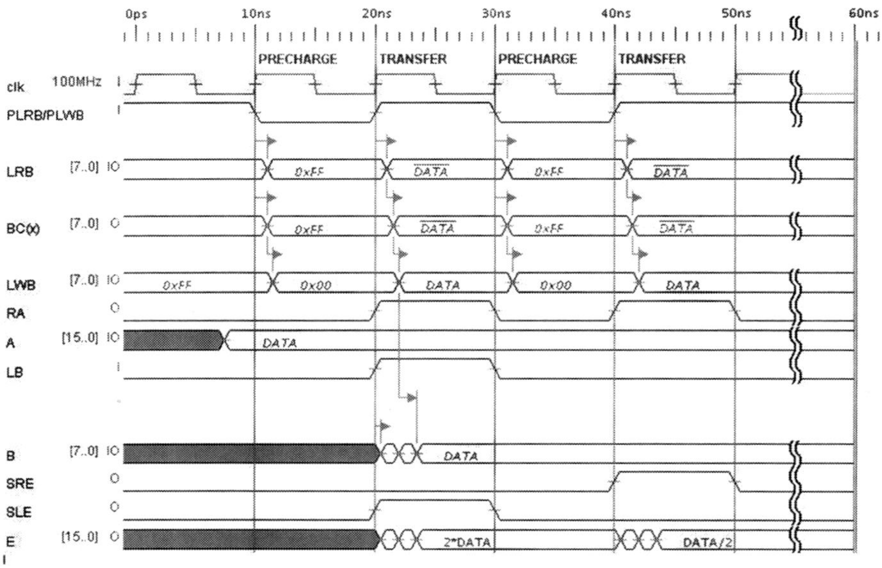


**Fig. 8** Timing diagram for two-phased register-transfer operation

as simultaneously so that the MSB of the register E is loaded into the LSB of the register F, and vice versa (Fig. 7). This feature enables efficient execution of global asynchronous distance transforms, unsigned multiplication and division.

The basic register-transfer operation comprises two clock cycles: *precharge* and *transfer*. The timing diagram in Fig. 8 explains the data transfer procedure based on simple program:

| **Time step** | **Code** | **Comment** | |
|---|---|---|---|
| 0: | mov B,A | //moves data from register A to register B | *Two commands can be combined because both have the same source* |
| 0: | sle E,A | //left shift data from reg. A to shift reg. E | |
| 1: | sre E,A | // right shift data from reg. A to shift reg. E | |

During the *precharge cycle*, both the LRB and an internal node *X* in the BC are precharged so that values on the LRB and the LWB are 0xFF and 0x00 correspondingly. During the *transfer* cycle, the data is read from specified register(s) or accumulator to the LRB, transferred through the BC to the LWB and then loaded into another register(s) or accumulator. At the end of the transfer cycle, when load signals are logic '0', the data is stored in the dynamic register. The communication between local neighbours is identical to local register-transfer operations and also performed within two clock cycles. If the data from one source needs to be transferred to a number of registers, these operations can be combined.

## 3.2 ALU

The ALU is designed to perform arithmetic and logic operations, so that every PE is capable of processing integer or fixed-point data. Restricted by area constraints, the ALU is based on bit-serial architecture. In addition, it can perform a logic XOR operation, which extends the set of logic operations executable in each cell.

The block-diagram of the ALU is presented in Fig. 9. It consists of an input multiplexer, full subtractor (SUB), two dynamic flip-flops (DFF) for storing *carry* (ACC_C) and temporary single bit result (ACC_R), demultiplexer and the accumulator for the final result (ACC). The input multiplexer is connected to the LWB and by applying control bit select signals ($bs_1 \ldots bs_7$), it is possible to choose, which bit is currently being processed. The operation of the ALU is based on binary subtraction. The basic binary subtraction, e.g. $A_i - B_i$ is executed in four clock cycles. It is very similar to two `mov` operations. During the first operation, the stored value in register `B` is read to the LWB. The appropriate $i$th bit is then selected and stored in the ACC_R and subsequently applied to the input `B` of the subtractor. The second operation reads value from register `A`, transfers it to the LWB, and similar to the previous one, multiplexes $i$th bit to the input A of the subtractor. The output of the subtractor will now hold the result of binary subtraction $A_i - B_i$ and on the next clock cycle it will be stored in the $i$th bit of the ACC.

In order to perform addition, a subtraction of one operand from zero is required to form a negative number. Multiplication and division could be realized by shift-and-add and shift-and-subtract algorithms, respectively. However, in cases when

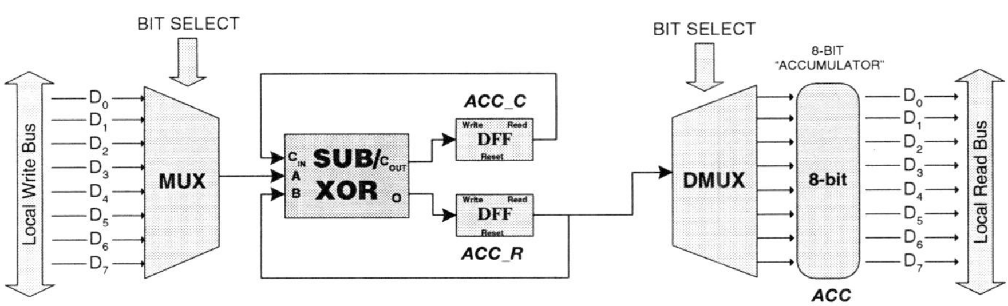


**Fig. 9** Bit-serial ALU

a register value is multiplied by a constant (e.g. $a \times B$, where $a$ – constant, $B$ – register), it is convenient to utilize the representation of this constant in binary format as $a = \sum a_i 2^i$, so that multiplication $a \times B = B \times \sum a_i 2^i$ is realized by a limited number of shift operations and additions or subtractions. Simple analysis shows that multiplication by any factor smaller that 20 takes a maximum of two arithmetic and six shift operations.

### *3.3 Flag*

The conditional execution of broadcast instructions is implemented by introducing an activity flag mechanism. Conditional operation is based on the fact that some instruction lines contain pass transistors, which are controlled by a flag value. If the PE has flag set to logic '0', the following instruction signals are 'ignored' until flag is reset back to logic '1'.

The schematic diagram of the FR is shown in Fig. 10. The flag value can be selected based on three inputs: carry (*C*), propagation (*P*) and binary output from the photo-sensor (*PIX*). These inputs are selected by corresponding signals *SFC*, *SFP* and *SFX*. The actual flag value is stored on the node $F_{stor}$, which is connected to the gates of pass transistors on flagged instruction lines. Two control signals *RIFL* and *RFL* define whether complement or direct value of the specified input is transferred to $F_{stor}$. During unconditional operation, $F_{stor}$ is continuously precharged by setting *SFG*, *RFL* and *RIFL* to logic '0'. If a conditional instruction such as `if C` (or `if !C`) is met, then *SFG, SFC, RFL* (or *RIFL*) are set to '1', so that value *C* (!*C*) appears at node $F_{stor}$. The FR is also used for random access pixel addressing. The pixel is selected when both $\text{SEL}_{\text{COL}}$ and $\text{SEL}_{\text{ROW}}$ are logic '0'.

For the purpose of design optimization, conditional execution is only applied to load operations (*LA, LB, LC, LD, LSE, LE, RSE, LFE, LF, RFE, LG*) and instructions

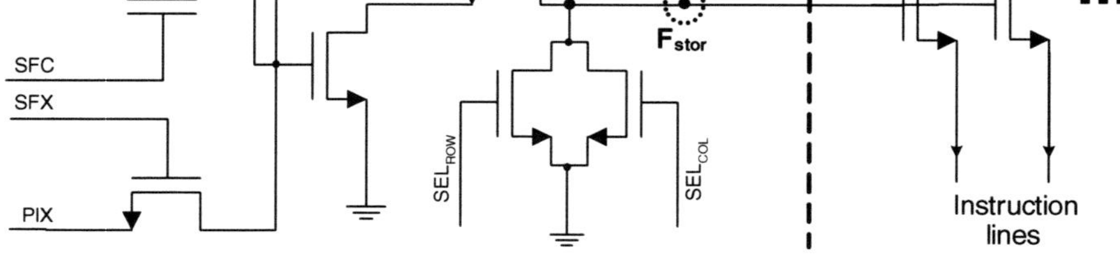


**Fig. 10** Flag register

controlling the flow of asynchronous propagations (*PFL, PL, PG*). In addition, it controls one register-read operation (*RD*) to enable division operations, flexible data transfers and data output. In other words, if a conditional instruction is met, pixels where $F_{stor}$ is logic '0' still execute all read and arithmetic operations, but ignore all memory write commands. Such approach reduces the number of 'pass' transistors and correspondingly the design size, yet at the price of limited power optimization.

Nested conditions can be realized only with complement flag values, i.e. `if !C` or `if !P`, because every such condition is expected to define a subspace of the space defined by the previous conditional operation. This is only possible when the node $F_{stor}$ is discharged since this operation can only reduce the number of previously flagged pixels.

The simple scenario of conditional operation looks as follows:

```
if C // ... set SFG, SF2 and RFL to logic '1'
  mov a,b
  add b,c
  ...
endif //... set SFG, SF2 and RFL to logic '0'
```

An example of nested conditional branching code is listed below. After the $F_{stor}$ value is set based on *P* flag, the following instructions are executed only in those pixels, where *P* is logic '0'. The second conditioning is performed on complement value of the *C* flag. If *C* is logic '0', the $F_{stor}$ node will keep the value set by previous operation `if !P`, otherwise if $C =$'1' the $F_{stor}$ will be discharged, the following instruction will be ignored. In other words, operations `sub g,d` and `shl e,g` will be performed only in pixels, where both *C* and *P* are logic '0' (e.g. `if !P && !C`).

```
if !P // ... set SFG,SF1 and RIFL to logic '1'
  mov a,b //set SF1,RIFL back to logic '0'
  add b,c
  ...
  if !C //SF2 and RIFL to logic '1'
    sub g,d
    shl e,g
    ...
endif //... set SFG,SF2 and RFL to logic '0'
```

## 3.4 Photo Sensor

A photo sensing circuit in every PE is based on a photodiode followed by a simple voltage comparator (Fig. 11). The photo detector works in an integration mode. The voltage across photodiode $V_{PD}$ is compared to the reference voltage $V_{ref}$. The inverted value of the comparator output is transferred to *PIX* input of the FR. Based on this flag indicator, a digital value that corresponds to light intensity can be written to

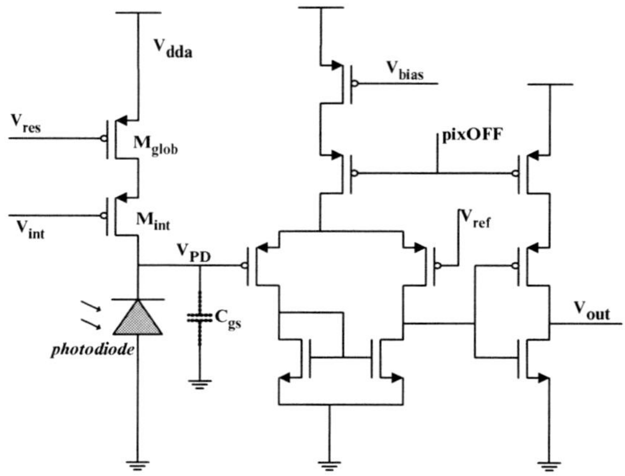


**Fig. 11** Photo sensor circuit

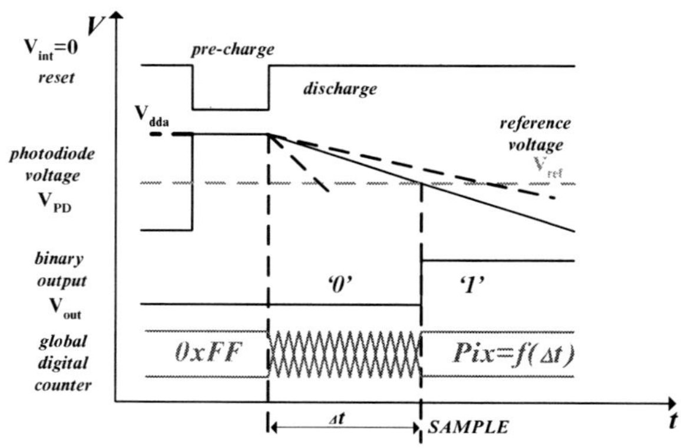


**Fig. 12** Light intensity to digital data conversion

a corresponding GPR. For grey-level conversion, the binary threshold is performed at different times (Fig. 12) and/or with multiple $V_{ref}$ levels. Signal *pixOFF* is used to turn off the biasing currents when the sensor is not accessed.

During reset phase, $V_{res}$ and $V_{int}$ are set into logic '0' so that switches $M_{glob}$ and $M_{int}$ are closed and capacitance $C_{PD}$ is charged with $V_{PD} = V_{dda}$. While $V_{res}$ is broadcast globally to all PEs at the same time, $V_{int}$ is a locally controlled signal connected to $LRB_7$. Such configuration enables local control over pixel reset/integration time. From a functionality perspective, it allows additional processing, e.g. adaptive sensing and data compression while sensing.

The relationship between light intensity, exposure time and photodiode voltage can be approximated by a linear equation (1). Here, $L$ is light intensity, $t$ is time after precharge stage, $a$ is a constant.

$$V_{PD} = V_{dda} - a \times L \times t. \tag{1}$$

The real relationship is not always linear (especially in corners of the range) due to non-linearity of the photodiode capacitance and photosensitivity; however, this problem can be addressed by appropriate calibration. Hence for a linear relation between the counter value and light intensity, a non-linear counter sequence must be used [35]. From the functionality perspective, the time-domain conversion allows wide dynamic range sensing and additional processing during photocurrent integration, which can be efficiently used for adaptive sensing.

## 3.5 Data I/O

The ASPA chip provides random access data I/O operations. The data to be output must be placed in the register dedicated for global read out. The difference between this register and others is that the read operation of this register is controlled by the flag, so that only addressed pixels output data, whereas other pixels outputs are in high impedance state. In practice, data from any GPR or accumulator can be output, but because they are output to the same bus, the result value would be logic OR of all the outputs. The procedure of outputting data to the global bus is identical to local register-transfer operation, but instead of writing to a local register, the $SEL_{GLOB}$ signal is set to logic '1' and data from the LRB is transferred to the precharged global data bus (GDB) by discharging it. In a similar way (through a precharged bus, the local flag value is output to the global flag bus and then to the central controller.

ASPA chip also supports a flexible random-access addressing mode [32] that provides access to individual pixels, blocks of pixels and pixel patterns. Column and row addresses are represented by two 8-bit words: 'Select Word' and 'Select X Word'. 'Select Word' is the conventional address, e.g. $COL_{SEL}$ ='0001' represents first column. 'Select Word X' represents *don't care* values or a part of the address that can be represented both by zeroes and ones. An example of address formation is illustrated in Fig. 13. As a result of the illustrated operation, a set of addresses representing first and fifth column or row is obtained.

The addressing can be therefore divided into three modes.

*Mode 1:* This mode is used for individual pixel addressing. In this case, none of the Select Word X signals is used. This mode is usually used for output and input of image data.

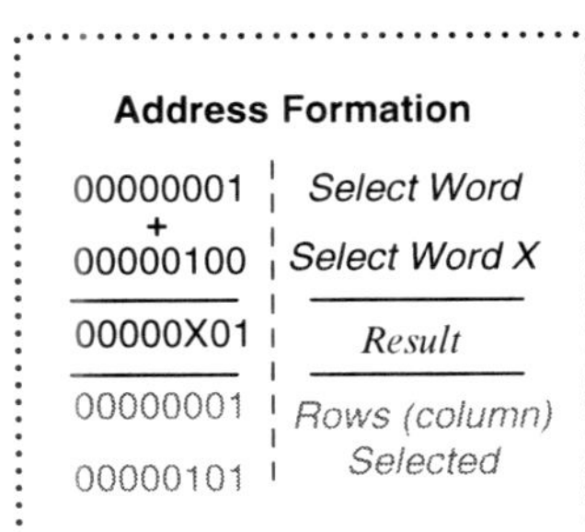


**Fig. 13** Formation of address word

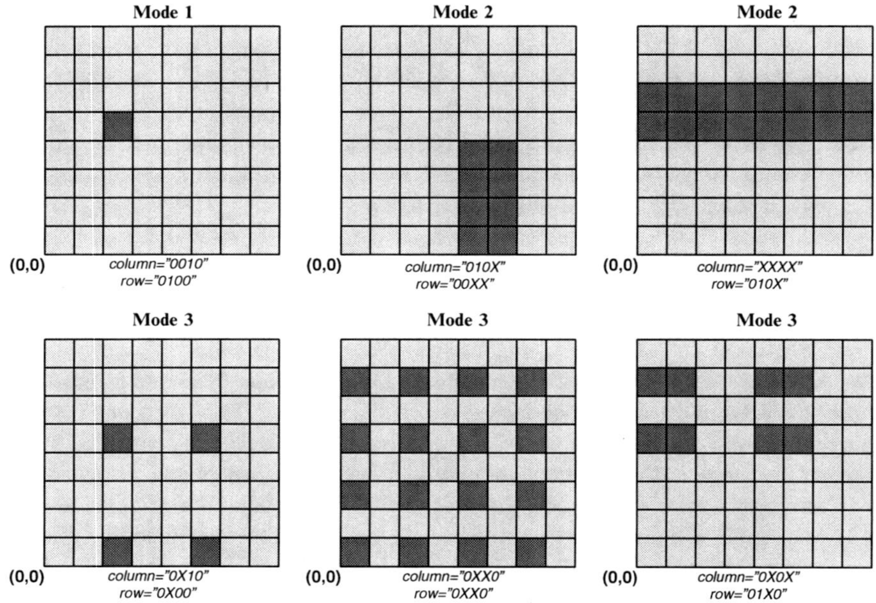


**Fig. 14** Flexible pixel addressing

*Mode 2:* This mode is used for pixel block selection. For such operation, the least significant bits of Select Word X are set to logic '1' (Fig. 14). In order to select the entire array Select Word X is set to '11111111', thus making the result address 'XXXXXXXX'. As follows from its description, this mode is used for block pixel processing.

*Mode 3:* This mode is used for periodic pixel selection. It is achieved by setting the most significant bits into a *don't care* value. Let us assume that the address is represented by an N-bit vector and $m$ most significant bits are set to a *don't care* value. Let us also assume that N–$m$ least significant bits represent the decimal value $k, k < 2^{N-m}$. Then this address word will correspond to the primitives (rows or columns) with following addresses: $k, 2^{N-m}+k, 2\times 2^{N-m}+k, 3\times 2^{N-m}+k, \ldots, (2^m-1)2^{N-m}+k$. Essentially, it represents the sequence of addresses with period $2^{N-m}$. This mode can be used for linear discrete transformation, 2D block data distribution, multiresolution image processing and processor clustering.

## 3.6 Periphery

One of the frequently used tasks in image processing is pixel search algorithm. The conventional approach to this problem is based on binary tree search algorithm, with the complexity of $O(\log_2 N)$. Being one of the fundamental operations

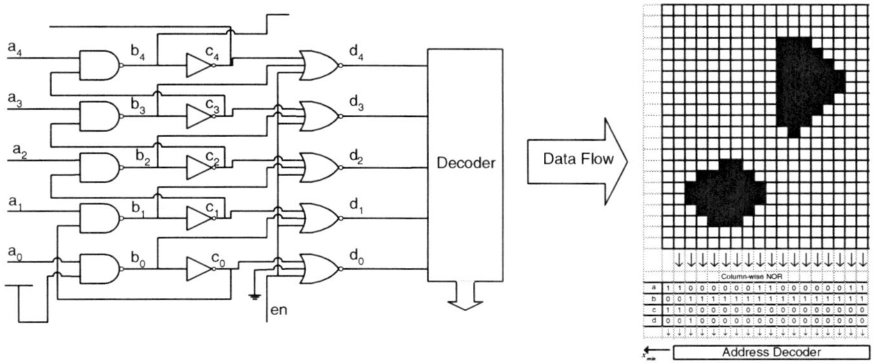


**Fig. 15** Asynchronous address extraction. The periphery circuits extracts $x_{\min}$ and $y_{\min}$ coordinates of segmented objects

in medium-level image analysis, it is beneficial to have this frequently used routine hardware optimized. A simple peripheral circuit that enables instant address extraction of pixels with smallest column and row indices previously marked by flag indicators is introduced in ASPA in order to enable AER readout mode.

The schematic diagram of the circuit and corresponding data flow for an example image are illustrated in Fig. 15. In the figure, $a_i$ represents the value of the $i$th column (row) flag bus. The data asynchronously transforms according to the following equations:

$$c_0 = \overline{b_i} = a_0$$

$$d_0 = \overline{en + c_0}$$

$$b_i = \overline{a_i \cdot c_{i-1}}$$

$$c_i = \overline{b_i}$$

$$d_i = \overline{en + b_{i-1} + c_i}.$$

The data transition is presented in Fig. 15. The output vector $d$ is then encoded to represent an address of the column/row with the minimum index among the marked pixels. The column can then be selected and followed by similar operation for rows to achieve $(x, y)$ coordinates of the object.

## 3.7 Instruction Set

Due to its SIMD architecture, the program for the ASPA chip effectively corresponds to the program for an individual cell, with some exceptions for global operations. The ASPA instruction set comprises register-transfer operations (including global transfers), conditional branching, arithmetic operations, global asynchronous operations and photo-sensing instructions. Presented in Table 1 are a number of most common instructions for an individual PE. It should be emphasized that it does not cover the full range of possible operations, as different operations can be combined together and form a new operation. For example, it is possible to run asynchronous propagations and perform the register-transfer operation at the same time.

### 3.8 Layout

The layout of a complete cell is presented in Fig. 16. The basic structures of the cell such as registers, BC, ALU and FR are marked.

The prototype design was fabricated in 0.35 μm 4-metal CMOS technology, which at the time of design was the most cost-effective way to fabricate the prototype ICs. Overall dimensions of the cell are $100 \times 117\,\mu m^2$ (thus providing cell density of 85 cells $mm^{-2}$) with $11.75 \times 15.7\,\mu m^2$ utilized as a photo-sensor. As the main purpose of this chip was to test the processing functionality of the architecture, the sensing part was given less priority. Therefore, the achieved fill-factor of ~2% is relatively small. Potentially, the area of the sensor could be extended underneath the wiring reaching $810\,\mu m^2$, which would provide fill-factor of more than 7%. Further improvement of the apparent fill-factor could be achieved by placing in-situ micro lens above each cell.

The PE circuitry consists of 460 transistors, 94% of which are $n$-type transistors (most of them have minimum size), which results in high transistor density, especially in memory elements. The basic memory cell occupies $3.6 \times 8.7\,\mu m^2$ so that 40 bits of internal memory (not including the shift registers and the accumulator) occupy $18 \times 70\,\mu m^2$ or only 10% of the pixel area. Since the memory elements are based on dynamic logic, the majority of charge storing nodes were shielded from the instruction lines by the ground plane to avoid cross coupling.

The microphotograph of the ASPA chip is provided in Fig. 17. A $19 \times 22$ PE array has been fabricated on a $9\,mm^2$ prototype chip. It should be emphasized that this architecture is scalable, i.e. array size may vary, although attention has to be paid to the periphery instruction drivers, since for larger arrays the load capacitance and resistance on instruction wires that go across entire array increases proportionally. Operating at 2.5 V, each cell consumes $63.04\,\mu W$, so that the entire array provides 38 GOPS/W for grey-scale operations. The extrapolated energy efficiency in asynchronous operations is approximately 2.5 TOPS/W (0.4 pJ/cell per propagation). Operating at 75 MHz, the ASPA chip delivers 15.6 GOPS for binary and 1 GOPS for grey-scale operations. In spite of the relatively large cell size and falling behind analogue designs in terms of power efficiency, a fully digital approach promises simple adaptation to much finer technology, resulting in small pitch and reduced power consumption.

While designing a compact processing cell based on digital architecture, one of the major challenges is the complicated routing of wires carrying control and data

**Table 1** Selected ASPA instructions

| Notation | Function | Microcodes | | | | | | | | |
|---|---|---|---|---|---|---|---|---|---|---|
| **nop** | No operation, default vector | RSTC, RSTACC, LGD, ld_c, ld_cx, ld_r, ld_rx, LADDR, RST_ADDR = 1, PFL, pixOFF = '1', all others are '0' | | | | | | | | |
| | | PLRB | PLWB | ... | LA | ... | RB | ... | INT_SEL | ... |
| **mov** A, B[a] | Load register A with value from register B: $A = B$ | 0 | 0 | ... | 0 | ... | 0 | ... | 0 | |
| | | 1 | 1 | | 1 | | 1 | | 1 | |
| | | PLRB | PLWB | ... | LA | ... | RB | RC | INT_SEL | ... |
| **or** A, B, C | Load register A with logic OR of B and C: $A = B \parallel C$ | 0 | 0 | ... | 0 | ... | 0 | 0 | 0 | ... |
| | | 1 | 1 | | 1 | | 1 | 1 | 1 | |
| | | PLRB | PLWB | ... | LSE | ... | RB | ... | INT_SEL | ... |
| **shl** E, B[b] | Shift left the value from register B and store in E: $E = B << 1$ | 0 | 0 | ... | 0 | ... | 0 | ... | 0 | ... |
| | | 1 | 1 | | 1 | | 1 | | 1 | |
| | | PLRB | PLWB | ... | RSE | ... | RB | ... | INT_SEL | ... |
| **shr** E, B | Shift right the value from register B and store in E: $E = B >> 1$ | 0 | 0 | ... | 0 | ... | 0 | ... | 0 | ... |
| | | 1 | 1 | | 1 | | 1 | | 1 | |
| | | PLRB | PLWB | ... | LSE | SHFT | RB | ... | INT_SEL | ... |
| **shlx** E, B | Shift left the value from register B, store in E and $LSB(E) = MSB(F)$ | 0 | 0 | ... | 0 | 0 | 0 | ... | 0 | ... |
| | | 1 | 1 | | 1 | 1 | 1 | | 1 | |
| | | PLRB | PLWB | ... | RSF | SHFT | RB | ... | INT_SEL | ... |
| **shrx** F,B | Shift right the value from register B, store in F and $MSB(F) = LSB(E)$ | 0 | 0 | ... | 0 | 0 | 0 | ... | 0 | ... |
| | | 1 | 1 | | 1 | 1 | 1 | | 1 | |
| **shl16** | Shift left registers E and F as one 16-bit register: $EF = EF << 1$; | **shlx** E,E<br>**shl** F,F | | | | | | | | |
| **shr16** | Shift right registers E and F as one 16-bit register: $EF = EF >> 1$; | **shrx** F,F<br>**shl** E,E | | | | | | | | |

(continued)

**Table 1** (continued)

| Notation | Function | Microcodes | | | | | | | | | | | |
|---|---|---|---|---|---|---|---|---|---|---|---|---|---|
| | | PLRB | PLWB | RA | RB | INT_SEL | RSTC | ReadC | LCarry | RSTACC | LACC | NLACC | $BS_i$ |
| **sub** A,B,i | Subtract *i*th bit of B from *i*th bit of A, result in $ACC_i : ACC_i = A_i - B_i$. The carry flag value is stored in a dedicated register ACC_C | 0 | 0 | 0 | 0 | 0 | 1 | 0 | 0 | 1 | 0 | 0 | 1 |
| | | 1 | 1 | 0 | 1 | 1 | 0 | 0 | 0 | 0 | 1 | 0 | 1 |
| | | 0 | 0 | 0 | 0 | 0 | 1 | 1 | 0 | 1 | 0 | 1 | 1 |
| | | 1 | 1 | 1 | 0 | 1 | 1 | 0 | 1 | 1 | 1 | 0 | 1 |
| **sub** A,B[c] | Subtract the value in register B from the value in register A, result in ACC: $ACC = A - B$ | **for** $(i = 0; i < 8; i++)$ **sub** A,B,i | | | | | | | | | | | |
| **add** A,B | Add register A to register B, result in register A: $A = A + B$ | **sub** 0x00, A **mov** A, ACC **sub** B,A **mov** A,ACC | | | | | | | | | | | |
| | | ReadC | … | SFG | RFL | … | SF2 | … | … | … | | | |
| **if** C[d] | Check whether the flag value is '1'. If $C = 1$, following instructions are executed | 1 | … | 0 | 0 | … | 0 | … | … | … | | | |
| | | 1 | | 1 | 1 | | 1 | | | | | | |
| | | ReadC | … | SFG | … | RIFL | … | SF2 | … | … | | | |
| **if** !C | Check whether the flag value is '0'. If $C = 0$ – following instructions are executed | 1 | | 0 | | 0 | | 1 | | | | | |
| | | 1 | | 1 | | 1 | | 1 | | | | | |
| | | $BS_0$ | … | SFG | … | … | … | PFL | PG | PL | | | |
| **prop** | Run asynchronous propagation, stores the result in prop. latch | 1 | | 1 | | | | 0 | 0 | 1 | | | |
| | | 0 | | 1 | | | | 0 | 1 | 1 | | | |

[a]Additional index can be used to indicate inter-pixel register-transfer, e.g. mov A,BN indicates that the value from register B is transferred to the register A of pixels south neighbour. An immediate operand is used for global data transfers only, e.g. mov A, 0xFF

[b]There are only two shift registers: E and F. Therefore, the first operand can be either E or F

[c]It is also possible to specify the bit range for this operation, i.e. sub A (4:7), B (4:7)

[d]In order to read the flag value globally (for example for global OR operation, coordinate extraction), the GLOB_SEL must be set to '1' during this operation

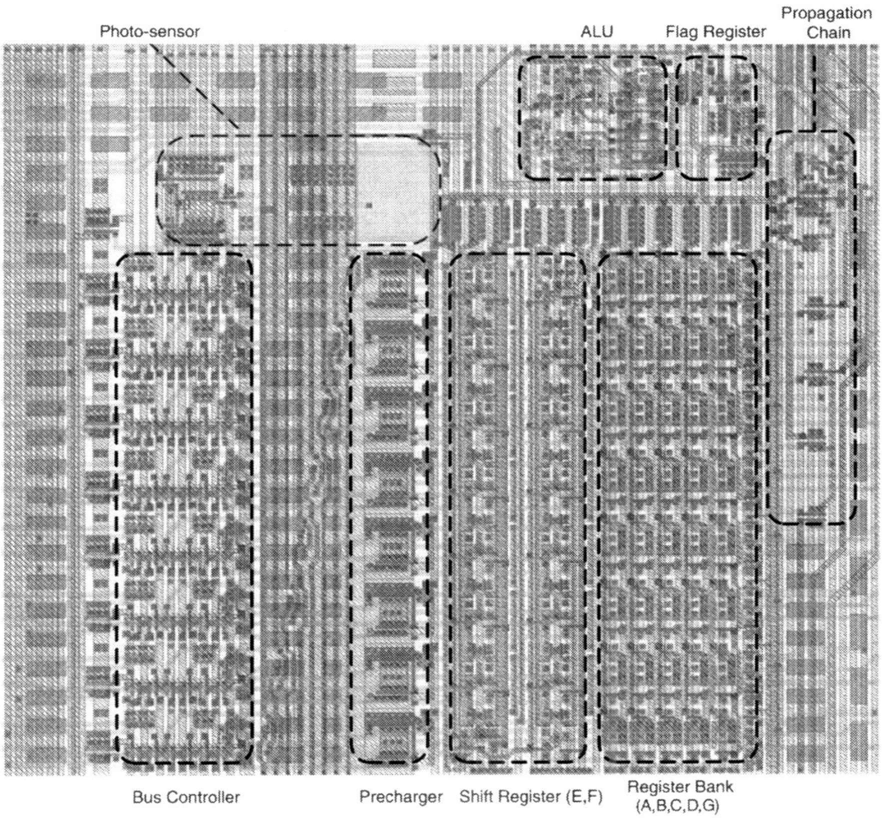


**Fig. 16** Layout of the PE in standard 0.35 μm CMOS technology

signals. The estimated area in the ASPA chip used only for routing (without any active logic) is approximately 32% of the cell area. Most of this area is used for routing of global and local data lines. It suggests that for a fabrication process with up to four metal layers, the area excess brought by digital bit-parallel architecture is significant and for an area effective solution one should look at either fully bit-serial or analogue architectures. However, our designs indicate that if we consider a fabrication process with six metal layers instead of four, the cell area can be reduced by more than 30%. Therefore, it can be concluded that the area utilization for digital design improves with advanced technology that provides more routing layers. We have recently designed a similar processor array [36] in 0.18 μm technology achieving 363.1 cells $mm^{-2}$. Further improvements in cell density are expected on finer technologies, considering 1,600 cells $mm^{-2}$ to be achievable on 65 nm.

Sub-100 nm technologies will require additional solutions to counter high leakage currents in dynamic registers and stronger crosstalks. However, small retention time can be addressed by a memory refresh routine, which may represent relatively small performance sacrifice compared to the achieved benefits.

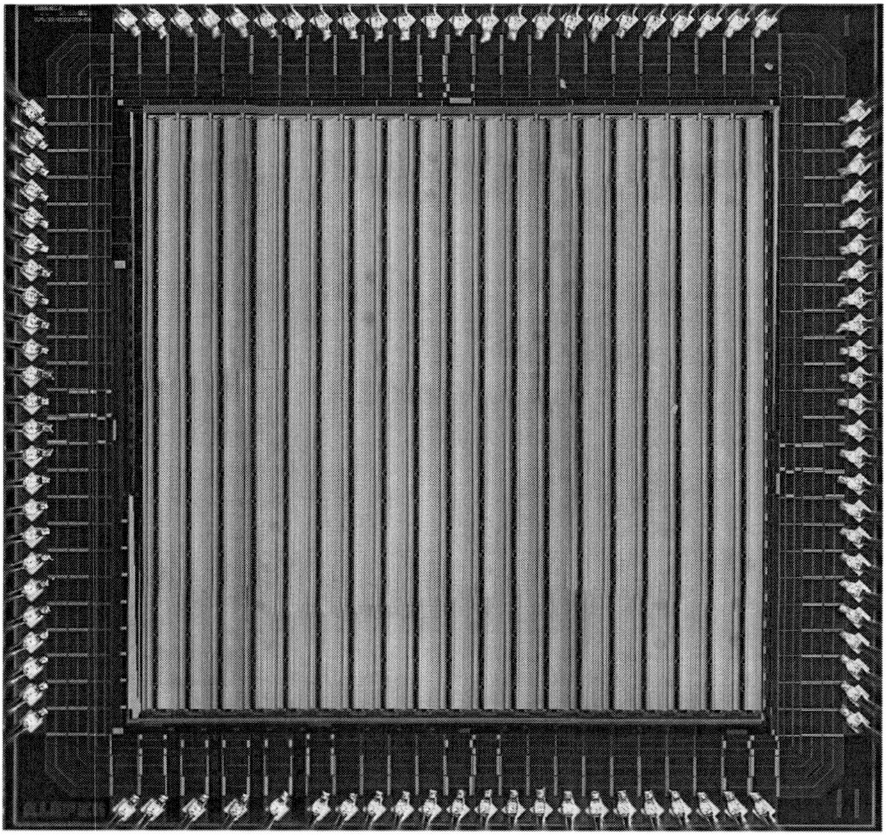

**Fig. 17** Microphotograph of the ASPA

# 4 Image Processing

## *4.1 Local Operations*

The main benefit of general-purpose vision chips is their software-programmable architecture, so that a wide range of image processing algorithms can be coded in software. The main difference between coding for conventional serial microprocessors and SIMD parallel processor arrays is that in the latter case the program is interpreted from the point of view of individual processing cell, with the exception of global operations. From this perspective, the approach to developing image processing code has to be adjusted. As an example, the comparison of code for basic binary threshold (with threshold level 128) on serial and pixel-parallel architectures is presented in Fig. 18.

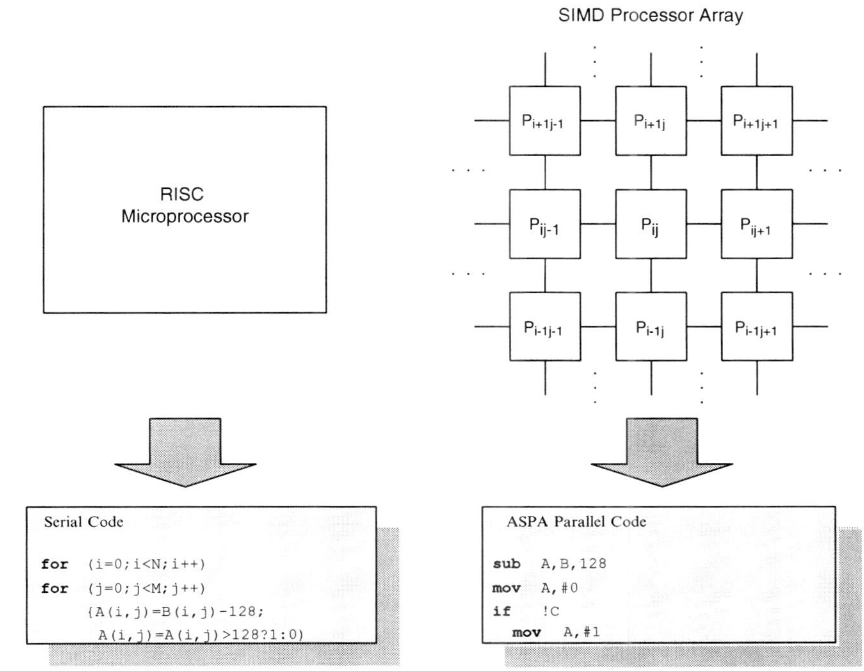


**Fig. 18** Different approach towards code development for serial and parallel architectures

The instruction set for ASPA vision chip consists of local instructions executed by every cell, and global instructions involving either periphery units or asynchronous data manipulation on the array. Every cell supports the set of basic operations necessary to implement convolutions and filtering. Some examples of basic image preprocessing with corresponding execution times are presented in Fig. 19.

Because all data is represented and processed in digital domain, it is straightforward to verify the results versus numeric simulations. A significant number of useful morphological image processing algorithms deal with images represented in binary format (binary dilation, erosion, skeletonization) and the ASPA can execute many of these tasks in an efficient way. Code examples for Sobel edge detection, smoothing (these are performed on grey-scale images in $3 \times 3$ neighbourhood) and closing operations are presented in Fig. 19a–d.

## 4.2 *Global Pixel-Parallel Operations*

### 4.2.1 Trigger-Wave Propagation

Apart from local operations, the ASPA chip supports global binary trigger-wave propagation, which is a basic global operation that can be used for geodesic

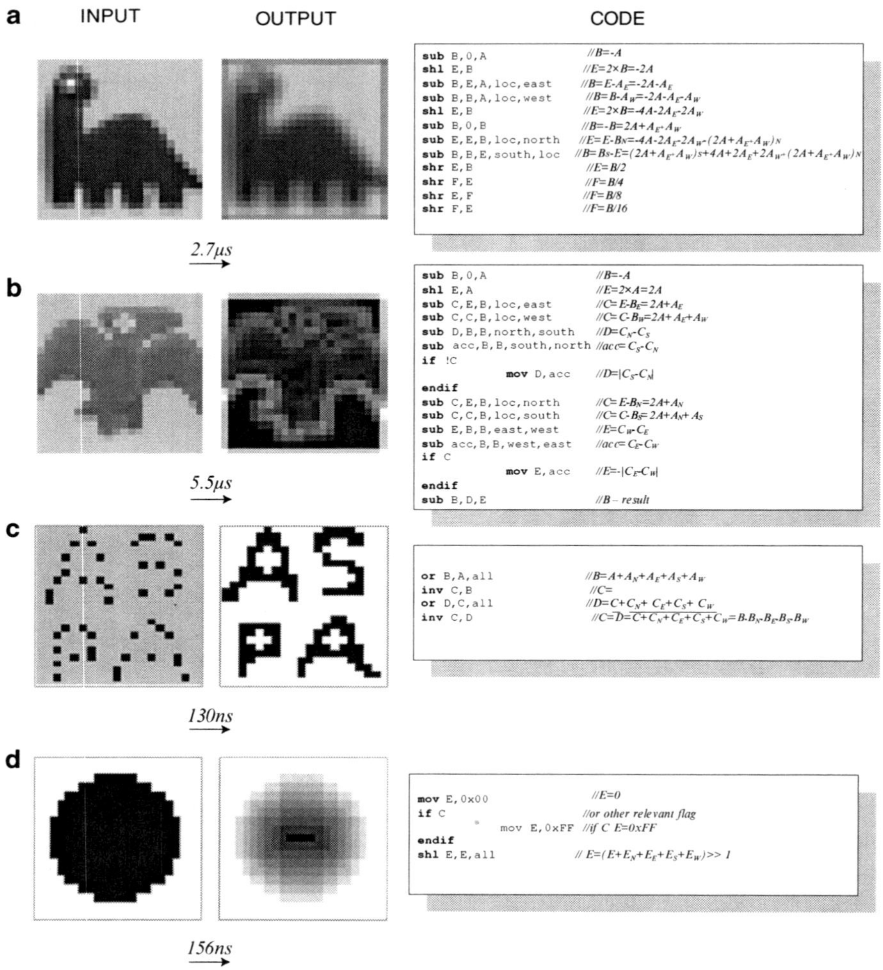


**Fig. 19** Low-level image processing in ASPA: (**a**) smoothing; (**b**) Sobel edge detection; (**c**) reconstruction by closing; (**d**) asynchronous distance transformation

reconstruction, hole filling (Fig. 20) and as part of more complex algorithms [37]. Although such operation can easily be implemented in synchronous fashion as an iterative process, such realization is inefficient in terms of power and overall performance, as the number of required iterations should cover 'worst case', e.g. $O(n \times m)$ for $n \times m$ image. With the propagation delay of 0.76 ns per pixel, the equivalent synchronous implementation on ASPA in an iterative manner would require operation at 2.7 GHz (two operations per cell).

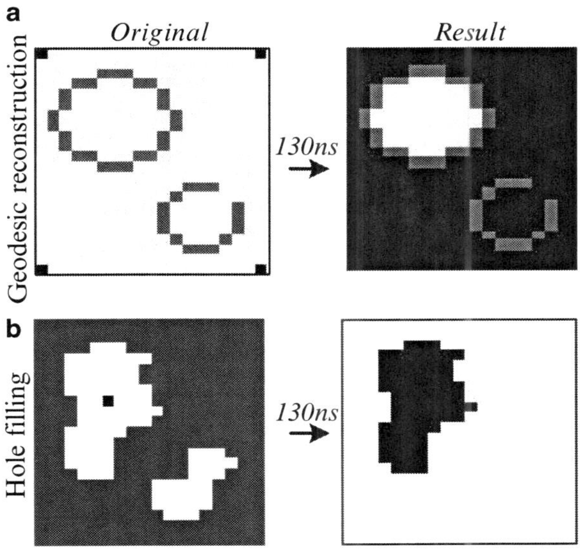


**Fig. 20** Global operations on the ASPA chip: (**a**) geodesic reconstruction; (**b**) asynchronous hole filling

### 4.2.2 Distance Transformation

By utilizing the features of the BC and shift registers, it is possible to perform simple processing while transferring data between cells. In particular, it is possible to perform a simplified version of the increment operation.

Let us consider the value '11111111' being stored in one of the two shift registers present in a PE. A shift left operation will translate this value to '11111110'. By applying this operation $m$ times ($m < 9$), we will achieve $m$ zeroes in least significant bits. Let us consider the chain of PEs with the shift register F of the processing cell $PE_{ik}$ loaded with '11111111' (Fig. 21). In other PEs, register F is loaded with '00000000'. Then, a simple distance transform can be computed by reading the shift register and loading it with data from the neighbours at the same time (**shl** $\mathrm{E}, \mathrm{E_N}, \mathrm{E_E}, \mathrm{E_S}, \mathrm{E_W}$). If necessary, an additional step can be employed to interpret the value as a binary number. The zero value indicates that processing cell $PE_{ik+N}$ is out of range of 7-pixel radius.

Taking into account that the BC executes a bitwise OR operation, it is easy to notice that this operation, when extended to a 2D array, provides a minimum function in the context of operated numbers (Fig. 22). The minimum is selected among the values of four neighbours and local data. Let us consider the processing of binary images (Fig. 18d). If we assign 0xFF to all the background pixels, 0x00 to foreground pixels and perform the operation described above, we will achieve a distance map using city block distance within 8 pixel range (0xFF corresponds to

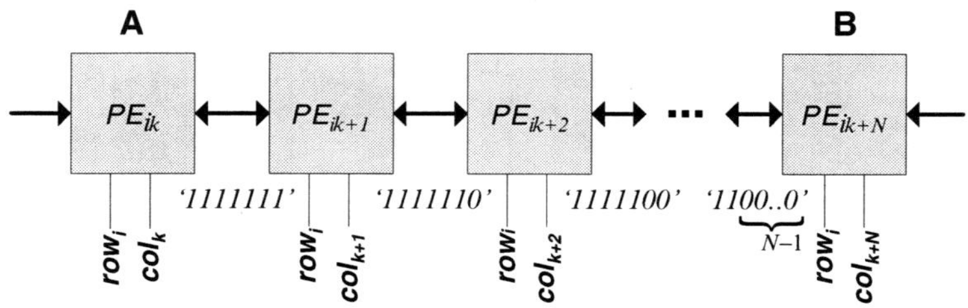


**Fig. 21** Asynchronous distance transform

| Register Data | | Interpreted Distance |
|---|---|---|
| HEX | BIN | |
| 0xFF | 11111111 | 0 |
| 0xFE | 11111110 | 1 |
| 0xFC | 11111100 | 2 |
| 0xF8 | 11111000 | 3 |
| 0xF0 | 11110000 | 4 |
| 0xE0 | 11100000 | 5 |
| 0xC0 | 11000000 | 6 |
| 0x80 | 10000000 | 7 |
| 0x00 | 00000000 | out of range |

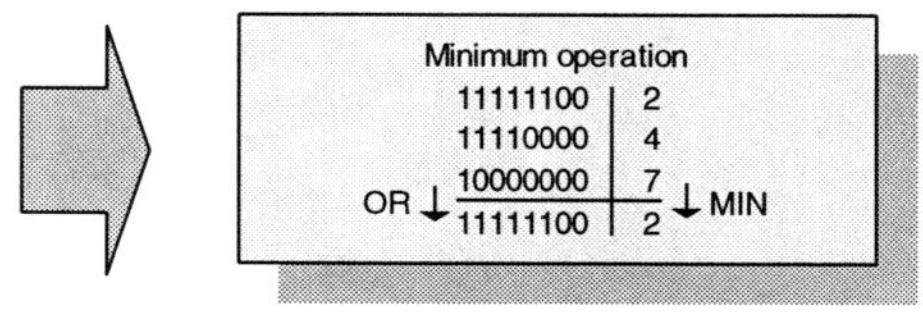


**Fig. 22** Binary data interpretation during asynchronous transitions

distance 0, 0x00 – to 8). If necessary, the full distance map can be calculated by applying the above procedure consequently $[R/8]+1$ times, where R is the maximum object radius.

### 4.2.3 Object Localization

There are a number of algorithms that require a fast location of the object, segmented from the background. Let us consider a simple example of address extraction with two segmented objects (Fig. 23a). In order to locate all objects in the image, it is necessary to execute the following steps:

1. Set flag for marked objects: 2 operations (Fig. 23a);
2. Read out flags and perform global OR operation: 1 operation;
3. Terminate if there are no marked pixels, i.e. OUT =0xFF(1 operation);
4. Read out extracted column-address(Fig. 23b): 1 operation; (addr = 3);
5. Select the propagation space (dark grey in Fig. 23c): 1 operation;
6. Select 3rd column (light grey in Fig. 23c): 1 operation;
7. Run propagation from selected pixels in the propagation space (black in Fig. 23d) and store propagation flag: 1 operation;
8. If propagation flag is '1', read out flags, i.e. extract the row-address (Fig. 23e): 1 operation;
9. If propagation flag is '1', remove the marker: 1 operation;
10. Go to 1;

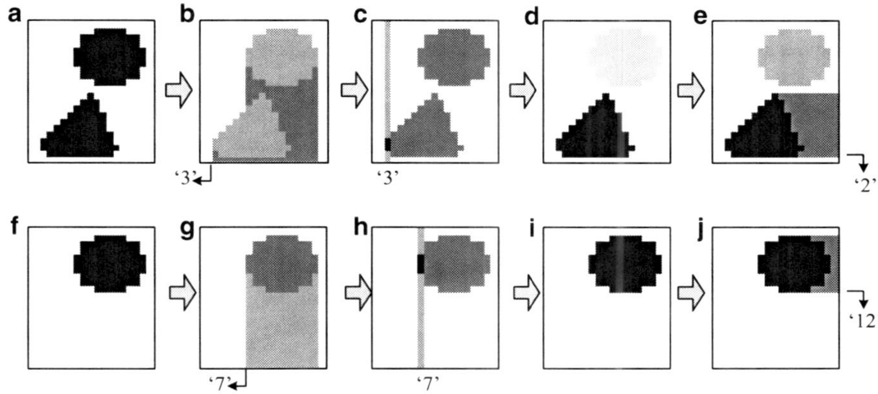


**Fig. 23** Coordinate extraction: (**a**) Two marked objects are selected; (**b**) The flags of selected objects are read globally, and the coordinate of the most left flagged pixel is extracted; (**c**) Objects are selected (*dark grey*) as a propagation space, third column is selected (*light grey*), propagation is initiated from the intersection of two selections (*black*); (**d**) Left object is selected; (**e**) Same as (**b**) for row-address; (**f**) Left object is unmarked; (**f–j**) are same as (**a–e**) but for the second object

In the presented example, this procedure is repeated for each object. At the end of this operation, coordinates of objects' borders are extracted. Extracting the boundary box of one binary object takes 18 clock cycles at 75 MHz clock or 240 ns. It should be noted that only one iteration is required for address extraction, as the flag values, which form the address value, are read in step 2 and there is enough time for signal to propagate across address extraction circuit.

The above-described basic global operations can be efficiently used as a part of computationally expensive algorithms, such as skeletonization, object tracking, recognition and watershed transformation [37].

# 5 Conclusions

The main challenge in early vision is high computational load, as the processing time is generally proportional to the image pixel resolution. Medium-level processing algorithms (e.g. object reconstruction, segmentation) pose further challenges. However, the high degree of data-parallelism motivates the design of an efficient yet compact massively parallel fine-grain processor-per-pixel architecture that fills the gap between image sensing and high-level processing. Such a device naturally maps onto parallel data structure and performs sensing, preprocessing and intermediate image analysis on a single die, outputting mainly essential image descriptors.

In order to achieve this goal, it is important to enable efficient execution of the global operations, which form the basis for many medium-level processing algorithms. The majority of complex algorithms can be efficiently decomposed into a small set of basic global operations, such as binary trigger-wave propagation, distance transformation, global summation, address extraction and global data transfer. The main benefit of these operations is that they can be implemented with relatively small hardware overhead and executed in asynchronous, single-step manner, thus significantly increasing overall performance. Yet, despite significant speed-up, it is important to estimate the area efficiency of asynchronous design, as global operations contribute only a part in overall preprocessing.

The solution presented in this chapter combines synchronous and asynchronous approaches in order to extend the range of vision chip applications from low- to medium-level image processing. The chip contains a $19 \times 22$ general-purpose processor-per-pixel array, whose functionality is determined by a program, broadcast by an off-chip controller. While serving as a proof-of-concept design, it demonstrates the feasibility of efficient global operations, executed in an asynchronous manner.

## References


1. G. M. Amdahl, Validity of the single-processor approach to achieving large-scale computing capabilities, in Proceedings of American Federation of Information Processing Societies Conference, pp. 483–485, 1967
2. M. J. B. Duff and D. M. Watson, The cellular logic array image processor, Computer Journal, 20, 68–72, 1977/02/ 1977
3. K. E. Batcher, Design of a massively parallel processor, IEEE Transactions on Computers, c-29, 836–840, 1980
4. K. Aizawa, Computational sensors-vision VLSI, IEICE Transactions on Information and Systems, E82-D, 580–588, 1999
5. A. Moini, Vision Chips or Seeing Silicon, Kluwer, Dordrecht, 1999
6. A. A. Abbo, R. P. Kleihorst, V. Choudhary, L. Sevat, P. Wielage, S. Mouy, B. Vermeulen, and M. Heijligers, Xetal-II: A 107 GOPS, 600 mW massively parallel processor for video scene analysis, IEEE Journal of Solid-State Circuits, 43, 192–201, 2008
7. L. Lindgren, J. Melander, R. Johansson, and B. Moller, A multiresolution 100-GOPS 4-Gpixels/s programmable smart vision sensor for multisense imaging, IEEE Journal of Solid-State Circuits, 40, 1350–1359, 2005
8. P. Dudek and P. J. Hicks, A general-purpose processor-per-pixel analogue SIMD vision chip, IEEE Transactions on Circuits and Systems I: Fundamental Theory and Applications, 52, 13–20, 2005/01/2005
9. T. Komuro, I. Ishii, M. Ishikawa, and A. Yoshida, A digital vision chip specialized for high-speed target tracking, IEEE Transactions on Electron Devices, 50, 191–199, 2003
10. G. C. Linan, A. Rodriguez-Vazquez, R. C. Galan, F. Jimenez-Garrido, S. Espejo, and R. Dominguez-Castro, A 1000 FPS at 128 x128 vision processor with 8-bit digitized I/O, IEEE Journal of Solid-State Circuits, 39, 1044–1055, 2004
11. A. Lopich and P. Dudek, Asynchronous cellular logic network as a co-processor for a general-purpose massively parallel array, International Journal of Circuit Theory and Applications, Article first published online: 29 Apr 2010, DOI: 10.1002/cta.679, 2011
12. R. Forchheimer and A. Odmark, A single chip linear array picture processor, Proceedings of the SPIE – The International Society for Optical Engineering, 397, 425–430, 1983
13. K. Chen, M. Afghahi, P. E. Danielsson, and C. Svensson, PASIC: A processor-A/D converter-sensor integrated circuit, in Proceedings – IEEE International Symposium on Circuits and Systems, New Orleans, LA, USA, 1990, pp. 1705–1708

14. K. Chen and C. Svensson, A 512-processor array chip for video/image processing, in From Pixels to Features II. Parallelism in Image Processing. Proceedings of a Workshop, Amsterdam, The Netherlands, 1991, pp. 187–199
15. M. Gokstorp and R. Forchheimer, Smart vision sensors, in IEEE International Conference on Image Processing, Chicago, IL, USA, 1998, pp. 479–482
16. D. Andrews, C. Kancler, and B. Wealand, An embedded real-time SIMD processor array for image processing, in Proceedings of the 4th International Workshop on Parallel and Distributed Real-Time Systems, Los Alamitos, USA, 1996, pp. 131–134
17. J. C. Gealow and C. G. Sodini, Pixel-parallel image processor using logic pitch-matched to dynamic memory, IEEE Journal of Solid-State Circuits, 34, 831–839, 1999
18. R. P. Kleihorst, A. A. Abbo, A. van der Avoird, M. J. R. Op de Beeck, L. Sevat, P. Wielage, R. van Veen, and H. van Herten, Xetal: A low-power high-performance smart camera processor, in IEEE International Symposium on Circuits and Systems, NJ, USA, 2001, pp. 215–218
19. J.-E. Eklund, C. Svensson, and A. Astrom, VLSI implementation of a focal plane image processor – a realization of the near-sensor image processing concept, IEEE Transaction on VLSI Systems, 4, 322–335, 1996
20. J. Poikonen, M. Laiho, and A. Paasio, MIPA4k: A 64 × 64 cell mixed-mode image processor array, in IEEE International Symposium on Circuits and Systems Taiwan, 2009, pp. 1927–1930
21. F. Paillet, D. Mercier, and T. M. Bernard, Second generation programmable artificial retina, in Twelfth Annual IEEE International ASIC/SOC Conference, 15–18 Sept. 1999, Washington, DC, USA, 1999, pp. 304–309
22. M. Wei, L. Qingyu, Z. Wancheng, and W. Nan-Jian, A programmable SIMD vision chip for real-time vision applications, IEEE Journal of Solid-State Circuits, 43, 1470–1479, 2008
23. T. Komuro, S. Kagami, and M. Ishikawa, A dynamically reconfigurable SIMD processor for a vision chip, IEEE Journal of Solid-State Circuits, 39, 265–268, 2004
24. V. Gies, T. M. Bernard, and A. Merigot, Convergent micro-pipelines: A versatile operator for mixed asynchronous-synchronous computations, in IEEE International Symposium on Circuits and Systems (ISCAS), NJ, USA, 2005, pp. 5242–5245
25. M. Arias-Estrada, M. Tremblay, and D. Poussart, A focal plane architecture for motion computation, Real-Time Imaging, 2, 351–360, 1996
26. B. Ducourthial and A. Merigot, Parallel asynchronous computations for image analysis, Proceedings of the IEEE, 90, 1218–1229, 2002
27. F. Robin, G. Privat, and M. Renaudin, Asynchronous relaxation of morphological operators: A joint algorithm-architecture perspective, International Journal of Pattern Recognition and Artificial Intelligence, 11, 1085–1094, 1997
28. F. Robin, M. Renaudin, and G. Privat, An asynchronous 16 × 16 pixel array-processor for morphological filtering of greyscale images, in European Solid-State Circuits Conference, Gif-sur-Yvette, France, 1996, pp. 188–191
29. A. Lopich and P. Dudek, Hardware implementation of skeletonization algorithm for parallel asynchronous image processing, Journal of Signal Processing Systems, 56, 91–103, 2009
30. B. Galilee, F. Mamalet, M. Renaudin, and P. Y. Coulon, Parallel asynchronous watershed algorithm-architecture, IEEE Transactions on Parallel and Distributed Systems, 18, 44–56, 2007
31. D. Noguet, Massively parallel implementation of the watershed based on cellular automata, in Proceedings of the International Conference on Application-Specific Systems, Architectures and Processors, Zurich, Switzerland, 1997, pp. 42–52
32. P. Dudek, A flexible global readout architecture for an analogue SIMD vision chip, in IEEE International Symposium on Circuits and Systems, Bangkok, Thailand, 2003, pp. 782–785
33. K. A. Zaghloul and K. Boahen, Optic nerve signals in a neuromorphic chip I&II, IEEE Transactions on Biomedical Engineering, 51, 657–666, 2004
34. D. L. Vilarino, V. M. Brea, D. Cabello, and J. M. Pardo, Discrete-time CNN for image segmentation by active contours, Pattern Recognition Letters, 19, 721–734, 1998
35. A. Kitchen, A. Bermak, and A. Bouzerdoum, A digital pixel sensor array with programmable dynamic range, IEEE Transactions on Electron Devices, 52, 2591–2601, 2005

36. A. Lopich and P. Dudek, An $80 \times 80$ general-purpose digital vision chip in 0.18 um CMOS technology, in IEEE International Symposium on Circuits and Systems, Paris, France, 2010, pp. 4257–4260
37. A. Lopich and P. Dudek, Global operations in SIMD cellular processor arrays employing functional asynchronism, in IEEE International Workshop on Computer Architecture for Machine Perception and Sensing, Montreal, Canada, 2007, pp. 16–23

# Focal-Plane Dynamic Texture Segmentation by Programmable Binning and Scale Extraction

**Jorge Fernández-Berni and Ricardo Carmona-Galán**

**Abstract** Dynamic textures are spatially repetitive time-varying visual patterns that present, however, some temporal stationarity within their constituting elements. In addition, their spatial and temporal extents are a priori unknown. This kind of pattern is very common in nature; therefore, dynamic texture segmentation is an important task for surveillance and monitoring. Conventional methods employ optic flow computation, though it represents a heavy computational load. Here, we describe texture segmentation based on focal-plane space-scale generation. The programmable size of the subimages to be analysed and the scales to be extracted encode sufficient information from the texture signature to warn its presence. A prototype smart imager has been designed and fabricated in 0.35 μm CMOS, featuring a very low-power scale-space representation of user-defined subimages.

## 1 Introduction

Dynamic textures (DTs) are visual patterns with spatial repeatability and a certain temporal stationarity. They are time varying, but some relations between their constituting elements are maintained through time. Because of this, we can talk about the frequency signature of the texture [1]. An additional feature of a DT is its indeterminate spatial and temporal extent. Smoke, waves, a flock of birds or tree leaves swaying in the wind are some examples. The detection, identification and tracking of DTs is essential in surveillance because they are very common in natural scenes. Amongst the different methods proposed for dynamic texture recognition, those based on optical flow are currently the most popular [2]. Optic flow is a computationally efficient and natural way to characterize the local dynamics of a temporal texture. This is the case for weak DTs, which become static when referred to a local coordinate system that moves across the scene. However, the recognition

J. Fernández-Berni (✉)
Institute of Microelectronics of Seville (IMSE-CNM-CSIC), Consejo Superior de Investigaciones Científicas, Universidad de Sevilla, C/Americo Vespucio s/n 41092 Seville, Spain
e-mail: jfberni@imse-cnm.csic.es

Á. Zarándy (ed.), *Focal-Plane Sensor-Processor Chips*, 
DOI 10.1007/978-1-4419-6475-5_5, 

of strong DTs implies a much greater computational effort. For these textures, which possess intrinsic dynamics, the brightness constancy assumption associated with standard optical flow algorithms cannot be applied. More complex approaches must be considered to overcome this problem. Recently, interesting results have been achieved by applying the so-called brightness conservation assumption [3]. However, this method means heavy computational load and the subsequent high energy consumption. For a particular type of artificial vision systems, a power-efficient implementation of dynamic texture recognition is mandatory. Wireless multimedia sensor networks [4] are an obvious example. These networks are composed of a large number of low-power sensors that are densely deployed throughout a region of interest (ROI) to capture and analyse video, audio and environmental data from their surroundings. The massive and scattered deployment of these sensors makes them quite difficult to service and maintain. Therefore, energy efficiency must be a major design goal to extend the lifetime of the batteries as much as possible.

We propose an approach that does not rely on heavy computation by a general-purpose processor, but on an adapted architecture in which the more tedious tasks are conveyed to the focal plane to be realized concurrently with the image capture. This results in a simplified scene representation that carries, nevertheless, all the necessary information. In this scheme, redundant spatial information is removed at the earlier stages of the processing by means of simple, flexible and power-efficient computation at the focal plane. This architecture encodes the major features of the DTs, in the sense that the spatial sampling rate and the spatial filter passband limits are programmed into the system. This permits to track textures of an expected spatial spread and frequency signature. The main processor operates then on a reduced representation of the original image obtained at the focal plane, thus its computational load is greatly alleviated.

## 2 Simplified Scene Representation

### 2.1 Programmable Binning and Filtering

In general, existing research on dynamic textures recognition is based on global features computed over the whole scene. A clear sign of this fact is that practically all the sequences composing the reference database *DynTex* [5] contain only close-up sequences. It does make sense, in these conditions, to apply strategies of global feature detection. However, in a different context, e.g., video-surveillance, textures can appear at any location of the scene. Local analysis is required for texture detection and tracking. One way of reducing the amount of data to be processed is to summarize the joint contribution of a group of pixels to the appropriate medium-size-feature index. Let us consider the picture in Fig. 1a, which depicts a flock of starlings. It is known that these flocks maintain an internal coherence based on local rules that place each individual bird at a distance prescribed by their wing span [6]. This is an example of self-organized collective behavior, whose emergent

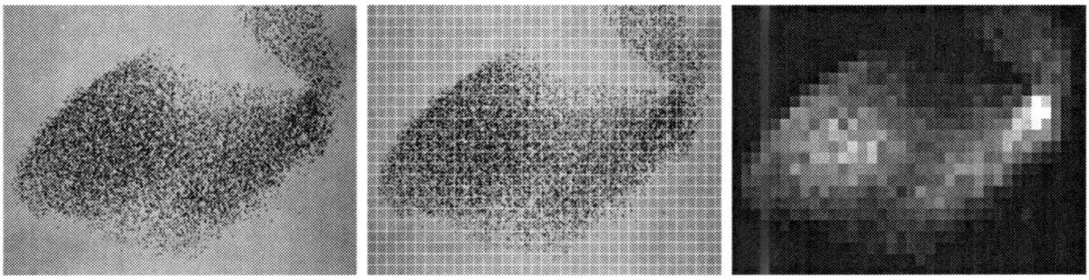

**Fig. 1** Binning and filtering applied to a scene containing a flock of starlings

features are characterized by a set of physical parameters such as flock density. We can estimate the density of the flock – more precisely, the density of its projected image – by conveniently encoding the nature of the object into the spatial sampling rate and the passband of the spatial filter selected to process the subimages.

The first step is then to subdivide the image into pixel groups that, for the sake of clarity, will be of the same size and regularly distributed along the two image dimensions (Fig. 1b). Then, if the image size is $M \times N$ pixels, and the image is divided into equally sized subimages, or bins, of size $m \times n$ pixels, we will end in a representation of the scene that is $1/R$ times smaller than the original image, being:

$$R = \left(\frac{m}{M}\right)\left(\frac{n}{N}\right). \tag{1}$$

The problem is conveyed now to finding a magnitude that summarizes the information contained in each $m \times n$-pixel bin and is still useful for our objective of texture detection. In the case of the starling flock density, a measure of the number of birds contained in a certain region of the image can be given by the high-frequency content of each bin. Notice that features of a low-spatial frequency do not represent any object of interest, i.e., bird, but details belonging to the background. Therefore, the value of each bin, represented in Fig. 1c, can be defined as the quotient:

$$B_{kl} = \frac{\sum_{\forall|\mathbf{k}|>0} E_{kl}(\mathbf{k})}{\sum_{\forall\mathbf{k}} E_{kl}(\mathbf{k})}, \tag{2}$$

where $k \in \{1,\ldots,M/n\}$ and $l \in \{1,\ldots,N/n\}$. Also, $\mathbf{k} = (u,v)$ represents the possible wave numbers and each summand $E_{kl}(\mathbf{k})$ is the energy associated with frequency $\mathbf{k}$, computed within the bin indexed by $(k,l)$. This is a spatial highpass filter normalized to the total energy associated with the image bin. If $V_{ij}$ is the value of the pixel located at the $i$th row and $j$th column of the bin, and also, $\hat{V}_{uv}$ is the component of the spatial DFT of the bin corresponding to $u$ and $v$ reciprocal lengths, the total image energy associated with the bin is:

$$\sum_{\forall\mathbf{k}} E_{kl}(\mathbf{k}) = \sum_{u=0}^{m-1}\sum_{v=0}^{n-1} |\hat{V}_{uv}|^2 = \sum_{i=1}^{m}\sum_{j=1}^{n} |V_{ij}|^2. \tag{3}$$

The result is an estimation of the bird density at a coarser grain than the full-size image, avoiding pixel-level analysis.

## 2.2 Linear Diffusion as a Scale-Space Generator

As already described, apart from the appropriate binning that sets the interesting feature size and geometrical ratio, a suitable filter family is needed to discriminate the spatial frequency components within each subimage. Let us consider that an image is a continuous function evaluated over the real plane, $V(x,y)$, that assigns a brightness value to each point in the plane. If this brightness is regarded as the concentration of a scalar property [7], the flux density is proportional to the gradient and oriented in the opposite direction. The diffusion equation:

$$\frac{\partial V(x,y,t)}{\partial t} = D\nabla^2 V(x,y,t) \tag{4}$$

follows from continuity considerations, i.e., no sources or sinks of brightness are present in the image plane. The original image is assumed to be the initial value at every point, $V(x,y,0)$. Then, applying the Fourier transform to this equation and solving in time:

$$\hat{V}(\mathbf{k},t) = \hat{V}(\mathbf{k},0)\mathrm{e}^{-4\pi^2|\mathbf{k}|^2 Dt} \tag{5}$$

what represents, in the reciprocal space, the transfer function of a Gaussian filter:

$$G(\mathbf{k};\sigma) = \mathrm{e}^{-2\pi^2\sigma^2|\mathbf{k}|^2} \tag{6}$$

whose smoothing factor is related to the diffusion duration $t$ through:

$$\sigma = \sqrt{2Dt}. \tag{7}$$

Hence, the larger the diffusion time, $t$, the larger the smoothing factor, $\sigma$, thus, the narrower the transfer function (Fig. 2), and so, the smoother the output image. Gaussian filters are equivalent to the convolution with Gaussian kernels, $g(x,y;\sigma)$, of the reciprocal widths:

$$g(x,y;\sigma) * V(x,y) = \frac{1}{2\pi\sigma^2}\int_{-\infty}^{\infty}\int_{-\infty}^{\infty} V(x-x',y-y')\mathrm{e}^{-\frac{x'^2+y'^2}{2\sigma^2}}\mathrm{d}x'\mathrm{d}y'. \tag{8}$$

These kernels hold the scale-space axioms: linearity, shift invariance, semi-group structure, and not enhancement of local extrema. This makes them unique for

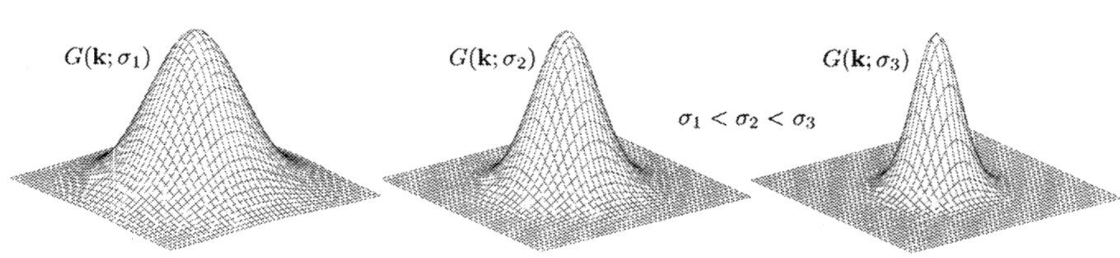


**Fig. 2** Gaussian filters of increasing $\sigma$

scale-space generation [8]. Scale-space is a framework for image processing [9] that makes use of the representation of the images at multiple scales. It is useful in the analysis of the image, e.g., to detect scale-invariant features that characterize the scene [10]. Different textures are noticeable at a particular scale, $\xi$, which is related to the smoothing factor:

$$\xi = \sigma^2 \tag{9}$$

However, convolving two large images or, alternatively, computing the FFT, and the inverse FFT, of a given image with a conventional processing architecture represents a very heavy computational load. We are interested in a low-power focal-plane operator able to deliver an image representation at the appropriate scale.

## 2.3 Spatial Filtering by a Resistor Network

Consider a grid composed of resistors like the one depicted in Fig. 3a. Let $V_{ij}(0)$ be the voltages stored at the grounded capacitors attached to each node of the resistor grid. If the network is allowed to evolve, at every time instant each node will satisfy:

$$\tau \frac{\mathrm{d}V_{ij}}{\mathrm{d}t} = -4V_{ij} + V_{i+1,j} + V_{i-1,j} + V_{i,j+1} + V_{i,j-1}, \tag{10}$$

where $\tau = RC$. Applying the spatial DFT, for a grid of $M \times N$ nodes, we arrive to:

$$\tau \frac{\mathrm{d}\hat{V}_{uv}}{\mathrm{d}t} = -4\left[\sin^2\left(\frac{\pi u}{M}\right) + \sin^2\left(\frac{\pi v}{N}\right)\right]\hat{V}_{uv}. \tag{11}$$

Notice that $\hat{V}_{uv}$ represents the discrete Fourier transform of $V_{ij}$, which is also discrete in space. Therefore, $u$ and $v$ take discrete values ranging from 0 to $M-1$ and $N-1$ respectively. Solving (11) in the time domain, we obtain:

$$\hat{V}_{uv}(t) = \hat{V}_{uv}(0)\mathrm{e}^{-\frac{4t}{\tau}\left[\sin^2\left(\frac{\pi u}{M}\right)+\sin^2\left(\frac{\pi v}{N}\right)\right]} \tag{12}$$

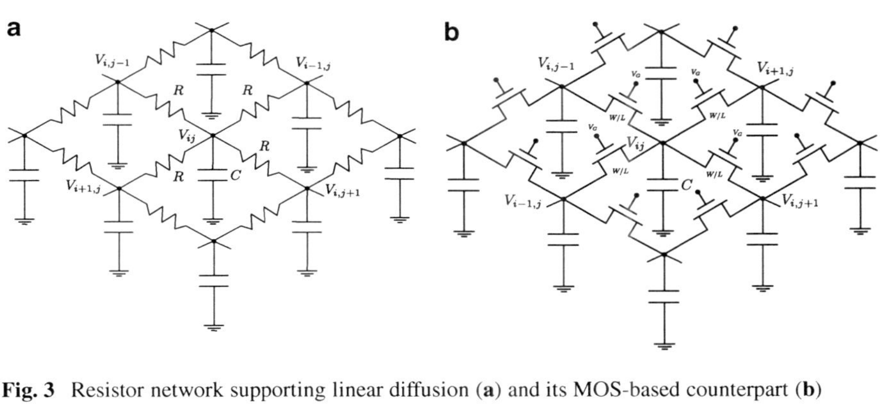


**Fig. 3** Resistor network supporting linear diffusion (**a**) and its MOS-based counterpart (**b**)

what defines a discrete-space version of the Gausian filter in (6) given by:

$$G_{uv}(\sigma) = \mathrm{e}^{-2\sigma^2\left[\sin^2\left(\frac{\pi u}{M}\right)+\sin^2\left(\frac{\pi v}{N}\right)\right]}, \tag{13}$$

where now $\sigma = \sqrt{2t/\tau}$. This function approximates quite well the continuous-space Gaussian filter at the lower frequencies and close to the principal axes of the reciprocal space. At higher frequencies, specially at the bisectrices of the axes, i.e., when $u$ and $v$ both become comparable to $M$ and $N$, respectively, isotropy degrades as the approximation of $\sin^2(\pi u/M)$ and $\sin^2(\pi v/N)$ by $(\pi u/M)^2$ and $(\pi v/N)^2$, respectively, becomes too coarse.

# 3 VLSI Implementation of Time-Controlled Diffusion

## 3.1 MOS-Resistor Grid Design

Resistive networks are, as we have already seen, massively parallel processing systems that can be employed to realize spatial filtering [11]. But a true linear resistive grid is difficult to implement in VLSI. The low sheet-resistance exhibited by the most resistive materials available in standard CMOS renders too large areas for the necessary resistances. A feasible alternative is to employ MOS transistors to replace the resistors one by one. They can achieve larger resistances with lesser area than resistors made of polysilicon or diffusion strips. In addition, by controlling the gate voltage, their resistance can be modified. They can also be operated as switches, thereby configuring the connectivity of the network. This substitution of resistors by MOS transistors, however, entails, amongst others, linearity problems. In [12], the linearity of the currents through resistive grids is achieved by using transistors in weak inversion. The value of the resistance associated with each transistor is directly controlled by the corresponding gate voltage. This property of current linearity is also applicable even if the transistors leave weak inversion as long as all of them share the same gate voltage [13]. Linearity is not so easy to achieve when signals are encoded by voltages as in Fig. 3a. The use of MOSFETs operating in the ohmic region instead of resistors is apparently the most simple option [14]. However, the intrinsic non-linearity in the I–V characteristic leads to more elaborated alternatives for the cancellation of the non-linear term [15], even to transconductor-based implementations [16].

For moderate accuracy requirements, though, the error committed by using MOS transistors in the ohmic region can be kept under a reasonable limit if the elementary resistor is adequately designed. For an estimation of the upper bound of this error, let us compare the circuits in Fig. 4. They represent a 2-node ideal resistor grid and its corresponding MOS-based implementation. The gate voltage $V_G$ is fixed and we will assume, without loss of generality, that the initial conditions of the capacitors fulfill $V_1(0) > V_2(0)$, being $V_1(0) = V_1'(0)$ and $V_2(0) = V_2'(0)$. We will also assume that the transistor is biased in the triode region for any voltage at the drain and source

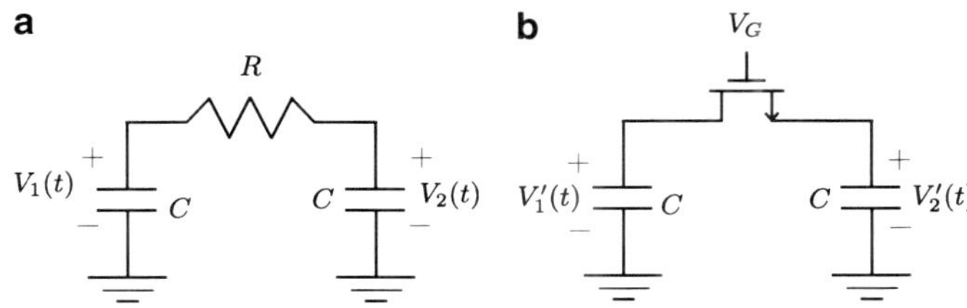


**Fig. 4** A 2-node ideal resistive grid (**a**) and its MOS-based implementation (**b**)

terminals, that will range from $V_{\min}$ to $V_{\max}$. The evolution of the circuit in Fig. 4a is described by this set of ODEs:

$$\begin{cases} C\frac{dV_1}{dt} = -\frac{V_1(t)-V_2(t)}{R} \\ C\frac{dV_2}{dt} = \frac{V_1(t)-V_2(t)}{R} \end{cases} \tag{14}$$

while the behavior of the circuit in Fig. 4b is described by:

$$\begin{cases} C\frac{dV_1'}{dt} = -G_{\mathrm{M}}\left[V_1'(t)-V_2'(t)\right] \\ C\frac{dV_2'}{dt} = G_{\mathrm{M}}\left[V_1'(t)-V_2'(t)\right] \end{cases} \tag{15}$$

by making use of the following transconductance:

$$G_{\mathrm{M}} = k_n S_n \left\{2\left(V_{\mathrm{G}}-V_{T_n}\right)-\left[V_1'(t)+V_2'(t)\right]\right\}, \tag{16}$$

where $k_n = \mu_n C_{ox}'/2$ and $S_n = W/L$. This transconductance remains constant during the network evolution, if we neglect the substrate and other second order effects, as the sum $V_1'(t)+V_2'(t)$ remains the same because of charge conservation. Therefore, and given that the charge extracted from one capacitor is ending up in the other, we can define the error in the corresponding node voltages as:

$$\begin{cases} V_1'(t) = V_1(t) + \varepsilon(t) \\ V_2'(t) = V_2(t) - \varepsilon(t) \end{cases} \tag{17}$$

or, equivalently:

$$\varepsilon(t) = \frac{V_1'(t)-V_2'(t)}{2} - \frac{V_1(t)-V_2(t)}{2}. \tag{18}$$

Because of our initial assumptions, $V_1(0) = V_1'(0)$ and $V_2(0) = V'2(0)$, we have that $\varepsilon(0) = 0$. Also, the stationary state, reached when $t \to \infty$, renders $\varepsilon(\infty) = 0$, as $V_1(\infty) = V_2(\infty)$ and $V_1'(\infty) = V_2'(\infty)$. Therefore, there must be at least one point in time, let us call it $t_{\mathrm{ext}}$ in which the error reaches an extreme value, either positive or negative. In any case, the time derivative of the error:

$$\frac{d\varepsilon}{dt} = \frac{1}{\tau}\left[V_1'(t)-V_2'(t)\right](1-G_{\mathrm{M}}R) - \frac{2\varepsilon(t)}{\tau} \tag{19}$$

must cancel in $t_{\text{ext}}$, resulting in an extreme error of:

$$\varepsilon(t_{\text{ext}}) = \frac{1}{2}\left[V_1'(t_{\text{ext}}) - V_2'(t_{\text{ext}})\right](1 - G_{\text{M}}R). \tag{20}$$

Notice that for $G_{\text{M}}R = 1$ the error is zero at any moment. This happens if the transistor aspect, $S_n$, is selected to match the resistance $R$ through:

$$S_n = \frac{1}{k_n R\left\{2(V_{\text{G}} - V_{T_n}) - \left[V_1'(0) + V_2'(0)\right]\right\}} \tag{21}$$

and the current values of $V_1'(0)$ and $V_2'(0)$ add up to the same $V_1'(0) + V_2'(0)$ with which $S_n$ was selected. Unfortunately, this will very seldom happen. And because we do not know a priori where within the interval $[V_{\text{min}}, V_{\text{max}}]$ are $V_1'(0)$ and $V_2'(0)$, neither $V_1'(t_{\text{ext}})$ and $V_2'(t_{\text{ext}})$, we are interested in a good estimate of the largest possible error within the triangle $\triangle ABC$ (Fig. 5), delimited by points $A : (V_{\text{min}}, V_{\text{min}})$, $B : (V_{\text{max}}, V_{\text{min}})$, $C : (V_{\text{max}}, V_{\text{max}})$. This is because one of our initial assumptions was that $V_1'(0) > V_2'(0)$, and this condition is maintained until these voltages identify at the steady state.

Let us express this error, $\varepsilon_x = \varepsilon(t_{\text{ext}})$, as a function of $V_{1_x}' = V_1'(t_{\text{ext}})$ and $V_{2_x}' = V_2'(t_{\text{ext}})$. It can be done because the sum $V_1'(t) + V_2'(t)$, present in the definition of $G_{\text{M}}$, is constant along the evolution of the network, therefore also at $t_{\text{ext}}$:

$$\varepsilon_x(V_{1_x}', V_{2_x}') = \frac{1}{2}\left(V_{1_x}' - V_{2_x}'\right)\left\{1 - k_n S_n R\left[2(V_{\text{G}} - V_{T_n}) - \left(V_{1_x}' + V_{2_x}'\right)\right]\right\}. \tag{22}$$

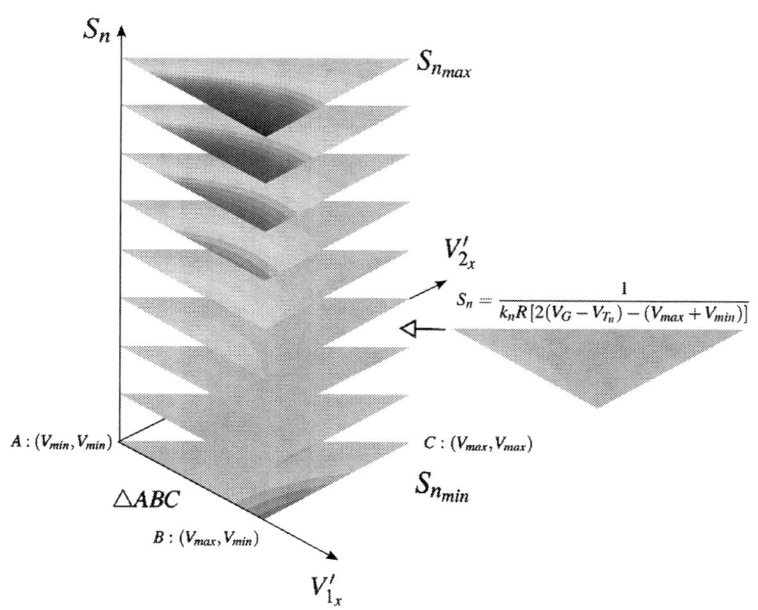


**Fig. 5** Error estimate for practical values of $S_n$

Then, any possible extreme of $\varepsilon_x(V'_{1_x}, V'_{2_x})$ must be at a critical point, i.e., a point in which $\nabla\varepsilon_x(V'_{1_x}, V'_{2_x}) = 0$. But the only one that can be found in $\triangle ABC$ is a saddle point. Therefore, we can only talk of absolute maxima or minima, and they will be at the borders of the triangle (Fig. 5). More precisely at sides $\overline{AB}$ and $\overline{BC}$, given that $\varepsilon_x \equiv 0$ along side $\overline{AC}$, at points:

$$\begin{cases} V'_{1_x} = V_{\max} & V'_{2_x} = V_{\mathrm{G}} - V_{T_n} - \frac{1}{2k_n S_n R} \\ V'_{1_x} = V_{\mathrm{G}} - V_{T_n} - \frac{1}{2k_n S_n R} & V'_{2_x} = V_{\min} \end{cases} \tag{23}$$

and their values are:

$$\begin{cases} \varepsilon_x|_{\max} &= \frac{1}{2}k_n S_n R\left(V_{\max} - V_{\mathrm{G}} + V_{T_n} + \frac{1}{2k_n S_n R}\right)^2 \\ \varepsilon_x|_{\min} &= -\frac{1}{2}k_n S_n R\left(V_{\min} - V_{\mathrm{G}} + V_{T_n} + \frac{1}{2k_n S_n R}\right)^2. \end{cases} \tag{24}$$

Notice that increasing or decreasing $S_n$ has antagonistic effects in the magnitude of $\varepsilon_x|_{\max}$ and $\varepsilon_x|_{\min}$ (Fig. 5). Therefore, the optimal design is obtained for:

$$S_n = \frac{1}{k_n R\left[2(V_{\mathrm{G}} - V_{T_n}) - (V_{\max} + V_{\min})\right]} \tag{25}$$

which minimizes the maximum error, rendering [17]:

$$\min\left(\max|\varepsilon_x|\right) = \frac{1}{16}\frac{(V_{\max} - V_{\min})^2}{V_{\mathrm{G}} - V_{T_n} - \frac{V_{\max}+V_{\min}}{2}} \tag{26}$$

Notice that the design space is limited by the extreme values of $S_n$, those beyond which the target resistance $R$ fall outside the interval of resistance values that can be implemented within the triangle $\triangle ABC$ that represents all the possible values that can be taken by $V'_{1_x}$ and $V'_{2_x}$. These extrema are:

$$\begin{cases} S_{n_{\min}} = \frac{1}{2k_n R\left(V_{\mathrm{G}} - V_{T_n} - V_{\min}\right)} \\ S_{n_{\max}} = \frac{1}{2k_n R\left(V_{\mathrm{G}} - V_{T_n} - V_{\max}\right)}. \end{cases} \tag{27}$$

Notice also that if one chooses to select $S_n = (S_{n_{\min}} + S_{n_{\max}})/2$ – led by the groundless intuition that it will render the smallest error as it is at equal distance from the two extremes of the design space –, this will yield suboptimal design, as the optimum $S_n$, expressed in (25), is notably below the midpoint. It is also worth taking into account that (26) represents a conservative upper bound of the maximum error that is going to be achieved by optimal design. This is because we have not considered the relation between $V'_1(t)$ and $V'_2(t)$ imposed by (15):

$$V'_1(t_{\mathrm{ext}}) - V'_2(t_{\mathrm{ext}}) = \left[V'_1(0) - V'_2(0)\right] \mathrm{e}^{-\frac{2G_{\mathrm{M}}}{C}t_{\mathrm{ext}}}. \tag{28}$$

This means that not all the possible values contained in $\triangle ABC$, defined on the $V'_{1_x} - V'_{2_x}$ plane, not $V'_1(0) - V'_2(0)$, will be covered by all the possible trajectories of the circuit. Thus, by equating (19) to zero after having substituted the trajectories by their actual waveforms, $t_{\text{ext}}$ is obtained:

$$t_{\text{ext}} = \frac{\tau}{2} \frac{\ln G_{\text{M}} R}{G_{\text{M}} R - 1} \tag{29}$$

and the exact value of the error in the extreme is then given by:

$$\varepsilon(t_{\text{ext}}) = \frac{V'_1(0) - V'_2(0)}{2} (1 - G_{\text{M}} R) (G_{\text{M}} R)^{-\left(\frac{G_{\text{M}} R}{G_{\text{M}} R - 1}\right)}. \tag{30}$$

Unfortunately, to obtain the position of the extreme error in a closed form from this equation is not possible, but numerically we have confirmed that the $S_n$ rendering the smallest error is the one expressed in (25).

## 3.2 Extrapolated Results for Larger Networks

The results described above have been applied to the design of a $64 \times 64$ MOS-based resistive grid. Simulations have been realized using standard 0.35 μm CMOS 3.3 V process transistor models in HSPICE. The signal range at the nodes is [0–1.5 V], wide enough to evidence any excessive influence of the MOSFET non-linearities in the spatial filtering. $V_{\text{G}}$ is established at 3.3 V. We aim to implement an array of capacitors interconnected by a network of resistors with a time constant of $\tau = 100$ ns. For that we will assume a resistor of $R = 100\,\text{k}\Omega$ and a capacitor of $C = 1$ pF. $S_n$ is obtained according to (25). But this equation does not take into account the substrate effect, or in other words, $S_n$ is not only depending on the sum $V_{\text{max}} + V_{\text{min}}$ but also in the specific values of the initial voltages at drain and source that render the same sum. For a specific value of $S_n$, and the same $V'_1(0) + V'_2(0)$, the resistance implemented can vary $\pm 5\%$. We have selected $W = 0.4$ μm and $L = 7.54$ μm,[1] which result in an average resistance of $100\,\text{k}\Omega$ for all the possible initial conditions rendering the optimum sum, i.e., $V_{\text{max}} + V_{\text{min}}$.

The initial voltage at the capacitors is proportional to the image intensity displayed at Fig. 6a. A MOS-based resistor network runs the diffusion of the initial voltages in parallel with an ideal linear resistor network. The deviation is measured via RMSE (Fig. 7) and reaches a maximum soon after the beginning of the diffusion process. The state of the corresponding nodes in both networks at this point, displayed in Fig. 6b, c, can be compared, Fig. 6d–f. The maximum observed RMSE for

[1] This transistor length lies out of the physical design grid, which fixes the minimum feature size to be 0.05 μm. We are using it here as illustrative of the design procedure.

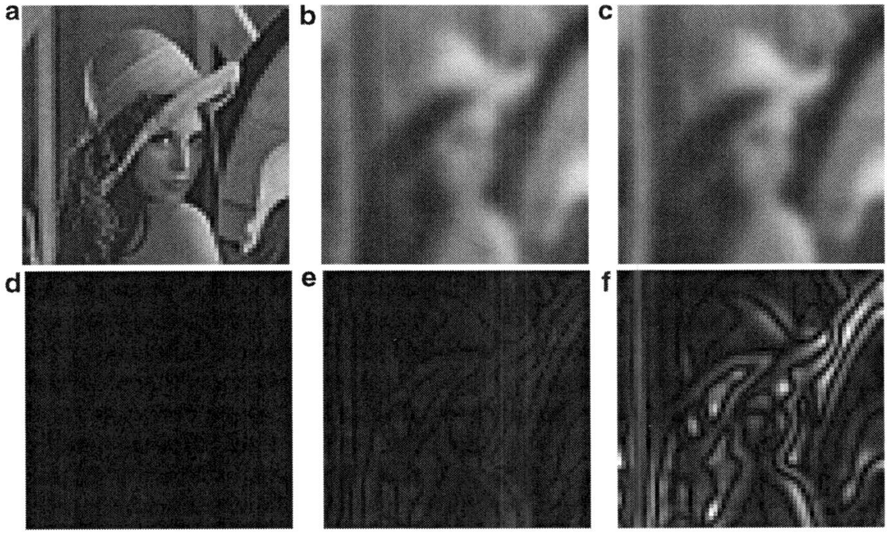


**Fig. 6** (**a**) Original image, (**b**) MOS-diffused image at instant of maximum error, (**c**) image diffused by resistor network, (**d**) absolute error, (**e**) absolute error multiplied by 10, (**f**) absolute error normalized to maximum individual pixel error

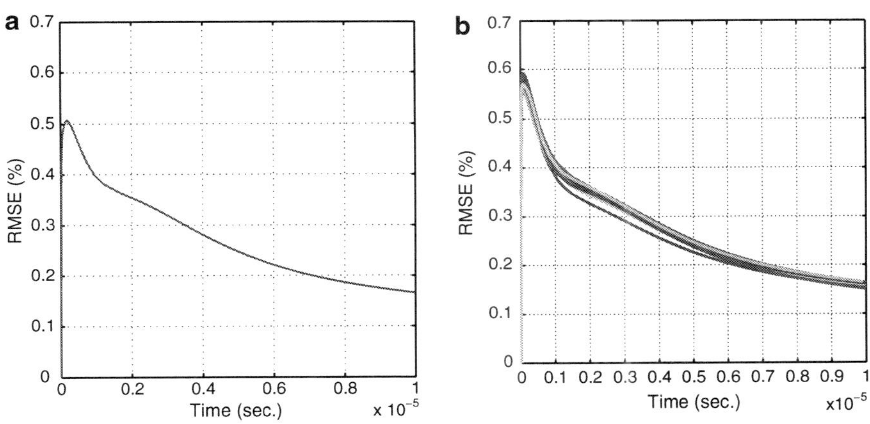


**Fig. 7** RMSE of the MOS-based grid state vs. resistor grid state: (**a**) w/o mismatch, (**b**) Monte Carlo with 10% mismatch

the complete image is 0.5%, while the maximum individual pixel error is 1.76%. This error remains below 0.6% even introducing an exaggerated mismatch (10%) in the transistors' $V_{T_{n0}}$ and $\mu_n$ (Fig. 7b). [2]

[2] Global deviations within the process parameters space have not been considered. In that case, the nominal resistance being implemented differs from the prescribed 100 kΩ.

### 3.3 Diffusion Duration Control

Implementing a Gaussian filter, (13), by means of dynamic diffusion requires a precise control of the diffusion duration, $t$, given that the scale, $\xi$, i.e., the square of the smoothing factor, is related to it through:

$$\xi = \frac{2t}{\tau}. \tag{31}$$

Therefore, to filter the image with a Gaussian function of scale $\xi$ means to let the diffusion run for $t = \tau\xi/2$ s. The actual value of $\tau$ is not important, as long as the necessary resolution of $t$ for the required $\xi$ is physically realizable. Actually, $\xi$ is not determined by $t$ itself, but by the quotient $t/\tau$. In other words, a different selection of $R$ and $C$ will still render the same set of $\xi$s as long as the necessary fraction of $\tau$ can be fine tuned. Implementing this fine control of the diffusion duration is not a trivial task when we are dealing with $\tau$s in the hundred-nanosecond range. To provide robust sub-$\tau$ control of the diffusion duration, the operation must rely on internally, i.e., on-chip, generated pulses. Otherwise, propagation delays will render this task very difficult to accomplish.

We propose a method for the fine control of $t$ based on an internal VCO. This method has been tested in the prototype that we describe later in this chapter. The first block of the diffusion duration control circuit is the VCO (Fig. 8a). It consists of a ring of pseudo-NMOS inverters in which the load current is controlled by $V_{\text{bias}}$, thus modifying the propagation delay of each stage. As the inverter ring is composed by an odd number of stages, the circuit will oscillate. A pull-up transistor, with a weak aspect ratio, has been introduced to avoid start-up problems. Also, a flip-flop is placed to render a 50% duty-cycle at the output. This circuit provides an internal clock that will be employed to time pulses that add up to the final diffusion duration.

The main block of the diffusion control is the 12-stage shift register. It will store a chain of 1s, indicating how many clock cycles the diffusion process will run. The clock employed for this will be either externally supplied or driven by the already described internal VCO.[3] The output signal, *diff_ctrl* in Fig. 8b, is a pulse with the desired duration of the diffusion, which is delivered to the gates of the MOS-resistors.

### 3.4 Image Energy Computation

For real-time detection and tracking of dynamic textures, we will be interested in a simplified representation of the scene. It can be constructed from the filtered image

[3] The aim of the internal VCO is to reach a better time resolution than an external clock. For programming the appropriate sequence into the SHR, an external, and slower, clock is usually preferred.

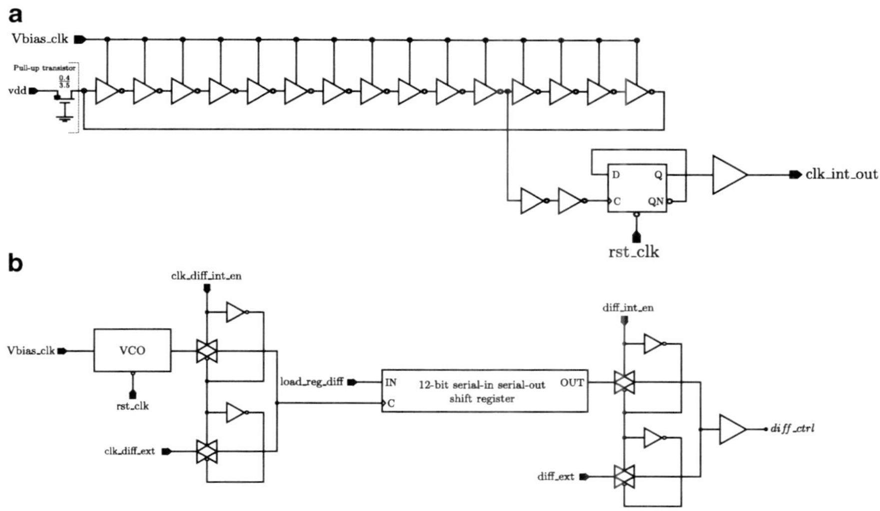


**Fig. 8** (**a**) 15-stage inverter ring VCO and (**b**) diffusion control logic

by first dividing it into subimages, usually of the same size. Each subimage is then represented by a number that encodes information related to the spatial frequency content of the subimage. This number is the image energy, as defined in (3). The energy of the image bin can be expressed as a function of time:

$$E_{kl}(t) = \sum_{\forall \mathbf{k}} E_{kl}(\mathbf{k};t) = \sum_{u=0}^{m-1}\sum_{v=0}^{n-1} |\hat{V}_{uv}(t)|^2 = \sum_{i=1}^{m}\sum_{j=1}^{n} |V_{ij}(t)|^2 \tag{32}$$

meaning that the energy of the image at time $t$ accounts for the frequency components that have not been filtered yet. In terms of the dynamics of the resistor grid, the total charge in the array of capacitors is conserved but, naturally, the system evolves towards the least energetic configuration. Therefore, the energy at time $t$ indicates how the subimage has been affected by the diffusion until that exact point in time. The longer $t$ the less of $E_{kl}(0)$ will remain in $E_{kl}(t)$. The energy lost between two consecutive points in time during the diffusion corresponds to that of the spatial frequencies filtered. Notice also that changing the reference level for the amplitude of the pixels does not have an effect beyond the dc component of the image spectrum. A constant value added to every pixel does not eliminate nor modify any of the spatial frequency components already present, apart from changing the dc component.

To analyse the presence of different spatial frequency components within a particular bin of the image, we would need to measure the energy of the bin pixels once filtered. Remember that for analysing a particular band of frequencies we will subtract two lowpass-filtered versions of the image. In this way, only the components of the targeted frequency band will remain. This will allow for tracking changes at that band without pixel-level analysis. The hardware employed to calculate the energy of

**Fig. 9** In-pixel energy computing circuit

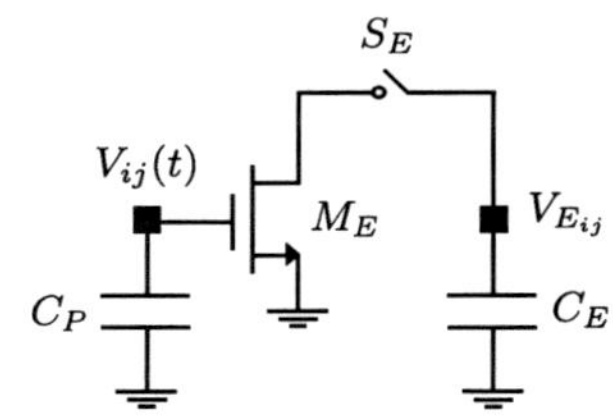


the bins at the pixel-level (Fig. 9) consists of a transistor, $M_{\mathrm{E}}$, that converts the pixel voltage to a current that is proportional to the square of the voltage, and a switch $S_{\mathrm{E}}$ to control the amount of charge that will be subtracted from the capacitor $C_{\mathrm{E}}$ that realizes charge-to-voltage conversion. All the $C_{\mathrm{E}}$s within the bin will be connected to the same node. At the beginning, all of them are pre-charged to $V_{\mathrm{REF}}$. Then, for a certain period of time, $T_{\mathrm{E}}$, the transistor $M_{\mathrm{E}}$ is allowed to discharge $C_{\mathrm{E}}$. But because the $m \times n$ capacitors of the bin are tied to the same node, the final voltage representing the bin energy after $t$ seconds of diffusion will be:

$$V_{E_{kl}}(t) = V_{\mathrm{REF}} - \frac{k_{\mathrm{E}} S_{\mathrm{E}} T_{\mathrm{E}}}{mnC_{\mathrm{E}}} \sum_{i=1}^{m} \sum_{j=1}^{n} [V_{ij}(t) - V_{T_{n0}}]^2. \tag{33}$$

We are assuming that all the $M_{\mathrm{E}}$s are nominally identical and operate in saturation. The offset introduced by $V_{T_{n0}}$ does not affect any spatial frequency other than the dc component. Deviations occur from pixel to pixel due to mismatch in the threshold voltage ($V_{T_{n0}}$), the transconductance parameter ($k_{\mathrm{E}}$), and the body-effect constant ($\gamma_{\mathrm{E}}$, not in this equation). These deviations are area dependent; therefore, transistor $M_{\mathrm{E}}$ is tailored to keep the error in the computation within the appropriate bounds. Also, mobility degradation contributes to the deviation from the behavior expressed in (33), which will ultimately limit the useful signal range.

# 4 Prototype Texture Segmentation Chip

## *4.1 Chip Description and Data*

The floorplan of the prototype chip is depicted in Fig. 10. It is composed of a mixed-signal processing array, concurrent with the photosensor array, intended to carry out Gaussian filtering of the appropriate scale, within user-defined image areas. In addition, the total energy associated with each image bin is calculated. On the periphery, there are circuits for bias and control of the operation. The outcome of the processing can be read out pixel-by-pixel in a random access fashion. The value of the pixel is buffered at the column bus and delivered to either an analog output pad or an on-chip 8-bit SAR ADC.

The elementary cell of the analog core (Fig. 11) was described in [18]. It contains a diffusion network built with p-type MOS transistors. The limits of the diffusion

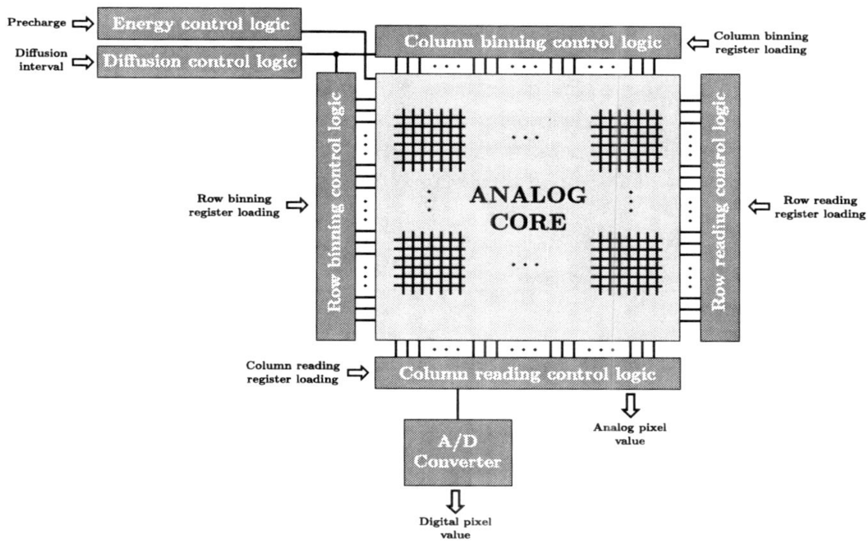


**Fig. 10** Floorplan of the prototype chip

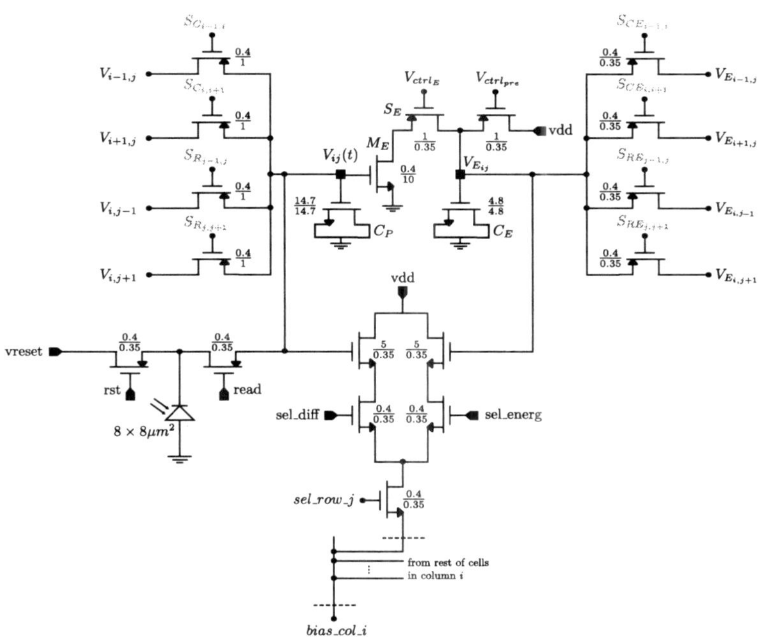


**Fig. 11** Schematic of the elementary cell

areas are column-wise and row-wise selected, enabled by the appropriate connection pattern. In this way, scale spaces can be generated at different user-defined areas of the scene. The pulse which determines the duration of the diffusion can either be put in externally or generated internally, as mentioned in Sect. 3.3. The main characteristics of the chip are summarized in Table 1. A microphotograph of the prototype chip with a close-up of the photosensors is shown in Fig. 12.

**Table 1** Prototype chip data

| | |
|---|---|
| Technology | 0.35 μm CMOS 2P4M |
| Vendor (process) | Austria Microsystems (C35OPTO) |
| Die size (with pads) | 7280.8 μm × 5780.8 μm |
| Pixel size | 34.07 μm × 29.13 μm |
| Fill factor | 6.45% |
| Resolution | QCIF: 176 × 144 px |
| Photodiode type | $n$-well/$p$-substrate |
| Power supply | 3.3 V |
| Power consumption (including ADC) | 1.5 mW |
| Internal clock freq. range | 10–150 MHz |
| ADC throughput (I/O limit) | 0.11 MSa s$^{-1}$ (9 μs Sa$^{-1}$) |
| Exposure time range | 100 μs–500 ms |

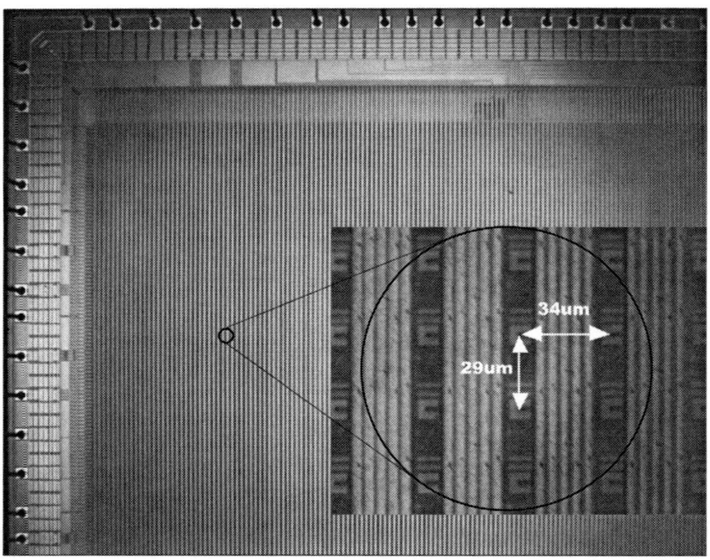


**Fig. 12** Microphotograph of the prototype chip

## 4.2 Linearity of the Scale-Space Representation

As expected from simulations, the use of MOS transistors instead of true linear resistors in the diffusion network (Fig. 3b) achieves moderate accuracy even under strong mismatch conditions. However, the value of the resistance is implemented by the transistors and therefore the value of the network time constant, $\tau$, is quite sensitive to process parameters. To have a precise estimation of the scale implemented by stopping the diffusion at different points in time – recall that $\xi = 2t/\tau$ – the actual $\tau$ needs to be measured. We have provided access to the extremes of the array and have characterized $\tau$ from the charge redistribution between two isolated pixels. The average $\tau$ measured is of 71.1 ns ($\pm 1.8\%$). Attending to the technology corners, the predicted range for $\tau$ was $[49, 148]$ ns. By reverse engineering the time constant, using (25), the best emulated resistance ($R_{eq}$) is obtained. Once $\tau$ is calibrated, any on-chip scale space can be compared to its ideal counterpart obtained by solving the spatially discretized diffusion equation corresponding to a network consisting of the same $C$s and resistors of value $R_{eq}$. To generate an on-chip scale space, a single image is captured. This image is converted to the digital domain and delivered through the output bus. It becomes the initial image of both the on-chip scale space and the ideal scale space calculated off-chip. The rest of the on-chip scale space is generated by applying successive diffusion steps to the original captured image. After every step, the image is converted to digital and delivered to the test instruments to be compared to the ideal image generated by MATLAB (Fig. 13). The duration of each

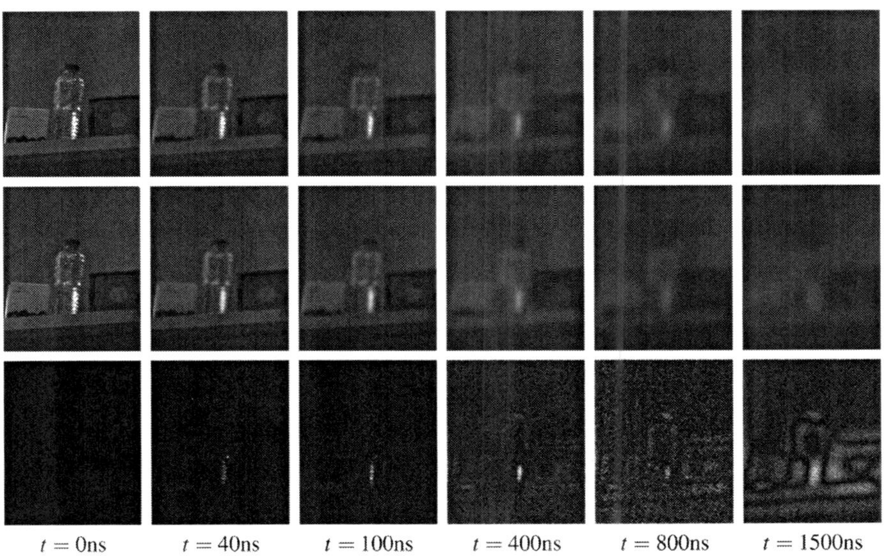


**Fig. 13** Comparative of scale spaces along the time. The first row corresponds to the on-chip scale space, the second one corresponds to the ideal scale space and finally the third one corresponds to their normalized difference

diffusion step is internally configured as sketched in Sect. 3.3. A total of 12 steps have been realized over the original captured image. Six of them are represented in Fig. 13 (first row) and compared to the ideal images (second row). The last row contains a pictorial representation of the error, normalized in each case to the highest measured error on individual pixels, which are 0%, 24.99%, 19.39%, 6.17%, 3.58% and 6.68%, respectively. It can be seen how FPN eventually becomes the dominant error at coarse scales. Keep in mind that this noise is present at the initial image of both scale spaces, but it is only added to each subsequent image of the on-chip scale space because of the readout mechanism. The key point here is that the error is kept under a reasonable level despite no FPN post-processing being carried out. This fact, together with the efficiency of the focal-plane operation, is crucial for artificial vision applications under strict power budgets. Two additional issues are worth to be mentioned. First of all, the accuracy of the processing predicted by simulation [17] is very close to that of the first images of the scale space, where fixed pattern noise is not dominant yet. Second, it has been confirmed for this and other scenes that the second major source of error in the scale space generation comes from uniform areas where the pixel values fall on the lowest extreme of the signal range. The reason is that the instantaneous resistance implemented by a transistor when its source and drain voltages coincide around the lowest extreme of the signal range presents the maximum possible deviation with respect to the equivalent resistance considered. The point is that, in such a situation, the charge diffused between the nodes involved is very small, keeping the error moderate despite the large deviation in the resistance.

## 4.3 Subsampling Modalities

Finally, as a glimpse into the possibilities of the prototype, independent scale spaces were generated within four subdivisions of the focal plane programmed into the chip (Fig. 14). Because of the flexible subdivision of the image, the chip can deliver from full-resolution digital images to different simplified representations of the scene,

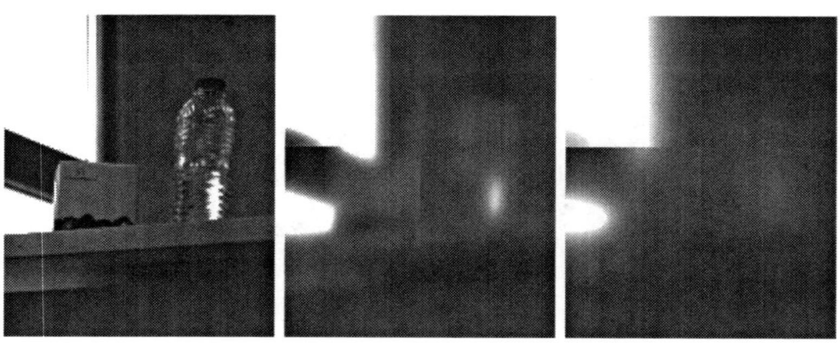

**Fig. 14** Independent scale spaces within four subdivisions of the focal plane

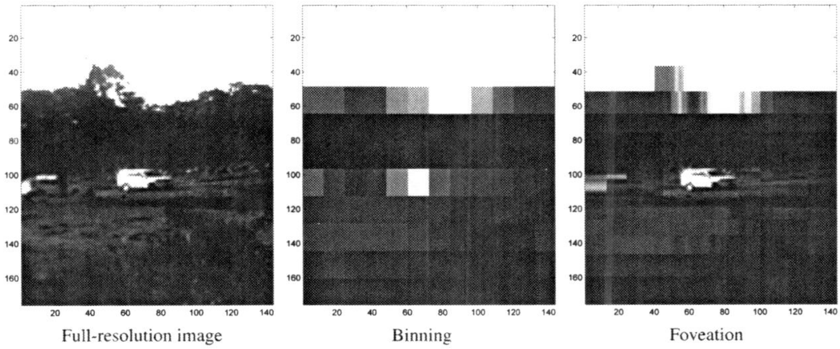


**Fig. 15** Example of the image sampling capabilities of the prototype

which can be reprogrammed in real time according to the results of their processing. As an example of the image subsampling capabilities of the chip, three different schemes are shown in Fig. 15. The first one corresponds to the full-resolution representation of the scene – it has been taken in the lab, by displaying a real outdoor sequence in a flat-panel monitor. The second one represents the same scene after applying some binning. In the third picture, the ROI, in the center, is at full resolution while the areas outside this region are binned together. The binning outside the ROI becomes progressively coarser. This represents a foveatized version of the scene in which the greater detail is only considered at the ROI, and then the grain increases as we reach further. From the computational point of view, this organization translates into important computing resources savings.

## 5 Conclusions

This chapter has presented a feasible alternative for focal-plane generation of a scale space. Its intention is to realize real-time dynamic textures detection and tracking. For that, focal-plane filtering with a resistor grid results in a very low-power implementation, while the appropriate image subdivision accommodated to the size of the targeted features also contributes to alleviate the computing load. A methodology for the design of the MOS-based resistor network is explained, leading to optimal design of the grid. Also, the means for a simplified representation of the scene are provided at the pixel level. These techniques have been applied to the design of a prototype smart CMOS imager. Some experimental results confirming the predicted behavior are shown.

**Acknowledgements** This work is funded by CICE/JA and MICINN (Spain) through projects 2006-TIC-2352 and TEC2009-11812, respectively.

## References

1. R. Nelson, R. Polana, CVGIP: Image Understanding **56**(1), 78 (1992)
2. D. Chetverikov, R. Péteri, in *International Conference on Computer Recognition Systems (CORES'05)* (Rydzyna Castle, Poland, 2005), pp. 17–26
3. T. Amiaz, S. Fazekas, D. Chetverikov, N. Kiryati, in *International Conference on Scale Space and Variational Methods in Computer Vision (SSVM'07)* (2007), pp. 848–859
4. I. Akyildiz, T. Melodia, K. Chowdhury, Computer Networks **51**(4), 921 (2007)
5. R. Péteri, M. Huskies, S. Fazekas, Dyntex: A comprehensive database of dynamic textures (2006). http://www.cwi.nl/projects/dyntex/
6. M. Ballerini, N. Cabibbo, R. Candelier, A. Cavagna, E. Cisbani, I. Giardina, A. Orlandi, G. Parisi, A. Procaccini, M. Viale, V. Zdravkovic, Animal Behaviour **76**(1), 201 (2008)
7. B. Jahne, H. Hau$\beta$ecker, P. Gei$\beta$ler, *Handbook of Computer Vision and Applications* (Academic, NY, 1999), vol. 2, chap. 4
8. J. Babaud, A.P. Witkin, M. Baudin, R.O. Duda, IEEE Transactions on Pattern Analysis and Machine Intelligence **8**(1), 26 (1986)
9. T. Lindeberg, International Journal of Computer Vision **30**(2), 79 (1998)
10. D.G. Lowe, International Journal of Computer Vision **60**(2), 91 (2004)
11. L. Raffo, S. Sabatini, G. Bo, G. Bisio, IEEE Transactions on Neural Networks **9**(6), 1483 (1998)
12. E. Vittoz, X. Arreguit, Electronic Letters **29**(3), 297 (1993)
13. K. Bult, J. Geelen, IEEE Journal of Solid-State Circuits **27**(12), 1730 (1992)
14. L. Vadasz, IEEE Transactions on Electron Devices **13**(5), 459 (1966)
15. H. Kobayashi, J. White, A. Abidi, IEEE Journal of Solid-State Circuits **26**(5), 738 (1991)
16. K. Hui, B. Shi, IEEE Transactions on Circuits and Systems-I **46**(10), 1161 (1999)
17. J. Fernández-Berni, R. Carmona-Galán, in *European Conference on Circuit Theory and Design (ECCTD'09)* (Antalya, Turkey, 2009)
18. J. Fernández-Berni, R. Carmona-Galán, in *12th International Workshop on Cellular Nanoscale Networks and Applications (CNNA)* (Berkeley, CA, 2010), pp. 453–458

# A Biomimetic Frame-Free Event-Driven Image Sensor

**Christoph Posch**

**Abstract** Conventional image sensors acquire the visual information time-quantized at a predetermined frame rate. Each frame carries the information from all pixels, regardless of whether or not this information has changed since the last frame had been acquired. If future artificial vision systems are to succeed in demanding applications such as autonomous robot navigation, high-speed motor control and visual feedback loops, they must exploit the power of the biological, asynchronous, frame-free approach to vision and leave behind the unnatural limitation of frames: These vision systems must be driven and controlled by events happening within the scene in view, and not by artificially created timing and control signals that have no relation whatsoever to the source of the visual information: the world. Translating the frameless paradigm of biological vision to artificial imaging systems implies that control over visual information acquisition is no longer being imposed externally to an array of pixels but the decision making is transferred to the single pixel that handles its own information individually. The notion of a frame has then completely disappeared and is replaced by a spatio-temporal volume of luminance-driven, asynchronous events. ATIS is the first optical sensor to combine several functionalities of the biological 'where'- and 'what'-systems of the human visual system. Following its biological role model, this sensor processes the visual information in a massively parallel fashion using energy-efficient, asynchronous event-driven methods.

## 1 Introduction

State-of-the-art image sensors suffer from severe limitations imposed by their very principle of operation. Nature suggests a different approach: Highly efficient biological vision systems are driven and controlled by events happening within the scene

C. Posch (✉)
AIT Austrian Institute of Technology, Vienna, Austria
e-mail: Christoph.Posch@ait.ac.at

Á. Zarándy (ed.), *Focal-Plane Sensor-Processor Chips*,
DOI 10.1007/978-1-4419-6475-5_6, 

in view, and not – like image sensors – by artificially created timing and control signals that have no relation whatsoever to the source of the visual information: the world. Translating the frameless paradigm of biological vision to artificial imaging systems implies that control over the acquisition of visual information is no longer being imposed externally to an array of pixels but the decision making is transferred to the single pixel that handles its own information individually.

## 1.1 Neuromorphic Information Processing

Despite all the impressive progress made during the last decades in the fields of information technology, microelectronics and computer science, artificial sensory and information processing systems are still much less effective in dealing with real-world tasks than their biological counterparts. Even small insects still outperform the most powerful computers in routine functions involving for example, real-time sensory data processing, perception tasks and motion control and are, most strikingly, orders of magnitude more energy efficient in completing these tasks. The reasons for the superior performance of biological systems are still only partly understood, but it is apparent that the hardware architecture and the style of neural computation are fundamentally different from what is state-of-the-art in artificial synchronous information processing. Very generally speaking, biological neural systems rely on a large number of relatively simple, slow and unreliable processing elements and obtain performance and robustness from a massively parallel principle of operation and a high level of redundancy, where the failure of single elements usually does not induce any observable system performance degradation.

The idea of applying computational principles of biological neural systems to artificial information processing exists since decades. The earliest work from the 1940s introduced a neuron model and showed that it was able to perform computation [1]. Around the same time, Donald Hebb developed the first models for learning and adaptation [2]. In the late 1980s, Carver Mead demonstrated [3–5] that modern silicon VLSI technology can be employed in implementing circuits that mimic neural functions and fabricate building blocks that work like their biological role models i.e., neurons, axons, ganglions, photoreceptors, etc., thus enabling the construction of biomimetic artifacts that combine the strengths of modern silicon VLSI technology with the processing abilities of brains. These observations revolutionized the frontier of computing and neurobiology to such an extent that a new engineering discipline emerged, whose goal is to design and build artificial neural systems and apply them for example, for vision systems, auditory processors or autonomous, roving robots. The field is referred to as "neuromorphic engineering." Neuromorphic systems, as the biological systems they model, are adaptive, fault tolerant and scalable and process information using energy-efficient, asynchronous, event-driven methods.

Representing a new paradigm for the processing of sensor signals, the greatest success of neuromorphic systems to date has been in the emulation of sensory signal acquisition and transduction, most notably in vision. Since the seminal attempt to build a "silicon retina" by Mahowald and Mead in the late 1980s [6], a variety of biomimetic vision devices has been proposed and implemented [7].

## 1.2 Biological Vision

In the field of imaging and vision, two remarkable observations are to be made: biology has no notion of a frame – and the world, the source of most of the visual information we are interested in acquiring, works asynchronously and in continuous time. The author is convinced that biomimetic asynchronous electronics and signal processing have the potential – also, and maybe especially, in fields that are historically dominated by synchronous approaches as it is the case for artificial vision, image sensing and image processing – to reach entirely new levels of performance and functionality, comparable to the ones found in biological systems. Furthermore, the author believes that future artificial vision systems, if they want to succeed in demanding applications for example, autonomous robot navigation, high-speed motor control, visual feedback loops. must exploit the power of the asynchronous, frame-free, biomimetic approach.

Studying biological vision, it has been noted that there exist two different types of retinal ganglion cells and corresponding retina-brain pathways in the, for example, human retina: X-cells or Parvo-cells and Y-cells or Magno-cells. The Y-cells are at the basis of what is named the transient channel or the Magno-cellular pathway. Y-cells are approximately evenly distributed over the retina. They have short latencies and use rapidly conducting axons. Y-cells have large receptive fields and respond transiently, especially when changes – movements, onsets, offsets – are involved. The X-cells are at the basis of what is called the sustained channel, or, the Parvo-cellular pathway. X-cells are mainly concentrated in the fovea, the center of the retina. The X-cells have longer latencies and the axons of X-cells conduct more slowly. They have smaller receptive fields and respond in a sustained way. X-cells are most probably involved in the transportation of detailed pattern, texture and color information [8].

It appears that these two parallel pathways in the visual system are specialized for certain types of visual perception:

- The Magno-cellular system is more oriented toward general detection or alerting and is referred to as the "where" system. It has high temporal resolution and is sensitive to changes and movements. Its biological role is seen in detecting for example, dangers that arise in the peripheral vision. Magno-cells are relatively evenly spaced across the retina at a rather low spatial resolution and are the predominant cell type in the retinal periphery.

- Once an object is detected, the detailed visual information (spatial details, color) seems to be carried primarily by the Parvo-system. It is hence called the "what" system. The "what" system is relatively slow, exhibiting low temporal, but high spatial resolution. Parvo-cells are concentrated in the fovea, the retinal center.

Practically, all conventional frame-based image sensors can functionally be attributed to the "what" system side, thus completely neglecting the dynamic information provided by the natural scene and perceived in nature by the Magno-cellular pathway, the "where"-system. Attempts to implementing the function of the Magno-cellular transient pathway in an artificial neuromorphic vision system has recently led to the development of the "Dynamic Vision Sensor" (DVS) [9–11]. This type of visual sensor is sensitive to the dynamic information present in a natural scene; however, it neglects the sustained information perceived by the Parvo-cellular "what"-system.

## 1.3 Learning from Nature

For solving real-world problems, from an engineering perspective, it is not strictly necessary to build functionally accurate copies of biological neural systems but devise neuromorphic systems that exploit key principles in the context of the available technology – which is very different from biology. In vision, the biomimetic parallel signal processing approach utilizes the possibility of integrating processing elements in each pixel of an imaging array and merging low-level signal processing operations, mainly some type of feature extraction, with the classical analog photoreceptor circuitry. Further exploiting the concepts of biological vision suggests a combination of the "where" and "what"-system functionalities in a biomimetic, asynchronous, event-driven style. A visual device implementing this paradigm could open up a whole new level of sensor functionality and performance, and inspire a new approach for image data processing. A first attempt towards this goal is described here. ATIS, an synchronous, time-based image sensor, is the first visual sensor that combines several functionalities of the biological "where" and "what" systems with multiple bio-inspired approaches, such as event-based time-domain imaging, temporal contrast dynamic vision and asynchronous, event-based information encoding and data communication.

Technically, the imager incorporates an array of asynchronous, fully autonomous pixels, each containing event-based change detection and pulse-width-modulation exposure measurement circuits. The operation principle ideally results in highly efficient lossless video compression through temporal redundancy suppression at the focal-plane, while the asynchronous, time-based exposure encoding yields exceptional dynamic range, signal-to-noise ratio, fixed-pattern noise performance and temporal resolution along with the possibility to flexibly optimize trade-offs in response to differing application demands.

## 2 Limitations to Solid-State Imaging

Continuous advances in deep-submicron CMOS process technology allow building high-performance single-chip cameras, combining image capture and advanced on-chip processing circuitry in the focal plane. Despite all progress, several problems with solid-state imaging remain unresolved and performance is limited, mostly due to physical constraints of fabrication technology and operating principles.

### *2.1 Temporal Redundancy*

Conventional image sensors acquire visual information in the form of image frames, time-quantized at a predetermined frame rate. Each frame conveys the information from all pixels, regardless of whether or not this information, or a part of it, has changed since the last frame had been acquired. This method obviously leads, depending on the dynamic contents of the scene, to a more or less high degree of redundancy in the acquired image data. Acquisition and handling of these dispensable data consume valuable resources and translate into high transmission power dissipation, increased channel bandwidth requirements, increased memory size and post-processing power demands.

One fundamental approach to dealing with temporal redundancy in video data is frame difference encoding. This simple form of video compression includes transmitting only pixel values that exceed a defined intensity change threshold from frame to frame after an initial key-frame. Frame differencing is naturally performed off-sensor at the first post-processing stage [12, 13]; yet, a number of image sensors with focal-plane frame differencing have been reported [14–17]. However, all these frame differencing imagers still rely on acquisition and processing of full frames of image data and are not able to self-consistently suppress temporal redundancy and provide real-time compressed video output. Furthermore, even when the processing and difference quantization is done at the pixel-level, the temporal resolution of the acquisition of the scene dynamics, as in all frame-based imaging devices, is still limited to the achievable frame rate and is time-quantized to this rate. One major obstacle for sensor-driven video compression lies in the necessity to combine a pixel identifier and the corresponding grayscale value and implement conditional readout using standard array scanning readout techniques.

A natural approach to autonomous suppression of temporal redundancy, and consequently sensor-driven video compression, is pixel-individual exposure on-demand, based on asynchronous, pixel-autonomous change detection. The problem of efficient, combined transmission of pixel addresses and intensity values can be resolved by using time-based exposure measurement and asynchronous, event-based information encoding and data communication [7, 18, 19]. This fully asynchronous operation, in addition, avoids unnatural time-quantization at all stages of image data acquisition and early processing.

The device described here implements this approach and achieves highly efficient sensor-driven ideally lossless video compression, delivering high-quality streaming video with compression factors depending essentially only on scene activity.

## 2.2 Dynamic Range and Signal-to-Noise Ratio

The dynamic range (DR) of an image sensor is defined as the ratio of the maximum processable signal and the noise floor under dark conditions. Conventional CMOS active pixel sensors (APS) are based on some variation on the 3T or 4T voltage-mode pixel. In the standard APS scheme, the exposure time and the integration capacitance are held constant for the pixel array. For any fixed integration time, the analog readout value has a limited signal swing that determines the maximum achievable DR as

$$\mathrm{DR} = 10\log\left(\frac{V_{\mathrm{sat}}^2}{V_{\mathrm{dark}}^2 + V_{\mathrm{reset}}^2 + V_{\mathrm{out}}^2}\right), \tag{1}$$

where $V_{\mathrm{sat}}$ is the maximum allowed voltage at the integration node and $V_{\mathrm{dark}}$, $V_{\mathrm{reset}}$ and $V_{\mathrm{out}}$ are darkcurrent, reset ($kTC$) and readout noise voltages, respectively.

Most voltage (and current) mode image sensors exhibit a saturating linear response with a DR limited to 60–70 dB. Both the signal saturation level and the noise floor are essentially constrained by the fabrication process. Light from natural scenes can span up to 140 dB of DR, ranging from 1 mlx up to 10 klx and more. According to notable experts in the field, it is clear that high DR imaging will dominate the market in the near future [20].

The signal-to-noise ratio (SNR) as an important criterion for image quality is defined as the quotient of the signal power and the average noise power:

$$\mathrm{SNR} = 10\log\left(\frac{V_{\mathrm{sig}}^2}{V_{\mathrm{dark}}^2 + V_{\mathrm{photo}}^2 + V_{\mathrm{reset}}^2 + V_{\mathrm{out}}^2}\right) \tag{2}$$

with $V_{\mathrm{photo}}$ representing the photocurrent shot noise. Since the photocurrent shot noise is the dominant noise source for moderate and high light illumination conditions, (2) can be approximated as:

$$\mathrm{SNR} \approx 10\log\left(\frac{C_{\mathrm{D}} \cdot V_{\mathrm{sig}}}{q}\right), \tag{3}$$

where $C_{\mathrm{D}}$ is the photodiode integration capacitance and $q$ the elementary charge. Because the SNR is proportional to the integration voltage $V_{\mathrm{sig}}$, in conventional APS image sensors with a fixed integration time for all pixels, the image quality strongly depends on the illuminance.

### 2.2.1 Time-Domain Imaging

To overcome the standard image sensor's DR and SNR limitations, several approaches have incorporated the dimension of time, in one form or another, as a system variable. While some designs use either variable integration times or time-dependent well capacities to increase DR [21–23], other designs are based on directly measuring the time it takes the photocurrent to produce a given voltage change at the sense node. This technique is commonly called time-domain or pulse modulation (PM) imaging.

In PM imaging, the incident light intensity is not encoded in amounts of charge, voltage, or current but in the timing of pulses or pulse edges. Dual to the voltage-mode pixel, the "integration-time"-mode pixel connects the sense node to a comparator which toggles state when $V_{int}$ goes beyond some reference value $V_{ref}$. The state is reflected in the binary signal $V_{out}$, which may be connected to an output bus and/or fed back to the reset transistor (Fig. 1). If an external signal $V_{reset}$ is used to reset the sense node, the pixel operates as a timer. If the loop is closed to connect the comparator output to the reset transistor, the pixel becomes an oscillator which generates pulses on the $V_{out}$ node at a frequency inversely related to the integration time. PM imaging can thus be coarsely classified into two basic techniques, namely pulse width modulation (PWM) encoding and pulse frequency modulation (PFM) encoding (Fig. 1). References [24–26] dating from 1996 onwards report early PWM image sensor implementations. The first PFM circuit was reported by Frohmader et al. [27] in 1982. The first PFM-based image sensor was proposed in 1993 [28] and demonstrated in 1994 [29]. Both schemes allow each pixel to autonomously choose its own integration time. By shifting performance constraints from the voltage domain into the time domain, DR is no longer limited by the integration voltage swing.

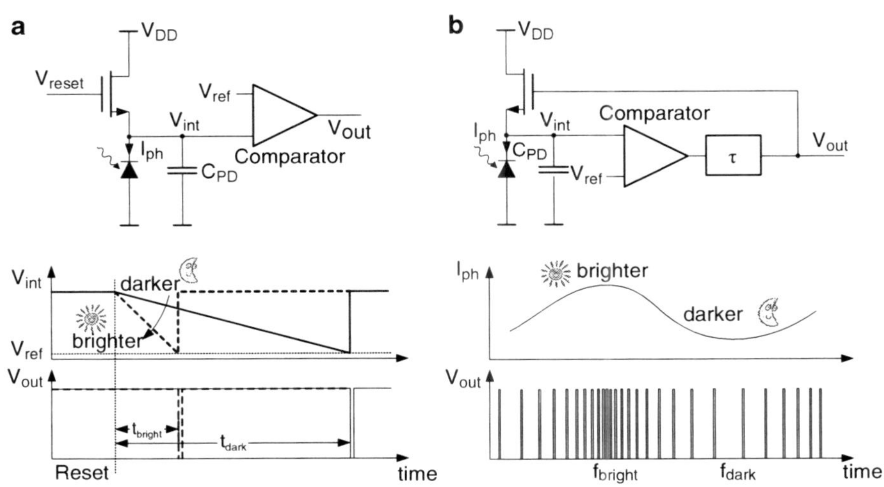


**Fig. 1** Time-domain encoding of pixel exposure information: PWM (**a**) and PFM (**b**)

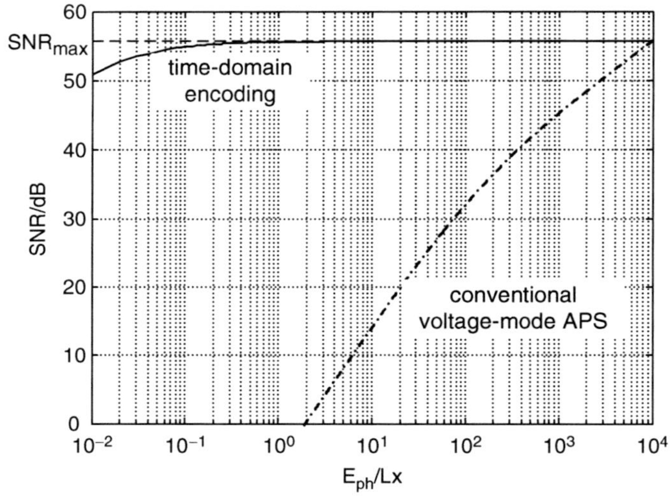


**Fig. 2** Comparison of the signal-to-noise ratio of a standard voltage-mode APS pixel and a time-domain pixel vs. illuminance

DR in PWM exposure encoding is given by the simple relation

$$\mathrm{DR} = 20\log\frac{I_{\mathrm{max}}}{I_{\mathrm{min}}} = 20\log\frac{t_{\mathrm{int,max}}}{t_{\mathrm{int,min}}}, \tag{4}$$

where the maximum integration time $t_{\mathrm{int,max}}$ is limited by the darkcurrent (typically seconds) and the shortest integration time by the maximum achievable photocurrent and the sense node capacitance (typically microseconds). DR values of the order of 100–120 dB have been reported for various PWM imagers [30, 31].

Also, the sensor's SNR benefits from the time-domain approach. In time-based image sensors, every pixel reaches the maximum integration voltage in every integration cycle. Consequently the achievable SNR is essentially independent of illuminance and photocurrent (compare (3)). Figure 2 plots SNR for a voltage mode APS and a time-domain pixel as a function of illuminance. The strong light dependency of APS SNR is apparent while the time-based pixel reaches full SNR already at low light conditions.

## 3 ATIS Imager Concept

As touched on above, the adverse effects of data redundancy, common to all frame-based image acquisition techniques, can be tackled in several different ways. The biggest conceivable gain, however, is achieved by simply not recording the

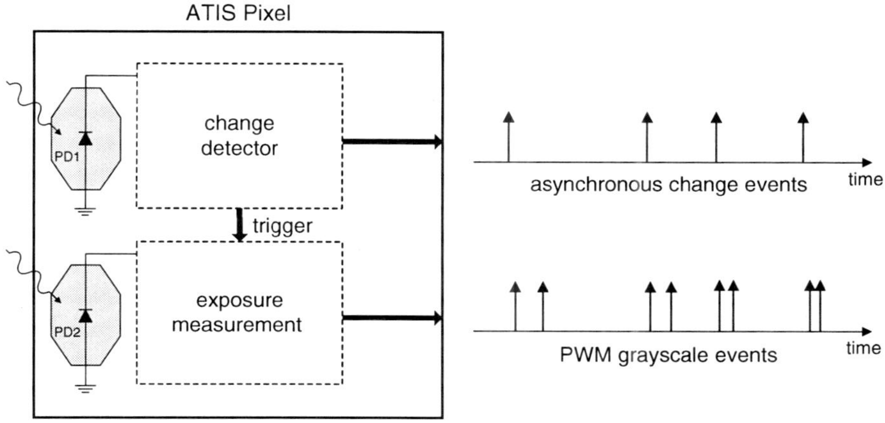


**Fig. 3** Functional diagram of an ATIS pixel. Two types of asynchronous "spike" events, encoding change and brightness information, are generated and transmitted individually by each pixel in the imaging array

redundant data in the first place, thus reducing energy, bandwidth/memory requirements and computing power in data transmission and processing. A fundamental solution for achieving complete temporal redundancy suppression is using an array of fully autonomous pixels that contain a *change detector* (CD) and a *conditional* exposure measurement (EM) device. The change detector individually and asynchronously initiates the measurement of a *new* exposure/grayscale value only if – and immediately after – a brightness change of a certain magnitude has been detected in the field-of-view of the respective pixel. Such a pixel does not rely on external timing signals and independently requests access to an (asynchronous and arbitrated) output channel only when it has a new grayscale value to communicate. Consequently, a pixel that is not stimulated visually does not produce output. In addition, the asynchronous operation avoids the time quantization of frame-based acquisition and scanning readout.

Pixels autonomously communicate change and grayscale events independently to the periphery in the form of asynchronous "spike" events. The events are arbitrated by asynchronous bus arbiters, furnished with the pixel's array address by an address encoder and sent out on an asynchronous bit-parallel bus (AER [7, 18, 19]). Figure 3 shows a functional diagram of the ATIS pixel.

## *3.1 Change Detection*

According to the German physicist and founder of psychophysics Gustav Theodor Fechner (1801–1887), the relationship between stimulus and perception is logarithmic (or the response is proportional to a relative change) [32]. For photopic vision, a relative change of one to two percent of light intensity is distinguishable by a human observer. To achieve ideal temporal redundancy suppression with respect to a human observer, the change detector must be able to sense a *relative* change of this magnitude or at least of one half of an LSB of the intended grayscale resolution over the full DR.

In an attempt to mimicking the Magno-cellular transient pathway of the human vision system, the DVS pixel circuit, a fast, continuous time logarithmic photoreceptor combined with asynchronous, event-based signal processing, that is sensitive to temporal contrast over 6 decades of illumination, has been developed [9–11]. This circuit is perfectly suited to serve as the sought-after low-latency, wide DR change detector. It combines an active, continuous-time, logarithmic photo-front end (*PD1*, $M_{\mathrm{fb}}$, *A1*), a well-matched, self-timed, self-balancing switched-capacitor amplifier (*C1, C2, A2*), and two threshold comparators for polarity-sensitive event generation (Fig. 4).

The photoreceptor responds logarithmically to intensity, thus implementing a gain control mechanism that is sensitive to temporal contrast or *relative* change. The circuit comprises a photodiode whose photocurrent is sourced by a saturated NMOS transistor $M_{\mathrm{fb}}$ operated in weak inversion. The gate of $M_{\mathrm{fb}}$ is connected to the output of an inverting amplifier whose input is connected to the photodiode. This transimpedance configuration converts the photocurrent logarithmically into a voltage and also holds the photodiode clamped at virtual ground [33]. As a result, the bandwidth of the photoreceptor is extended by the factor of the loop gain in comparison with a simple passive logarithmic photoreceptor. At low-light conditions, the bandwidth of photoreceptor is limited by the photocurrent and can be approximated by a first-order low-pass filter with corner frequency

$$f_{3\mathrm{dB}} = \frac{1}{2\pi C_{\mathrm{DMfb}}} \frac{I_{\mathrm{ph}}}{U_{\mathrm{T}}}, \tag{5}$$

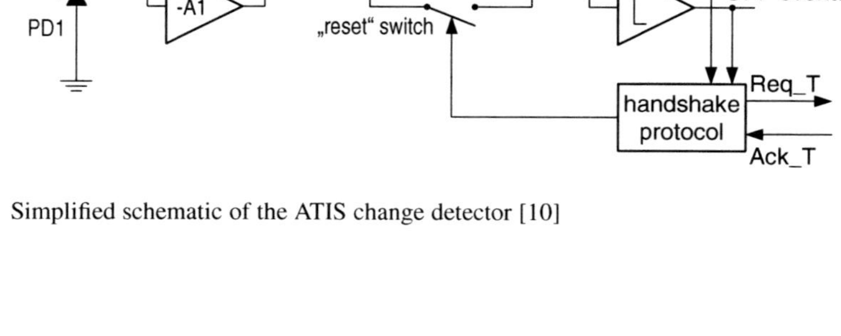


**Fig. 4** Simplified schematic of the ATIS change detector [10]

where $C_{\mathrm{DMfb}}$ is the gate-drain capacitance of the feedback transistor $M_{\mathrm{fb}}$ and $U_{\mathrm{T}}$ is thermal voltage. The bandwidth increase of the feedback configuration effects a corresponding reduction in SNR, which is given by:

$$\mathrm{SNR} \approx 10 \log \left( \frac{C_{\mathrm{DMfb}} \cdot U_{\mathrm{T}}}{\kappa_{\mathrm{Mfb}} \cdot q} \right), \tag{6}$$

where $\kappa_{\mathrm{Mfb}}$ is the subthreshold slope factor of transistor $\mathrm{M_{fb}}$. SNR levels of about 30dB can be reached with this configuration.

The DR of the continuous-time logarithmic photoreceptor is given by the same expression as one of the time-based image sensors (4), where $I_{\mathrm{max}}$ is the photocurrent at maximum illuminance and $I_{\mathrm{min}}$ is the darkcurrent. Assuming equal darkcurrent densities in both photodiodes, both pixel circuits, CD and EM, exhibit a very similar DR.

The photoreceptor output is buffered by a source follower and then differentiated by capacitive coupling to a floating node at the input of a common-source amplifier stage with switched capacitor feedback. The source follower isolates the sensitive photoreceptor from the rapid transients in the differencing amplifier. The amplifier is balanced using a reset switch that shorts input and output, yielding a reset voltage level depending on the amplifier operating point. Transients sensed by the photoreceptor circuit appear as an amplified deviation from this reset voltage at the output of the inverting amplifier. The event generation circuitry, composed of two threshold comparators, responds with pulse events of different polarity to positive and negative gradients of the photocurrent. Consequently, the rate of change is encoded in the inter-event intervals. Each of these change events is used to trigger and send a "start-integration" signal to the EM part. The polarity information contained in the change events is not required for the conditional EM functionality but is useful in various machine vision applications that rely on high temporal resolution event-based change information [34, 35].

## 3.2 Exposure Measurement

The EM part of the pixel is realized as a time-based PWM circuit. The time-domain approach to exposure measurement has been chosen for reasons of DR and SNR performance and its affinity to event-based information encoding and data communication.

For the photocurrent integrator circuit, an n-well/p-sub photodiode with PMOS reset transistor is used (Fig. 3a). The standard CMOS mixed-mode/RF fabrication process allows realizing the reset transistor $\mathrm{M_{Rst}}$ as *p*-type, thus maximizing integration swing. The sense node is directly coupled to the voltage comparator input. A differential slope sampling scheme based on two global integration thresholds ($V_{\mathrm{refH/refL}}$) for the first time implements true time-domain correlated double sampling (TCDS) [36]. The true differential operation within one integration cycle

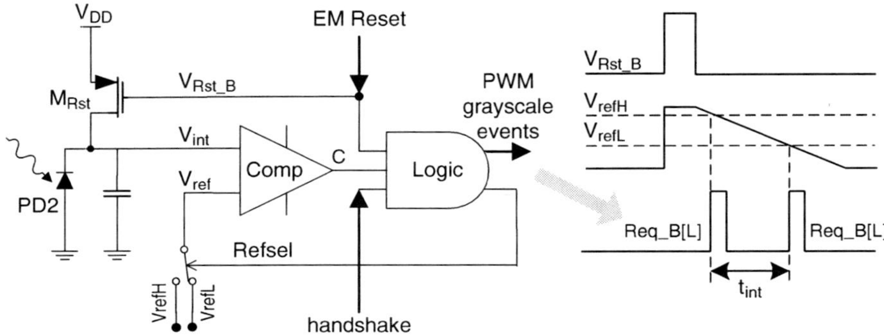


**Fig. 5** PWM dual-threshold exposure measurement

eliminates both comparator offset and reset *kTC* noise (Fig. 5) and comes at the cost of additional state and control logic in the pixel. Two 1-bit SRAM cells store the instantaneous pixel state and control the reference switch accordingly. Furthermore, the asynchronous digital logic is responsible for the event-based communication with the AER arbiter.

In the following, one cycle of transient change detection and exposure measurement is explained and illustrated by typical signal waveforms, taken from transistor-level simulation results (Fig. 6) [37]. The CD responds to a relative change in illumination by triggering the transmission of an address-event and simultaneously delivers a pulse on the reset line *Rst_B* (via row and column reset/mode-control circuits), which initiates an exposure-measurement cycle. The *Rst_B* signal briefly closes the switch $M_{\mathrm{Rst}}$, connecting the sense node to $V_{\mathrm{DD}}$. The pixel state control logic ensures that, at this point, the higher threshold voltage $V_{\mathrm{refH}}$ is connected as the reference voltage by setting the *RefSel* signal accordingly. By releasing the *Rst_B* signal, the integration process starts and the voltage $V_{\mathrm{int}}$ on the photodiode decreases proportionally to the photocurrent and thus proportionally to the illumination at the photodiode. When the photodiode voltage $V_{\mathrm{int}}$ reaches $V_{\mathrm{ref}}$, the comparator output $C$ toggles, causing the state logic to trigger the transmission of an event by activating the *Req_B[H]* signal, and to toggle the *RefSel* signal – now $V_{\mathrm{refL}}$ is set as the voltage reference and the comparator output $C$ toggles back. The integration continues in the meantime. $V_{\mathrm{int}}$ reaching $V_{\mathrm{refL}}$ marks the end of the measurement cycle, $C$ toggles again and the logic releases another event by sending a *Req_B[L]* signal. The time between the two events, triggered by *Req_B[H]* and *Req_B[L]*, is inversely proportional to the average pixel illumination during the integration.

CD and EM operation, once started, are completely detached and do not influence each other (in particular do not share a common output channel), with one important exception: If the CD senses another change before the integration process has finished, the current measurement cycle is aborted and the integration is restarted. In this case, the *Req_B[H]* event is discarded by the post-processor (which it detects in receiving two consecutive *Req_B[H]* events from the same pixel address). This behavior is intentional and does not imply information loss (depending on

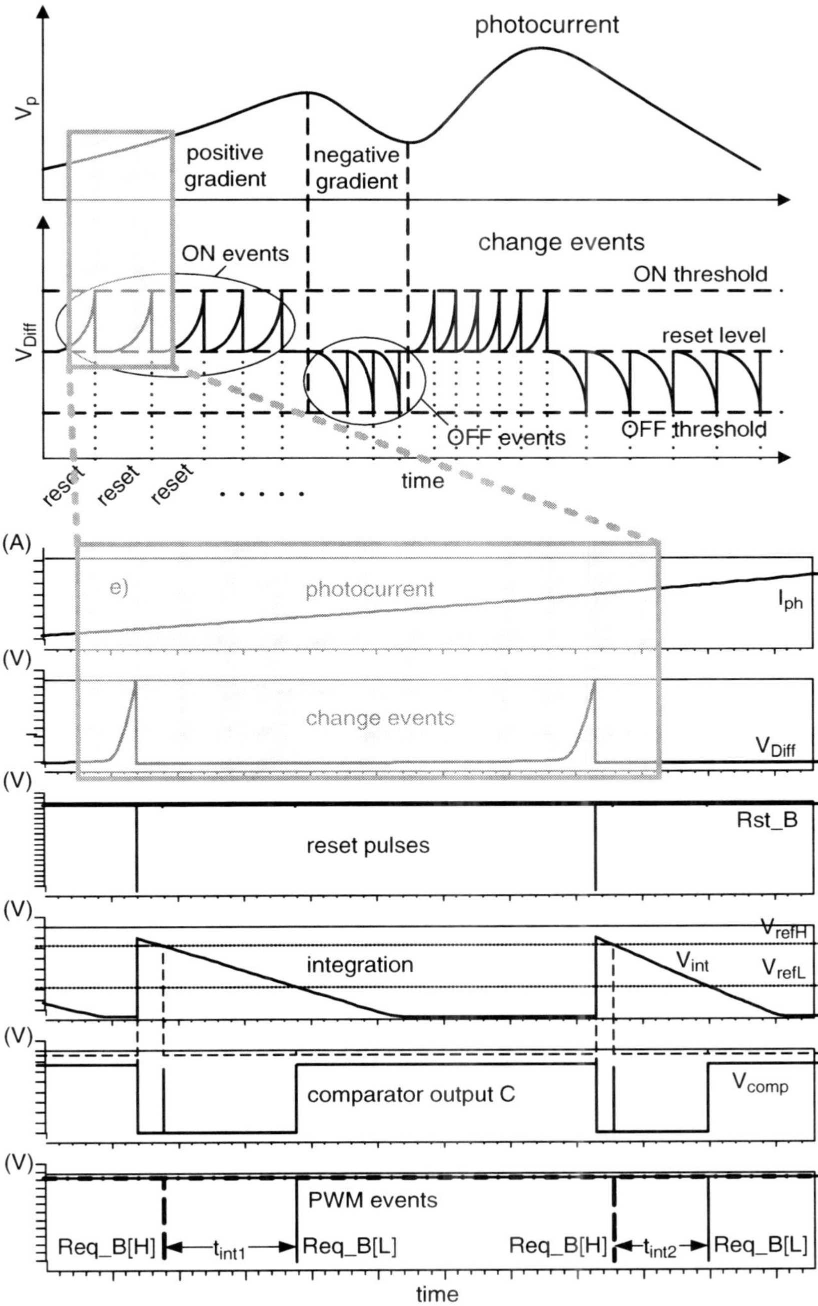


**Fig. 6** Two exposure measurement cycles triggered by change events. The upper plot shows typical signal waveforms of the change detector circuit. The upper trace represents an arbitrary voltage waveform at the node $V_p$ tracking the photocurrent through PD1. Signals from the PWM exposure measurement circuit are shown in the lower part of the figure

the observation time-scale), because, as a further change in illumination had taken place, the initial exposure result would have been already obsolete. This conduct ensures that each transmitted exposure result is as accurate and recent as possible.

## 3.3 System Considerations

The asynchronous change detection and the time-based exposure measurement approach harmonize remarkably well, mainly for two reasons: On the one hand, because both reach a DR of $>120$ dB – the first is able to detect small relative changes over the full range, the latter is able to resolve the associated grayscales independently of the initial light intensity. On the other hand, because both circuits operate event-based, namely the events of detecting illumination or reflectance changes and the events of pixel integration voltages reaching reference thresholds. Consequently, an asynchronous, event-based communication scheme (Address Event Representation, AER [7, 18, 19]) is used in order to provide efficient allocation of the transmission channel bandwidth to the active pixels and enable sensor-driven video compression output. Along with the pixel array address, the relevant information is inherently encoded in the event timing. Time-to-digital conversion of the event timings and the calculation of grayscale values from integration times are done off-chip.

The ATIS dynamic vision and image sensor is built around a QVGA ($304 \times 240$) pixel array and uses separate bus arbiters and event-parallel AER channels for communicating change events and grayscale encoding events independently and in parallel. Furthermore, the sensor features a flexible column/line-wise reset/trigger scheme for various modes of operation. Besides the (default) self-triggered mode, there are for example, external trigger modes for "snapshot" frame acquisition with "time-to-first-spike" (TTFS) encoding [38], or column-parallel relay readout [39]. Change detection and externally triggered imager operation can be fully decoupled and used independently and concurrently. Programmable regions-of-(non)-interest (ROI/RONI) are available and can be applied independently to CD and EM circuits.

### 3.3.1 Modes of Operation

The handshake signals for both CD and EM blocks are generated row and columnwise (and not pixel wise) in order to save chip area. The per-pixel *Rst_B* signals are generated by combinatorial logic from the row and column reset signals. In addition, the reset control logic can be configured through a digital serial interface to trigger (ROI) or ignore (RONI) selected regions of interest. The ROI/RONI selection of individual pixels can be configured independently for CD and EM and can be combined with a wide variety of trigger modes:

- In *normal operation mode (ATIS mode),* the start of the exposure measurement of one pixel is triggered from the event signal of the CD of the same pixel.

- In *global shutter mode,* groups of pixels or the full array, defined by the ROI, are reset simultaneously by a global control signal. This snapshot mode, essentially implementing a TTFS scheme [38], can run concurrently to the normal operation mode, and allows the quick acquisition of a reference frame during normal operation or an overlaid synchronous video mode.
- In the *asynchronous column-parallel readout (ACPR) sequential mode*, the integration start of one pixel (N) in one column is triggered by the preceding pixel (N–1). After externally starting for example the pixels of the first (top) row of the pixel array, the trigger runs in parallel along the columns, each pixel triggering its bottom neighbor when it has reached its first integration threshold $V_{\mathrm{refH}}$. This asynchronous "rolling shutter" mode is intended to avoid a familiar problem of event collisions in TTFS imagers (having many pixels finishing integration and trying to send their event at the same time when seeing highly uniform scenes) at the cost of slower image acquisition by spreading data readout in time. Multiple pixel rows across the array (e.g. rows 1 and 121, or 1, 61, 121, 181) can simultaneously be selected as starting rows to decrease frame acquisition time at the cost of higher event collision probability. Also this mode can run concurrently to the normal operation (ATIS) mode. The ACPR mode has been described in detail and analyzed in [39].

### 3.3.2 Data Readout

The pixels in the sensor array communicate with column and row arbiters via 4-phase AER handshaking as described in detail for example, in [10]. The 18-bit pixel addresses (8 bits row address, 9 bits column address, 1 polarity/threshold bit) are determined by row and column address encoders. The row signals yReq and yAck are shared by pixels along rows and the signals xReq and xAck are shared along columns. The peripheral AER circuits communicate without event loss. Bus collisions are resolved by delaying the transmission of events essentially on a "first-come-first-served" basis.

The self-timed communication cycle starts with a pixel (or a set of pixels in a row) pulling a row request (yReq) low against a global pull-up (wired OR). As soon as the row address encoder encodes the y-address and the row arbiter acknowledges the row (yAck), the pixel pulls down xReq. If other pixels in the row also have participated in the row request their column requests are serviced within the same row request cycle ("burst-mode" arbiter [38]). Now, the column address encoder encodes the x-address(es) and the complete pixel address(es) is/are available at the asynchronous parallel address bus. Assuming successful transmission and acknowledgment via the Ack_ext signal by an external data receiver, the Ack_col signal is asserted by the column handshake circuit. The conjunction of xAck and yAck signals generates control signals for the pixel that either (a) reset the transient amplifier in the CD part and eventually take away the pixel request, or (b) control the state logic in the EM part, respectively. The self-timed logic circuits ensure that all required ordering conditions are met. This asynchronous event-based communication works in an identical way both for CD and for EM events. Two completely separate and independent communication channels are used for the two types of events.

## 3.4 Layout

The chip has been implemented in a standard 0.18-μm 1P6M mixed-mode/RF CMOS process. Figure 7 shows the layout of the pixel with the main circuit parts annotated. The square pixel covers 900 $\mu m^2$ of silicon area (30 μm pixel pitch). The two photodiodes for continuous time operation of the CD (PD2) and integrating PWM exposure measurement (PD1) are placed side by side at the top edge of the pixel area. The fill factor of the pixel is 10% of total pixel area for the CD and 20% of total pixel area for the EM part.

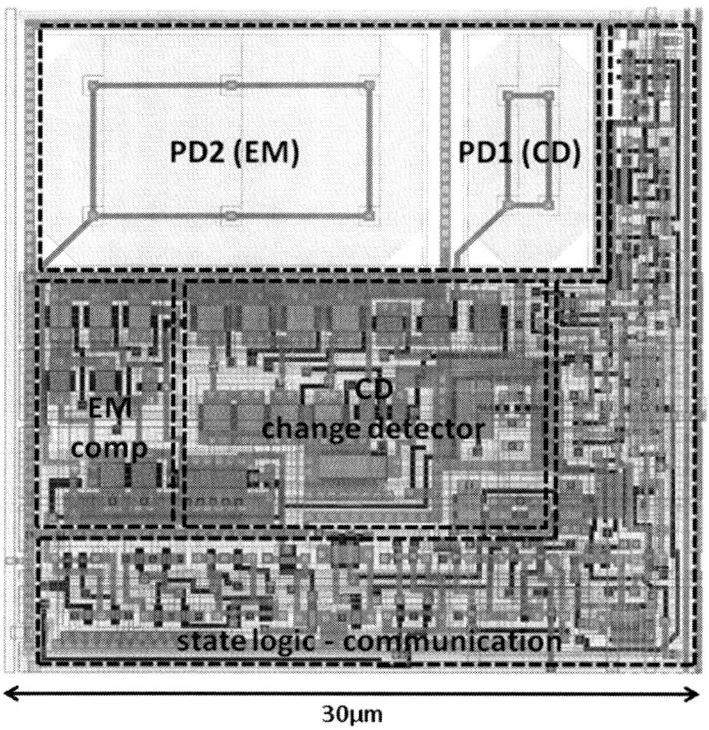


**Fig. 7** Pixel layout with the analog and digital circuit parts annotated. Two separate photodiodes – for continuous time operation of the change detector and reset-and-integrate PWM exposure measurement – are used in each pixel

## 4 Image Sensor Performance

The charge capacity of the integration node depends on operation voltages and approaches ~450,000 electrons at maximum (2V) integration swing $\Delta V_{\text{th}}$. Sense node capacitance is 36fF, yielding a conversion gain of 4.4 $\mu$V/e$^-$. Photodiode darkcurrent has been measured at ~3 fA, darkcurrent shot noise is 470e$^-$ corresponding to 2.1 mV r.m.s. for an integration swing of 1 V.

### *4.1 PWM Imaging: Transfer Function*

What is exposure time to conventional voltage-mode image sensors is the integration voltage swing in time-domain imaging. Figure 8 plots measured integration times for integration swings $\Delta V_{\text{th}}$ between 0.5 V and 2 V as a function of pixel illumination. The theoretically asserted 1/x-relation is accurately satisfied. Integration times range from for example, 10 ms @ 10lx to 10 $\mu$s @ 10klx for an integration swing of 500 mV.

### *4.2 Signal-to-Noise Ratio*

In Fig. 9, measured imager SNR as a function of integration swing for different light intensity is shown. SNR is >56 dB at an integration swing $\Delta V_{\text{th}}$ of 2 V and

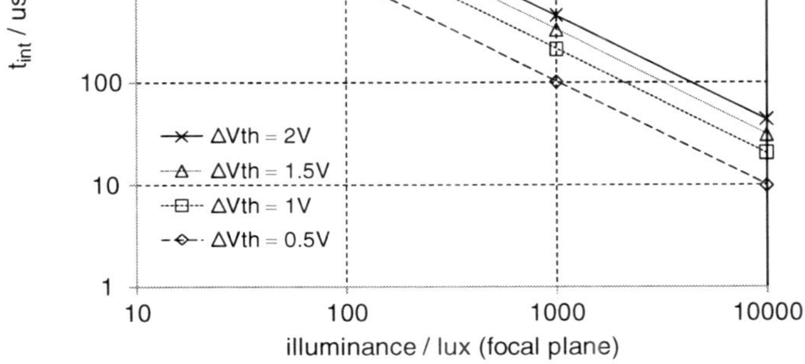


**Fig. 8** Measured PWM transfer function (integration time vs. lux) for four different values of integration swing $\Delta V_{\text{th}}$

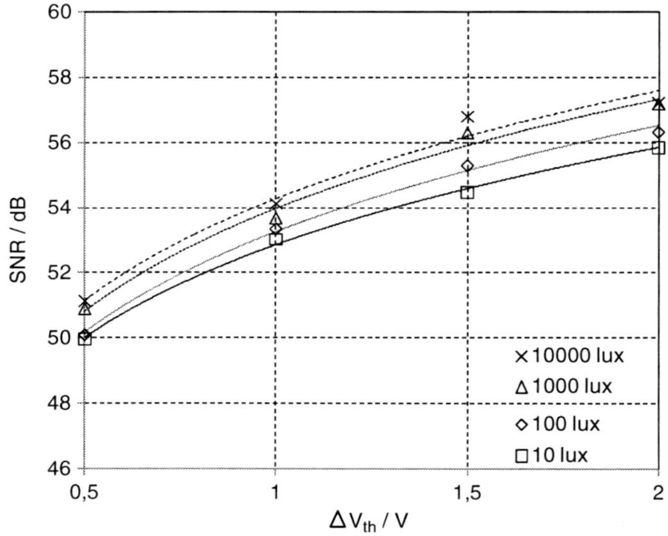


**Fig. 9** Measured SNR as functions of integration swing $\Delta V_{th}$ and light intensity. SNR is >56 dB for an integration swing of 2 V for light levels above 10lx. For $\Delta V_{th} = 100\,\text{mV}$ and 10lx SNR is still at 42.3 dB

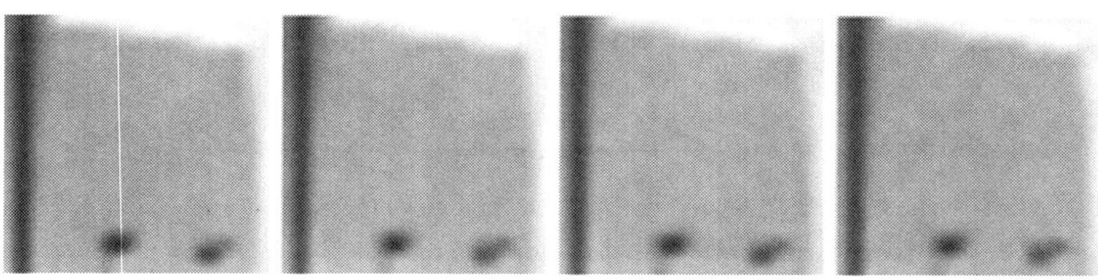

**Fig. 10** Detail of the scene in Fig. 11, acquired at four different integration swings: 2 V, 0.75 V, 0.25 V and 0.1 V. The slight decrease in SNR is observable

light levels above 1lx. Standard 8-bit grayscale resolution (48 dB) is achieved for very low illuminations and small integration swings. For $\Delta V_{th} = 100\,\text{mV}$ and 10lx, SNR is still at 42 dB, allowing for 7-bit resolution imaging at very short integration times (<2 ms @ 10lx). The result is 500 fps equivalent temporal resolution imaging and video at low-light conditions. The weak dependence of SNR on illuminance for time-based image sensors, as illustrated in Fig. 2, is well observable in the measured data. Figure 10 illustrates the slight increase in image noise for decreasing integration swings from 2 V down to 100 mV.

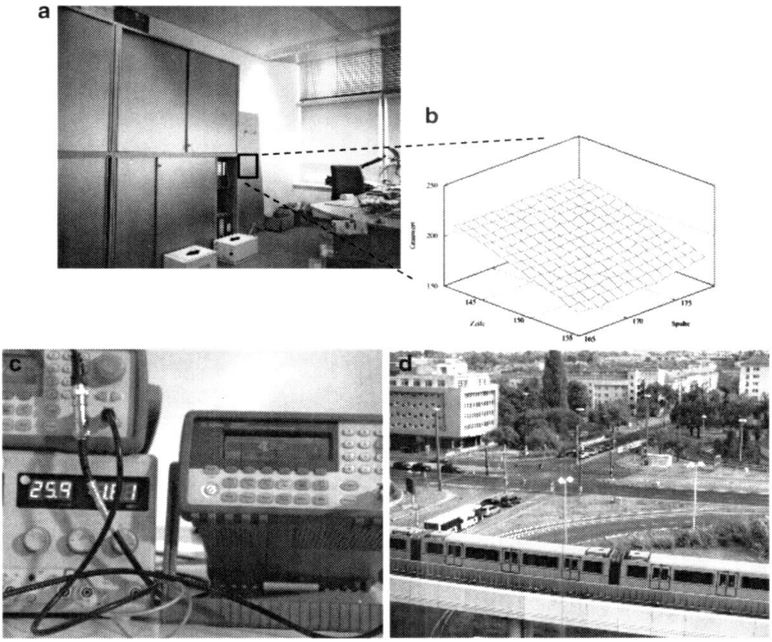


**Fig. 11** (**a**) Indoor scene at ~100lx, (**b**) mesh plot of grayscale values of the gradient in the marked area of (**a**); (**c**) is an example image of an indoor scene at ~200lx with fluorescent tube illumination and (**d**) an outdoor scene at ~3,000lx daylight

## *4.3 Fixed-Pattern Noise*

Fixed-pattern noise (FPN) is the temporally constant spatial nonuniformity of pixel response in an array due to device mismatch in the pixel circuits. An upper bound to the imager FPN of 0.25% was established from evaluating different homogenous parts of recorded image series, similar to the one shown in Fig. 11a.

Figure 11b shows grayscale values of an approximately planar gradient taken from the image in Fig. 11a displayed as a mesh plot for illustrating pixel response uniformity. Also owing to the SNR of 56 dB (9.3 bit), grayscale gradients are resolved smoothly without visible artifacts. Figure 11c, d show example images of an indoor and an outdoor scene at different illumination conditions.

## *4.4 Dynamic Range*

The achievable image sensor DR shows a trade-off with the temporal resolution required to capture scene dynamics and is limited by maximum allowable integration time at the dark end and AER communication channel data throughput at the bright end. A static scene DR of 143 dB at a maximum integration time of 4.5 s has been measured according to (4).

This temporal resolution, however, is inadequate for applications involving fast changing scenes or motion. To increase the temporal resolution without trading-off much of the DR, a method that is complementary to the multiple-exposure technique used for DR improvement in standard voltage-mode imagers is introduced. Owing to the presence of two independent thresholds, otherwise used for differential TCDS, it is possible to apply *two* integration swings (the complementary parameter to exposure time) during *one* image acquisition. With DR still at 143 dB, the longest integration times can so be reduced by a factor of about 20 (to $<\sim$200 ms, equivalent to 5 fps) if the time between *change event* and *upper* TCDS threshold is used to determine pixel exposure in the dark parts of the scene. Consequently for a temporal resolution of 33 ms (30 fps video speed equivalent temporal resolution), a DR of 125 dB is achieved. The penalty to pay for this high DR at high-speed operation is reduced SNR and FPN (no TCDS) for pixels that reach only the first threshold. From a system operation point of view, each pixel, depending on individual illumination, practically chooses for itself which threshold to use. The second threshold event is either simply ignored by the post-processor when arriving too late or will never appear since a new exposure has been started before.

Figure 12 shows image data acquired with the ATIS sensor from a real-world high-DR scene in one exposure. The first three images, top left, top right and bottom left, show different scalings of the exposure data, each revealing different details of

**Fig. 12** High DR imaging – three scalings of the same exposure image data and composite image

the scene outside the window and in the room, while bottom right is an illustrative attempt to generate a composite image by equalizing the data using a histogram equalization method.

### 4.5 Video Compression

The temporal redundancy suppression of the ATIS change-detector controlled operation ideally yields lossless focal-plane video compression with compression factors depending only on scene dynamics. Theoretically approaching infinity for static scenes, in practice, due to sensor nonidealities producing background noise events, the achievable compression factor is limited and appears to be of the order of 1,000 for bright static scenes as compared to a conventional, frame-based imager of the same resolution delivering raw 8-bit grayscale data at video speed of 30 fps.

Figure 13 shows a typical surveillance scene generating a 2.5–50 k events $s^{-1}$ @ 18-bit/event continuous-time video stream. The actual event rate depends on instantaneous scene activity. Comparing corresponding bit rates – 45–900 k bit $s^{-1}$ – to the raw data rate of a QVGA 8-bit grayscale sensor at 30 fps of 18 Mbit $s^{-1}$ demonstrates lossless video compression with compression factors between 20 and 400 for this example scene. Figure 13a contains a still frame taken from a continuous-time video sequence, Fig. 13b shows the same frame assuming video transmission has started from an empty image. The effect of objects triggering exposure measurement in the pixel they hit while moving across the focal plane becomes apparent (e.g., a white car moving from the bottom-left corner of the image toward the center).

Static background does not produce data apart from the odd CD noise event, also triggering exposure measurement in the respective pixel. On the one hand, this effect reduces the achievable video compression factor to about 1,000 for bright, static scenes (from infinity in an ideal, noise-free world). On the other hand, this effect is useful for capturing very slow changes in the scene (e.g., varying scene illumination from sun light due to passing clouds) through a continuous statistically distributed slow update of the entire image. The noise events are not perceivable in the video as usually a noise-triggered pixel replaces its grayscale value with an identical new one. Typical background noise activity is of the order of 1–3 k pixels $s^{-1}$, about 3 orders of magnitude below raw data rate from a conventional, frame-based sensor of same array size running at 30 fps. Figure 13c shows CD events collected during a time slice of 33 ms (ON events in white and OFF events in black) and Fig. 13d the grayscale data generated in response to the change events in c).

The compression gain is even larger when the temporal resolution (e.g., 500 fps equivalent for scenes $>$10lx) of ATIS is taken into account. Assuming the CDs to work 100% reliably and according to the requirements concerning contrast sensitivity stated earlier, the video data delivered by the ATIS sensor contain exactly the same information as a video stream from a conventional, frame-based imager – at a fraction of the data rate and hundreds-of-frames per second equivalent temporal resolution.

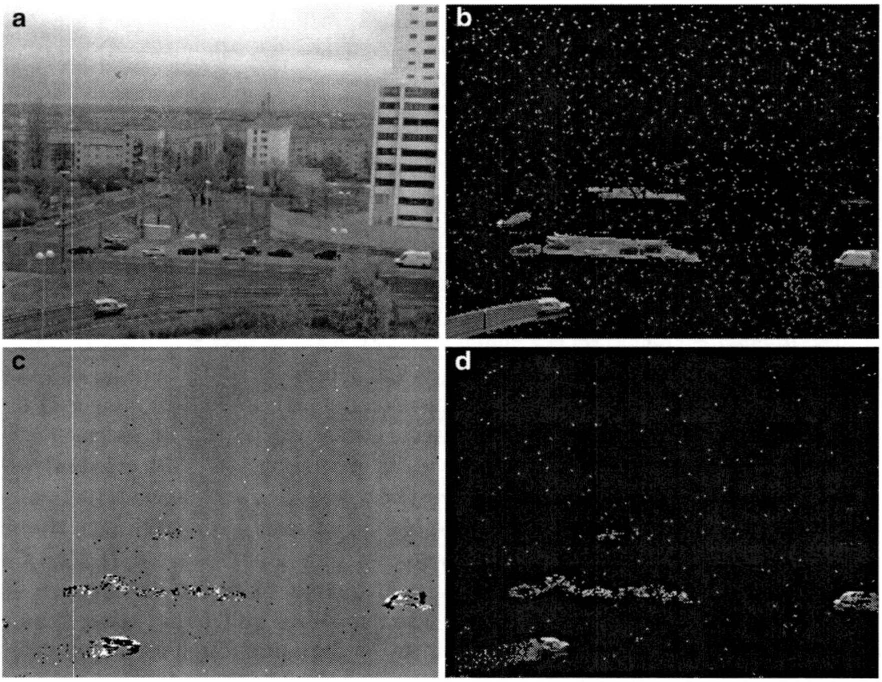


**Fig. 13** Traffic scene generating between 2.5 k and 50 k events $s^{-1}$, depending on instantaneous scene activity. The video compression factor w.r.t. raw data from a QVGA 30 fps 8-bit grayscale sensor was measured to be 20–400 for this example scene. (**a**) shows a still frame taken from a continuous-time video sequence recorded with the ATIS sensor; (**b**) shows the same frame assuming video transmission has started from an empty image. The effect of objects triggering exposure measurement in the pixel they hit while moving across the focal plane becomes apparent; (**c**) contains CD events recorded during a time slice of 33 ms (depicted as white/black pixels according to the event polarity – ON/OFF), and (**d**) shows the new pixel grayscale values measured in response to these detected changes. The video compression is essentially lossless, no dynamic image errors or artifacts are visible in the video (**a**)

# 5 Conclusion

ATIS is a biomimetic frame-free, high-DR, high-temporal resolution vision and image sensor with focal-plane data processing and compression. The sensor comprises an array of autonomous pixels that individually detect illumination changes and asynchronously encode in inter-event intervals the instantaneous pixel illumination after each detected change, ideally realizing optimal lossless pixel-level video compression through temporal redundancy suppression. Familiar deficiencies of time-based imagers have been remedied (a) using a novel time-domain correlated double sampling technique [36] and (b) by realizing illumination-dependent readout load spreading. Intra-scene DRs of 143 dB static and 125 dB @ $t_{int}$ <30 ms have been achieved. Target application areas are high-speed/high temporal-resolution

**Table 1** Summary table of sensor characteristics

| Parameter | Value |
|---|---|
| Fabrication process | UMC L180 MM/RF 1P6M Standard CMOS |
| Supply voltage | 3.3 V (analog), 1.8 V (digital) |
| Chip size | $9.9 \times 8.2\,\text{mm}^2$ |
| Optical format | 2/3″ |
| Array size | QVGA (304 × 240) |
| Pixel size | 30 μm × 30 μm |
| Pixel complexity | 77T, 3C, 2PD |
| Fill factor | 30% (20% EM, 10% CD) |
| Integration swing $\Delta V_{\text{th}}$ | 100 mV–2.3 V (adjustable) |
| SNR typ. | >56 dB (9.3bit) @ $\Delta V_{\text{th}} = 2\,\text{V}$, >10lx |
| SNR low | 42.3 dB (7bit) @ $\Delta V_{\text{th min}}$ (100 mV), 10lx |
| $t_{\text{int}}$ @ $\Delta V_{\text{th min}}$ (100 mV) | 2 ms @ 10lx (500 fps equ. temp. resolution) |
| Temporal resolution EM | 500 fps equ. (@ 10lx),<br>50 kfps equ. (@ 1,000lx) |
| Temporal resolution CD | 100 kfps equ. (@ >100lx) |
| DR (static) | 143 dB |
| DR (30 fps equivalent) | 125 dB |
| FPN | <0.25% @ 10lx (with TCDS) |
| Sense node cap | 36 fF |
| Conversion gain | $4.4\,\mu\text{V}/\text{e}^-$ |
| Darkcurrent | $1.6\,\text{nA}\,\text{cm}^{-2}$ (@25°C) |
| Power consumption | 50 mW (static), 175 mW (high activity) |
| Readout format | Asynchronous address-events (AER), 2 ×18bit-parallel |

dynamic machine vision, low-data rate video for wireless or TCP-based applications and high-DR, high-quality, high-temporal resolution imaging and video for, for example, scientific applications. Table 1 summarizes the main sensor specifications.

## References


1. W. McCulloch, W. Pitts, A logical calculus of the ideas immanent in nervous activity, *Bull. Math. Bio.*, no. 5, pp. 115–133, 1943
2. D. Hebb, *The organization of behavior*. New York, NY, Wiley, 1949
3. C. Mead, *Analog VLSI and neural systems*. NY, Addison-Wesley, 1989
4. C. Mead, Neuromorphic electronic systems, *Proc. IEEE*, vol. 78, no. 10 pp. 1629–1636, 1990
5. M.A.C. Maher, S.P. Deweerth, M.A. Mahowald, C.A. Mead, Implementing neural architectures using analog VLSI circuits, *IEEE Trans. Circ. Syst.*, vol. 36, no. 5, pp. 643–652, 1989
6. M.A. Mahowald, C.A. Mead, The Silicon Retina, *Scientific American*, May 1991
7. K. Boahen, Neuromorphic Microchips, *Sci. Am.*, vol. 292, pp. 55–63, 2005
8. A.H.C. Van Der Heijden, *Selective attention in vision*, ISBN: 0415061059, New York, Routledge, 1992
9. P. Lichtsteiner, T. Delbruck, A 64 × 64 AER logarithmic temporal derivative silicon retina, *Research in Microelectronics and Electronics, 2005 PhD*, vol. 2, pp. 202–205, 25–28 July 2005

10. P. Lichtsteiner, C. Posch, T. Delbruck, A 128 × 128 120 dB 15 μs latency asynchronous temporal contrast vision sensor, *IEEE J. Solid-State Circ.*, vol. 43, no. 2, pp. 566–576, 2008
11. P. Lichtsteiner, C. Posch, T. Delbruck, A 128 × 128 120dB 30mW asynchronous vision sensor that responds to relative intensity change, *ISSCC, 2006, Dig. of Tech. Papers*, pp. 2060–2069, 6–9 Feb 2006
12. F.W. Mounts, A video coding system with conditional picture-element replenishment, *BSTJ*, pp. 2545–2554, 1969
13. Y. Chin, T. Berger, A software-only videocodec using pixelwise conditional differential replenishment and perceptual enhancements, *IEEE Trans. Circ. Syst. Video Tech.*, vol. 9, no. 3, pp. 438–450, 1999
14. K. Aizawa, Y. Egi, T. Hamamoto, M. Hatori, M. Abe, H. Maruyama, H. Otake, Computational image sensor for on sensor compression, *IEEE Trans. Electr. Devices*, vol. 44, no. 10, pp. 1724–1730, 1997
15. V. Gruev, R. Etienne-Cummings, A pipelined temporal difference imager, *IEEE J Solid-State Circ.*, vol. 39, no. 3, pp. 538–543, 2004
16. J. Yuan, Y.C. Ho , S.W. Fung, B. Liu, An activity-triggered 95.3 dB DR – 75.6 dB THD CMOS imaging sensor with digital calibration, *IEEE J. Solid-State Circ.*, vol. 44, no. 10, pp. 2834–2843, 2009
17. Y.M. Chi, U. Mallik, M.A. Clapp, E. Choi, G. Cauwenberghs, R., Etienne-Cummings, CMOS camera with in-pixel temporal change detection and ADC, *IEEE J. Solid-State Circ.*, vol. 42, no. 10, pp. 2187–2196, 2007
18. K. Boahen, A burst-mode word-serial address-event link-I: transmitter design, *IEEE Trans. Circ. Syst. I*, vol. 51, no. 7, pp. 1269–1280, 2004
19. K. Boahen, Point-to-point connectivity between neuromorphic chips using address events, *IEEE Trans. Circ. Syst. II*, vol. 47, no. 5, pp. 416–434, 2000
20. G. Ward, The hopeful future of high dynamic range imaging, *2007 SID International Symposium*, 22–25 May 2007
21. S.J. Decker, R.D. McGrath, K. Brehmer, C.G. Sodini, A 256 × 256 CMOS imaging array with wide dynamic range pixels and column-parallel digital output, *IEEE J. Solid State Circ.*, vol. 33, pp. 2081–2091, 1998
22. T. Lulé, B. Schneider, M. Böhm, Design and fabrication of a high dynamic range image sensor in TFA technology, *IEEE J. Solid State Circ.*, vol. 34, pp. 704–711, 1999
23. D.X.D. Yang, A. El Gamal, B. Fowler, H. Tian, A 640 × 512 CMOS image sensor with ultrawide dynamic range floating-point pixellevel ADC, *IEEE J. Solid State Circ.*, vol. 34, pp. 1821–1834, 1999
24. V. Brajovic, T. Kanade, A VLSI sorting image sensor: Global massively parallel intensity-to-time processing for low-latency adaptive vision, *IEEE Trans. Robot. Autom.*, vol. 15, no. 1, 67–75, 1999
25. J.-E. Eklund, C. Svensson, A. Astrom, VLSI implementation of a focal plane image processor-a realization of the near-sensor image processing concept, *IEEE Trans. VLSI*, vol. 4, no. 3, pp. 322–335, 1996
26. M. Nagata, J. Funakoshi, A. Iwata, A PWM signal processing core circuit based on a switched current integration technique, *IEEE J. Solid-State Circ.*, vol. 33, no. 1, pp. 53–60, 1998
27. K. Frohmader, A novel MOS compatible light intensity-to-frequency converter suited for monolithic integration, *IEEE J. Solid-State Circ.*, vol. 17, no. 3, pp. 588–591, 1982
28. K. Tanaka, et al., Novel digital photosensor cell in GaAs IC using conversion of light Intensity to pulse frequency, *Jpn. J. Appl. Phys.*, vol. 32, no. 11A, pp. 5002–5007, 1993
29. W. Yang, A wide-dynamic-range, low-power photosensor array, *ISSCC 1994, Dig. of Tech. Papers*, pp. 230–231, 1994
30. A. Kitchen, A. Bermak, A. Bouzerdoum, A digital pixel sensor array with programmable dynamic range, *IEEE Trans. Electr. Devices*, vol. 52, no. 12, pp. 2591–2601, 2005
31. Q. Luo, J. Harris, A time-based CMOS image sensor, *IEEE International Symposium on Circuits and Systems, ISCAS 2004*, vol. IV, pp. 840–843, 2004
32. G.T. Fechner, *Elemente der Psychophysik*, 2. Bände, Leipzig, 1860

33. T. Delbruck, C. A. Mead, Analog VLSI adaptive logarithmic wide dynamic-range photoreceptor, *IEEE International Symposium on Circuits and Systems, ISCAS 1994*, vol. 4, pp. 339–342, 1994
34. D. Bauer, et al., Embedded vehicle speed estimation system using an asynchronous temporal contrast vision sensor, *EURASIP J. Embedded Syst.*, vol. 2007, doi:10.1155/2007/82174, 2007
35. C. Posch, M. Hofstätter, D. Matolin, et al., A dual-line optical transient sensor with on-chip precision time-stamp generation, *ISSCC, 2007, Dig. of Tech. Papers*, pp. 500–501, 11–15 Feb, 2007
36. D. Matolin, C. Posch, R. Wohlgenannt, True correlated double sampling and comparator design for time-based image sensors, *IEEE International Symposium on Circuits and Systems, ISCAS 2009*, pp. 1269–1272, 24–27 May 2009
37. C. Posch, D. Matolin, R. Wohlgenannt, An asynchronous time-based image sensor, *IEEE International Symposium on Circuits and Systems, ISCAS 2008*. pp. 2130–2133, 2008
38. X. Guo, X. Qi, J. Harris, A time-to-first-spike CMOS image sensor, *IEEE Sensors J.*, vol. 7, no. 8, pp. 1165–1175, 2007
39. D. Matolin, R. Wohlgenannt, M. Litzenberger, C. Posch, A load-balancing readout method for large event-based PWM imaging arrays, *IEEE International Symposium on Circuits and Systems, ISCAS 2010*. May 2010

# A Focal Plane Processor for Continuous-Time 1-D Optical Correlation Applications

**Gustavo Liñán-Cembrano, Luis Carranza, Betsaida Alexandre, Ángel Rodríguez-Vázquez, Pablo de la Fuente, and Tomás Morlanes**

**Abstract** This chapter describes a 1-D Focal Plane Processor, which has been designed to run continuous-time optical correlation applications. The chip contains 200 sensory processing elements, which acquire light patterns through a 2 mm $\times$ 10.9 $\mu$m photodiode. The photogenerated current is scaled at the pixel level by five independent 3-bit programmable-gain current scaling blocks. The correlation patterns are defined as five sets of two hundred 3-bit numbers (from 0 to 7), which are provided to the chip through a standard $I^2C$ interface. Correlation outputs are provided in current form through 8-bit programmable gain amplifiers (PGA), whose configurations are also defined via $I^2C$. The chip contains a mounting alignment help, which consists of three rows of 100 conventional active pixel sensors (APS) inserted at the top, middle and bottom part of the main photodiode array. The chip has been fabricated in a standard 0.35 $\mu$m CMOS technology and its maximum power consumption is below 30 mW. Experimental results demonstrate that the chip is able to process interference patterns moving at an equivalent frequency of up to 500 kHz.

## 1 Introduction

This chapter presents an *application-specific focal plane processor* (ASFPP) with dedicated architecture, sensory front-end, computing resources and external interface. The chip has been developed in the framework of an Industrial R+D project, whose aim is to design a One-Dimension (1D) programmable opto-electronic device able to acquire the light fringes produced in optical encoders applications [1–4] and transform them into a set of electrical signals that can be employed to determine the relative, or absolute, movement of the object that the encoder is attached to.

G. Liñán-Cembrano (✉)
Instituto de Microelectrónica de Sevilla CNM-CSIC, Universidad de Sevilla, Americo Vespucio s/n 41092 Seville, Spain
e-mail: Gustavo.Linan@imse.cnm.es

Á. Zarándy (ed.), *Focal-Plane Sensor-Processor Chips*,
DOI 10.1007/978-1-4419-6475-5_7, 

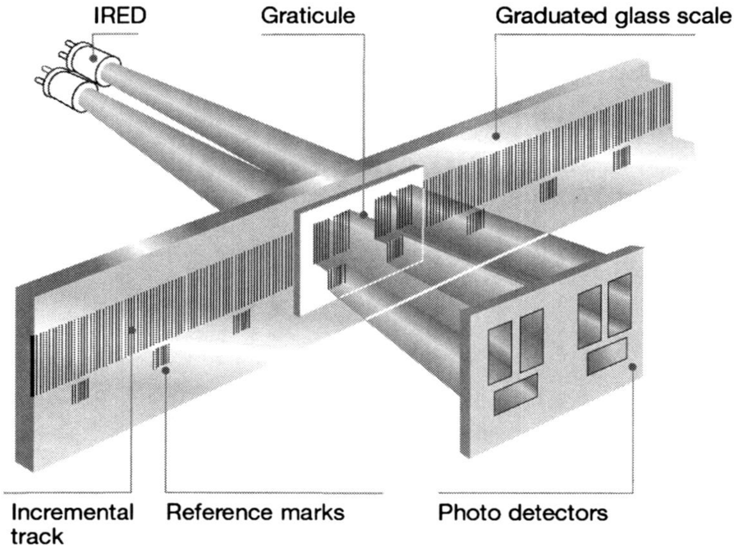


**Fig. 1** Typical distribution of elements in an optical encoder (from [1]. Printed with permission)

Figure 1 shows the typical distribution of elements in a transmission-type optical encoder [1]. Basically, four main blocks are identified. The first component is a near infrared[1] (NIR) LED, which acts as light source. Light from this diode reaches a glass (second component) – also known as *scale* – which is fixed to the axis where the movement is to be detected. This glass contains a pattern of bars that allow/avoid the transmission of the light to the next element in the system. Third and fourth elements are mounted on the encoder's head, which is attached to the moving object. The third element is a regular scanning grid (again a pattern of bars for light transmission or oclusion) whose period is different from that on the *scale*. Finally, the fourth element is an opto-electronic device which must acquire the light fringes produced by either Moiré, Talbot, or Lau effects [5–7], and transform them into suitable length-measuring electrical signals.

Just for illustration purposes, let us show a real example of how these fringes look like (Fig. 2). In this example, the period of the fringes over the photodiodes is 509.5 μm and, as one can easily see, what we obtain is not the typical ON/OFF pattern observed in simple relative encoders with only one scale – whose period is hundredths of micrometers – but a graded interference pattern instead. If we get slices of this image over the $x$ axis (see Fig. 3 left), we observe the profile of the fringes along the axis where the movement is to be detected. This image has been captured using a relatively low-pitch (7.2 μm) CCD camera in a real environment, this is why the lines (we are only plotting those corresponding to top and bottom rows of pixels) look so noisy. If we average pixel information along the $y$-axis and

[1] $\lambda = 880$ nm in our case.

**Fig. 2** Fringe patterns observed at the sensor's plane in a commercial encoder (acquired with a 640 × 480 7.2 μm CCD camera). Notice that it is not the typical light ON/OFF pattern

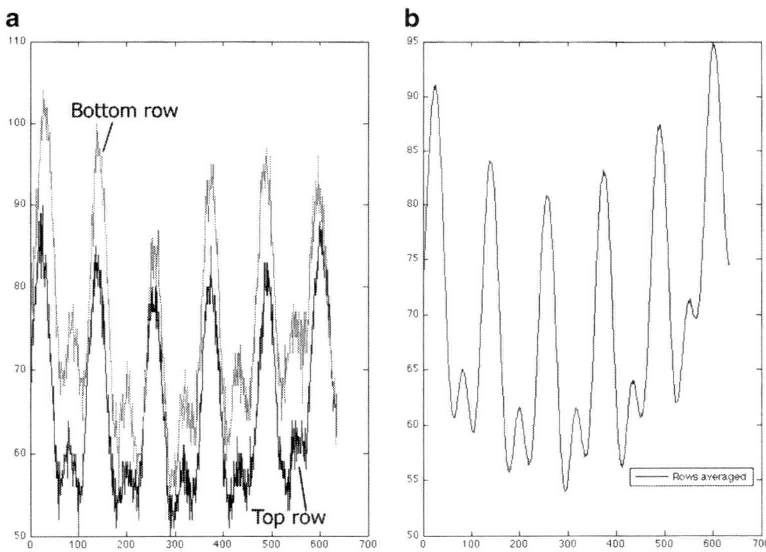


**Fig. 3** (**a**) Fringe patterns observed in first and last row on the CCD. (**b**) Result from averaging over the $y$ axis

plot the resulting mean fringe pattern, we obtain a cleaner view of how the fringes look like. Figure 3 (right) shows this result.[2] One can easily identify a repeating pattern in the fringes whose period, for this particular application, is 509.5 μm. We also observe the influence of using a single LED light source in the noticeable curvature appearing in the peaks of the fringe pattern. Obviously, when the object – with the encoder's head attached to it – suffers a displacement, these fringes also move on the focal plane[3] and, thus, the overall task of our optoelectronic device is to provide

[2] Which is approximately equivalent – neglecting partial suppression of reset noise due to averaging – to acquire the image with a 1D-CCD whose pixels have the same height as the array.

[3] Indeed, movement is amplified at the focal plane due to the optical setup of the system.

a few signals from which this movement can be measured precisely.[4] According to [4], the intensity of the light fringes produced over the detector's area along the $x$-axis can be compactly expressed as:

$$f(x,\theta) = A + B\sum_{n\geq 1}^{\infty} a_n \cos\left[\frac{2\pi n}{p}(x+\theta)\right], \tag{1}$$

where $x$ is the axis of movement of the object, $p$ is the period of the fringes, $\theta$ is the relative displacement between *scale* and *scanning* gratings,[5] and the $\{a_n\}$ Fourier coefficients depend on both the optical and the physical parameters of the system. The goal in this application consists of producing two quadrature signals $A$ and $B$:

$$A = k_A \sin\left(\frac{2\pi}{p}\theta\right)$$
$$B = k_B \cos\left(\frac{2\pi}{p}\theta\right) \tag{2}$$

from which one can obtain the relative displacement $\theta$ using:

$$\theta = \frac{p}{2\pi}\arctan\left(\frac{k_B A}{k_A B}\right). \tag{3}$$

Obviously, the more accurate we are in the generation of $A$ and $B$ signals, the more precise will be the measurement of the displacement, since the better would be the interpolation over the Lissajous plot – which ideally should result in a perfect circle. However, generating precise quadrature signals from this pattern of fringes is far from being easy. As demonstrated in [3, 4], *sine* and *cosine* functions can be obtained from such light fringes pattern by using the following $g_A(x)$ and $g_B(x)$ *kernel* functions,

$$g_A = \frac{1}{\pi x}\left[\sin\left(\frac{2\pi}{p_1}x\right) - \sin\left(\frac{2\pi}{p_2}x\right)\right]$$
$$g_B = \frac{1}{\pi x}\left[\cos\left(\frac{2\pi}{p_1}x\right) - \cos\left(\frac{2\pi}{p_2}x\right)\right], \tag{4}$$

where $p_1$ and $p_2$ are the periods of the two gratings. Then,

$$k_A \sin\left(\frac{2\pi}{p}\theta\right) = \int_{-\infty}^{+\infty} g_A(x) f(x,\theta)\mathrm{d}x$$
$$k_B \cos\left(\frac{2\pi}{p}\theta\right) = \int_{-\infty}^{+\infty} g_B(x) f(x,\theta)\mathrm{d}x, \tag{5}$$

where $f(x,\theta)$ is the light intensity of the fringe patterns on the focal plane. Kernel functions $g_A(x)$ and $g_B(x)$ are displayed in Fig. 4.

[4] *Precisely* means errors below 1 μm per meter in this case.

[5] Which indeed corresponds to the displacement of the object.

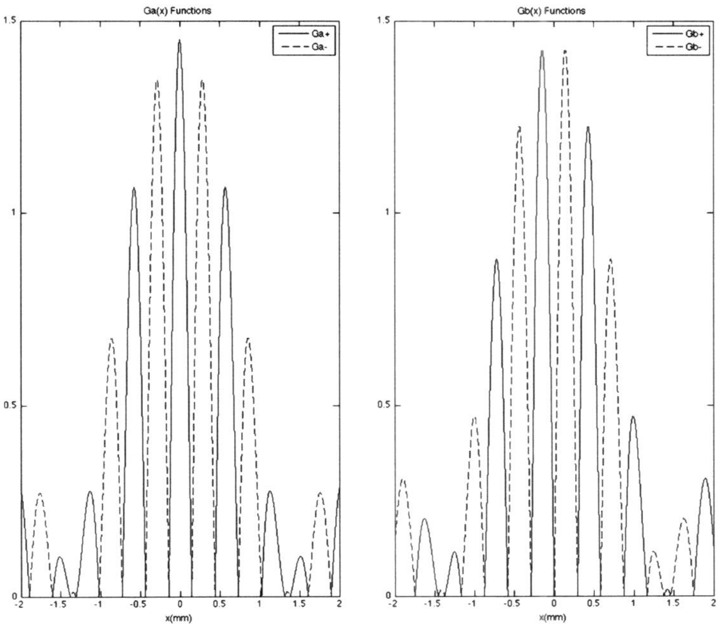


**Fig. 4** $g_B(x)$ (*left*) and $g_A(x)$ signals (*right*) – negative part of the functions are sign-inverted and plotted in *dashed line*

## 2 Description of the Operation and Design Specifications

Obviously, when mapping these equations into an electronic circuit one must consider different simplifications. First of all, we do not have the light intensity function $f(x,\theta)$ as input, but the current which is generated by discrete photodiodes, which are uniformly distributed along the focal plane. Hence, each of these diodes is providing a kind of averaging of the impinging light intensity over its area of influence – which, in the simpler case, will correspond to its area. Furthermore, we will not implement functions $g_A(x)$ and $g_B(x)$ in their continuous form but use a discrete approximation – with an equivalent number of bits. Finally, for implementation and versatility purposes, it is preferable to have separate outputs for the positive and negative parts of the $A$ and $B$ quadrature signals. Thus, instead of $g_A(x)$ and $g_B(x)$, we will have $g_{A_+}(x)$, $g_{A_-}(x)$, $g_{B_+}(x)$ and $g_{B_-}(x)$, let us simply denote them by $g_k(x)$ where the $k$ index specifies whether it is an $A$ or $B$ function and whether it is the positive or the negative part, $\{A_+, A_-, B_+, B_-\}$. Each of the outputs, $O_k$, of the chip is calculated as:

$$O_k(\theta) = \sum_{j=1}^{N_{\text{pixels}}} I_{\text{photo}_j}(\theta) m_{k,j}, \tag{6}$$

where each $m_{k,j}$ coefficient corresponds to the mean value of function $g_k(x)$ within the area of influence of diode *j*th,

$$m_{k,j} = \int_{X_L}^{X_R} g_k(x)\mathrm{d}x. \tag{7}$$

In the chip, we have implemented a 3-bit representation of these coefficients – after an analysis where performance and area occupation were balanced – which results in the functions in Fig. 5. Besides, we have added the possibility to adjust the global gain of each output channel by means of a Current-Mode Programmable-Gain Amplifier (CM-PGA). Therefore, *Sine* and *Cosine* outputs are obtained as:

$$\begin{aligned} k_A \sin\left(\frac{2\pi}{p}\theta\right) &= \alpha_1 O_{A_+} - \alpha_2 O_{A_-} \\ k_B \cos\left(\frac{2\pi}{p}\theta\right) &= \alpha_3 O_{B_+} - \alpha_4 O_{B_-}. \end{aligned} \tag{8}$$

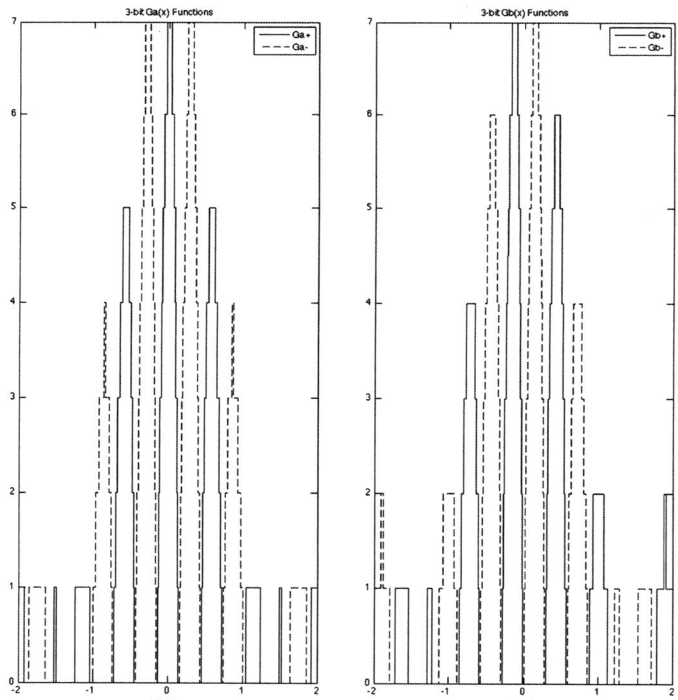


**Fig. 5** A 3-bit representation of $g_B(x)$ (*left*) and $g_A(x)$ signals (*right*) – negative part of the functions are sign-inverted and plotted in *dashed line*

Finally, the chip also provides a fifth output $O_R$ calculated as,

$$O_R = \alpha_5 \sum_{j=1}^{N_{\text{pixels}}} I_{\text{photo}_j} \cdot R_j; \qquad R_j = [0, 1, 2, 3, 4, 5, 6, 7], \tag{9}$$

which can be used to different purposes. Most commonly, it will be employed as a mechanism to obtain average illumination over the chip – by programming all $R_j$ coefficients to the same value –, and adjust, in a feedback loop, the current through the NIR LED to guarantee that the amplitude of *Sine* and *Cosine* outputs remains within an appropriate margin. Another possible use could be as a way to read absolute reference positions in double-chip configurations on the same head, where one chip is given the task of incremental displacement measurements, whereas the second chip performs readings of absolute marks.

## 2.1 Physical Information and System Requirements

The chip has been implemented using a 0.35 μm 4M-2P technology available through Europractice. This technology offers *Nwell–Psubstrate* photodiodes with a sensitivity around 0.3 A/W@880 nm. The following list summarizes the most relevant information and constraints for our design.

- Fringe period is 509.5 μm.
- Monochromatic light; $\lambda = 880$ nm.
- Incident light power at the focal plane between [5.5, 55] μW mm$^{-2}$. Hence, photogenerated current density will vary between [1.65, 16.5] μA mm$^{-2}$.
- Maximum frequency of fringes, due to head movement, at the focal plane is 500 kHz – head moving at 20 m s$^{-1}$.
- Fringe contrast is 12%.
- Diodes pitch and fringe period must be relative prime numbers.
- Sensing part must allocate at least four periods of fringes – array lenght ≥2.04 mm.[6]
- Minimum diode's height is 1.5 mm.
- Power consumption below 100 mW – using 3.3 V single power supply.
- Interfacing through I$^2$C 100 kbps standard only.
- *Continuous time* operation – conventional reset-exposure-readout operation is not allowed.

[6] Using four periods of fringes and having relative prime numbers in diode's pitch and fringe period improves the quality of the interpolation process over the Lissajous plot when measuring the displacement.

## 3 Architecture of the Chip

Figure 6 shows a simplified block diagram of the chip. As it has been mentioned, the main functionality of the chip is to provide five output currents,

$$O_k(t,\theta) = \alpha_k \cdot \sum_{j=1}^{N_{\text{pixels}}} I_{\text{photo}_j}(t,\theta) \cdot m_{k,j}, \tag{10}$$

where $\{m_{k,j}\}$ are positive integer numbers in the range $[0,7]$ and the $\alpha_k$ coefficients are positive programmable gains, which are defined as the quotient between any two 4-bit numbers,

$$\alpha_k = \frac{\sum_{n=0}^{3} 2^n}{\sum_{m=0}^{3} 2^m}. \tag{11}$$

The chip includes the following main blocks:

- A standard $I^2C$ interface which is the only mechanism to transmit commands and configuration setups to the chip. The chip address in the $I^2C$ bus is determined by a built-in defined constant, which is completed (LSB) with a bit provided through an external pin. Thus, two chips can be connected simultaneously to the same bus without conflicting. Readers with previous experience with the $I^2C$ bus will notice that the chip uses SDA_IN and SDA_OUT lines (top right corner in the

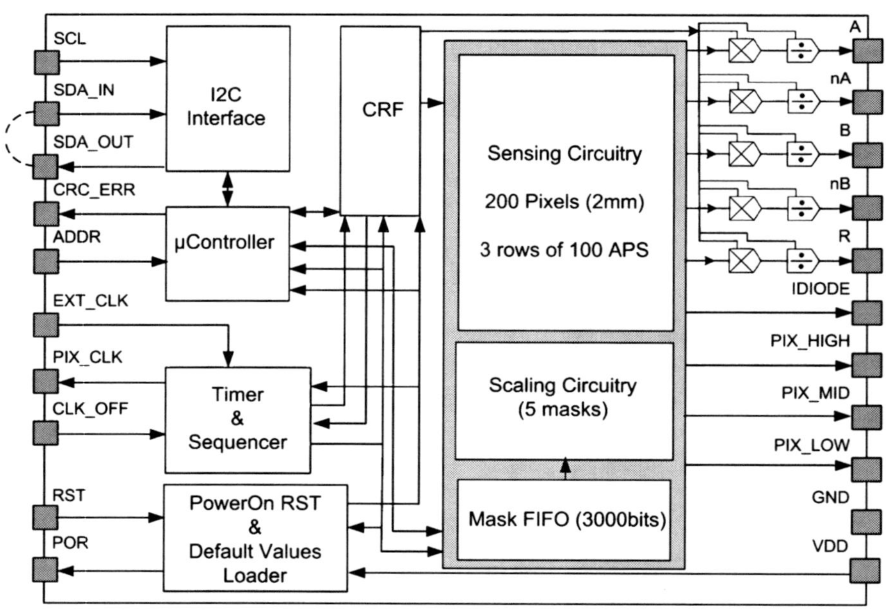


**Fig. 6** Block diagram of the chip

block diagram) instead of the conventional single SDA line. This is due to a lack of the proper bidirectional pad in the selected technology. SDA_IN uses a conventional input pad whether SDA_OUT uses a conventional open-drain pull-down output pad. Both signals are connected together at the board to the SDA data line of the $I^2C$ bus.

- A customized CISC microcontroller which implements five instructions – more details in Sect. 4.
- A block of timers and prescalers which includes a 2-bit configurable (through binary divisions $\times 1$, $\times 1/2$, $\times 1/4$, $\times 1/8$) 55 MHz oscillator.[7]
- A configuration register file (CRF), which comprises 26 bytes, that defines the state of the different programmable modules of the chip.
- A Power-On-Reset and Bootloader unit, which, on the one hand, detects the events of power-up/down and executes a system reset during these events, and, on the other hand, loads the CRF with its by-default information.
- Five independent current-mode gain-programmable amplifiers (CM-PGA), which implement the $\alpha_k$ coefficient in (10).
- The main array, with 200 SPEs. Basically, each pixel contains a photodiode and five 3-bit programmable output branches of a current mirror. The array also contains, inserted at the top, bottom, and centre, three rows of conventional APS sensors (100 pixels/row) that can be used during mounting of the head to help in the alignment process, or, in operation, to acquire the profile of the fringes being projected at a given time.
- A 3000-bit FIFO, which stores the two hundred 3-bit coefficients that define each of the five – one per output – sets $\{m_{k,j}\}$.

## 4 Digital Part

The digital block of the chip has been designed around a specific purpose microcontroller which is in charge of chip control and configuration. Once programmed, the chip can operate autonomously as required in most applications. In addition, the controller contains access resources to the FIFO data memory, a set of timers and prescalers with sequencers plus a CRC calculation unit. The microcontroller architecture follows the CISC paradigm and incorporates five simple instructions, which are summarized in Table 1.

The system only receives information through its $I^2C$ interface and, as mentioned above, can be configured through an external pin to have two different positions in the bus, allowing two chips to be simultaneously connected to the same bus – which is very important in advanced heads containing two sensors. The design of

[7] Nominal frequency of the designed token ring oscillator; process corners, mismatching, power supply variations, and temperature affect this frequency which might move $\pm 50\%$.

**Table 1** Customized microcontroller's instruction set

| CMD | CODE | 1st Arg. | 2nd Arg. | Data |
|---|---|---|---|---|
| WriCRF | $0 \times 01$ | [StartAddr] | N | {N bytes} |
| ReadCRF | $0 \times 02$ | [StartAddr] | N | {N bytes} read from chip |
| TIPS[a] | $0 \times 04$ | | | |
| WriFIFO | $0 \times 08$ | | | {K[b] bytes + CRC} |
| ReadFIFO | $0 \times 10$ | | | {CRC + 1,000 bytes} |

[a]Stands for trigger integrating pixels sequence
[b]There are two modes of FIFO writing: (1) Complete transmission of FIFOLEN bytes – FIFOLEN is defined as a 16-bit number (in two CRF registers) which by default takes the value of 1,000 (2) Marking the MSB of last byte to be transmitted with a logic 1. The specific mode during a transmission is defined in one of the registers in the CRF

the microcontroller has taken into account the analogue nature of the continuous time processing being performed by the system, this includes:

- To have low switching activity.
- To maintain its performance when clocked with a low-precision frequency oscillator – internal token ring oscillator – provided that its period remains between 12 ns and 140 ns.
- To remain *almost* idle during normal operation of the chip. The only modules that stay active during the operation of the chip are the one which detects if the chip has been addressed after any $I^2C$ start condition, and the sequencer of the Reset-Integrate-Readout process for the integration pixels – which indeed is not usually employed during normal use of the chip when making displacement measurements.

## *4.1 The Configuration Register File (CRF)*

The Configuration Register File, or simply CRF, is a 26-byte memory composed of single write port and double read port 8-bit registers (details in Table 2). The information stored in this register defines the status of all configurable options in the chip and it is written to the (safe) default value after a power-on reset or during a so-called *warm*[8] reset. The information stored in this register file can be divided into five groups of logic elements, namely:

1. The first group contains registers that inform about the last command code executed by the microcontroller, FIFO data related information (the last datum written in the FIFO together with the last datum read from the FIFO) and the CRC value corresponding to the data stored within the FIFO. Although this

[8] A reset commanded by the user by pulling down the *RST* pin of the chip.

**Table 2** The configuration register file

| Position | Name | Description |
|---|---|---|
| 0 | LASTCMD | Last command received |
| 1 | CRC | Last calculated CRC |
| 2 | LASTREAD | Last byte read from the chip |
| 3 | LASTWRI | Last byte written to the chip (no commands) |
| 4 | RCFP | Bit-masked configuration of the FIFO write/read and integration pixels. Some bits mask the operation of signals related to the operation of the integration pixels. This includes whether to reset the pixels, whether to reset the pointer that addresses the pixels during readout, whether to activate the prescaler that defines the duration of reset time, exposure time, and readout time per pixel, and whether the reset-integrate-readout process is to be executed continuously. Besides it also defines if FIFO write is controlled by the parameter FIFOLEN or by marking the last byte to be transmitted, whether FIFO readout (for test purposes) is destructive or not, and which clock has to be used during FIFO access operations (internal/SCL $I^2C$ clock) |
| 5 | TRST_MSB | Definition of RST time for integration pixels (2-byte variable) |
| 6 | TRST_LSB | |
| 7 | TPIX_MSB | Definition of output time per pixel (2-byte variable) |
| 8 | TPIX_LSB | |
| 9 | LASTPIX | Definition of the position of the last integration pixel that must be read (100) by default |
| 10 | TEXP_MSB | Definition of exposure time (3-byte variable) |
| 11 | TEXP_CSB | |
| 12 | TEXP_LSB | |
| 13 | PRESPIX_MSB | Definition of prescaler for integration pixel clock (2-byte variable) |
| 14 | PRESPIX_LSB | |
| 15 | PRESFIFO_MSB | Definition of prescaler for the FIFO (3-byte variable) |
| 16 | PRESFIFO_CSB | |
| 17 | PRESFIFO_LSB | |
| 18 | FIFOLEN_MSB | Number of bytes to be written to the FIFO (2-byte variable; 1,000 by default) |
| 19 | FIFOLEN_LSB | |
| 20 | RCNA | Bit masked configuration of analogue blocks in the SPE. One bit defines whether the analogue section of the chip is ON or OFF. A set of 5 bits define whether or not to activate the different additional functionalities on the pixel. Finally, two bits define the division ($\times 1$, $\times 0.5$, $\times 0.25$, $\times 0.125$) to apply to the on-chip built-in oscillator |
| 21 | GAIN_A | Gain of the CM-PGA in channel $A_+$ (by default it is set to $0\times$FF, which means that the implemented gain is 15/15) |
| 22 | GAIN_nA | Gain of the CM-PGA in channel $A_-$ (by default it is set to $0\times$FF, which means that the implemented gain is 15/15) |
| 23 | GAIN_B | Gain of the CM-PGA in channel $B_+$ (by default it is set to $0\times$FF, which means that the implemented gain is 15/15) |
| 24 | GAIN_nB | Gain of the CM-PGA in channel $B_-$ (by default it is set to $0\times$FF, which means that the implemented gain is 15/15) |
| 25 | GAIN_R | Gain of the CM-PGA in channel R (by default it is set to $0\times$FF, which means that the implemented gain is 15/15) |

information is not strictly necessary, it has been included for debugging and supervision purposes.
2. The second group contains the arguments that define the behavior of the FIFO accessing instructions.
3. The third group stores the configuration that controls the APS Reset-Exposure-Read timing sequence.
4. The fourth group is a single register that controls (bit-masked) the operation of the analog core and the configuration of the built-in clock divider.
5. The fifth group, composed of five registers, defines the gain of each output channel $\{A_+, A_-, B_+, B_-, R\}$.

## 4.2 The APS Sequencer

The APS sequencer generates timing and control signals that command the Reset-Exposure-Readout cycle. Its operation can be cyclic, activating a specific bit in CRF4, or not cyclic using the special single command instruction *TIPS* (see Table 1). The sequencer is fully programmable and completely idle when not in use.
Figure 7 shows the control signals generated by the sequencer. POINTRST initializes the circuitry that addresses the pixels during read time, while PIXRST is used to simultaneously initialize the photodiodes in all APS pixels. POINTRST and PIXRST are complementary signals and their duration is controlled by a programmable 16-bit timer (CRF5,CRF6), which consequently defines the reset time. PIXCLK drives the consecutive connection of each APS to its corresponding output node, defining the read time. The number of PIXCLK cycles is programmable (CRF9), and can be any number between 1 and 100. The duration of the PIXCLK

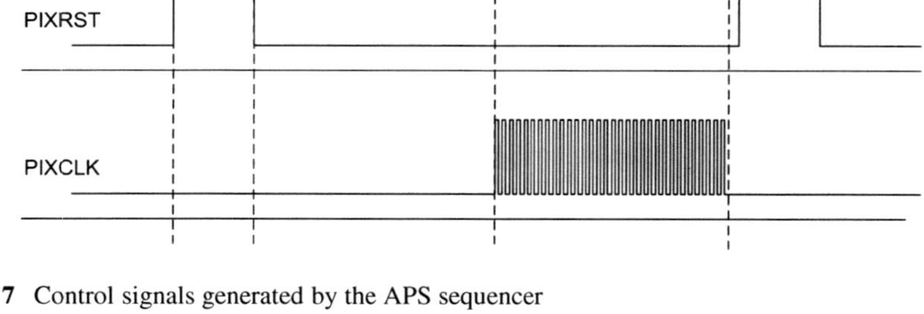


**Fig. 7** Control signals generated by the APS sequencer

cycle is programmable as well, and it is controlled by a 16-bit timer (CRF7, CRF8). The exposure time, defined as the time period between the end of reset time and the beginning of read time, is controlled by a programmable 24-bit timer (CRF10, CRF11, CRF12). It is also possible to extend these time intervals using a 16-bit prescaler (CRF13, CRF14), which can be activated if needed (by asserting a bit in CRF4). The sequencer includes a programmable bit mask option that allows the user to deactivate (masking) any of the control signals generated by its circuitry (POINTRST, PIXRST, and PIXCLK).

## *4.3 Accessing the FIFO*

The microcontroller has configuration options, which make the access to the correlation pattern FIFO memory more flexible. All the accessing options are available by programming the CRF registers adequately. The FIFO registers main clock (used during read and write operations) is selectable; the user can extract it from the I$^2$C Serial Clock or use an internal programmable 16-bit clock timer which employs the built-in token ring oscillator.

Write operations consist of storing the data transmitted by and external source in the FIFO. The data must always be followed by the corresponding CRC. Users can mark the end of transmission (EOT) in two ways, on the one hand, specifying the number of FIFO registers to be send, on the other hand, marking the EOT by asserting the MSB in the last byte to be transmitted (both options allow partial or total FIFO write operations). To verify the integrity of the received data, a CRC calculation unit performs CRC calculation during FIFO write operations. After the reception of the last datum, the controller compares the received and computed values and informs about matching status through a dedicated pin. As described above, users can also read the computed CRC value by downloading the information in CRF1.

Read operations involve the transmission to an external receptor of the total or partial content of the FIFO memory. The number of FIFO registers to be read is specified in variable FIFOLEN (CRF18, CRF19), although the external receptor can interrupt the transmission at any time by creating an I$^2$C stop condition. Read operations can be either destructive or non-destructive (by asserting a bit in CRF4). In the former case, every read and transmitted datum is eliminated. In the latter, during read operations, the controller interconnects the input and output ports of the FIFO. In this configuration, the FIFO is arranged in a ring structure, hence data are simply shifted circularly during read operations, therefore a complete non-destructive read operation of the FIFO memory leaves its registers unchanged. The microcontroller oversees the FIFO configuration to avoid writing operations while the FIFO is disposed in ring structure. Therefore, even if the user accidentally leaves the non-destructive FIFO read access option established, write operations can always be executed.

# 5 The Mixed-Signal Processing Core

The computing core of the chip is an 1-D array of 200 programmable sensory processing elements (SPEs). These SPEs transform the incident light (fringes) into a photo-generated current and scale this current to produce five independent versions of it (one per output channel). Scaling coefficients are integer numbers in the range [0–7] and are locally stored within each SPE in a 15-bit shift register. Registers in physically adjacent SPEs are connected in series (output from left-side to input of right side) in such a way that a 3000-bit ($15 \times 200$) shift register is formed – the previously described FIFO –, thus making the process of programming the coefficients quite straightforward. The SPE includes different configurable modules – whose state is defined in CRF20 – that allow for optimizing power consumption and accuracy according to the needs of the application at a given time. Thus, for instance, frequency response of the system can be modified – at the expense of power consumption – to allow processing fast moving fringes ($20\,\mathrm{m\,s^{-1}}$). In addition to the main array, the mixed-signal core of the chip contains three rows of 100 APS pixels inserted at the top, middle and bottom of the main diodes array. Finally, the mixed-signal core also contains five *Current-Mode* 8-bit *Programmable-Gain Amplifiers*, which generate the output of the chip as expressed in (10). The following subsections describe, in detail, the different modules in this mixed-signal processing core.

## 5.1 The Sensory-Processing Element

Figure 8 shows a block diagram – including a transistor-level representation of the Current-to-Voltage conversion unit – of the SPE. The blue-shaded area corresponds to a biasing unit which is shared by all SPEs in the chip and that is located at the periphery of the array. All biasing currents in the chip are obtained as scaled-up(/down)

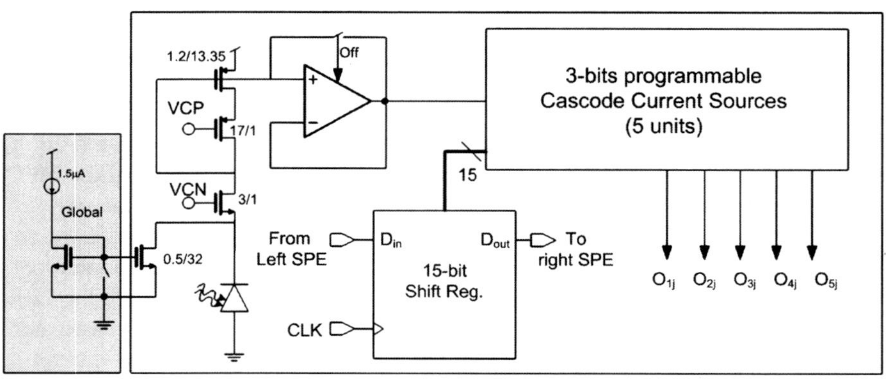


**Fig. 8** Block diagram of the SPE

versions of a single 15 μA source, which is generated by an internal band-gap circuit, thus, the 1.5 μA external source in Fig. 8 is obtained from the 15 μA source and a $\times 10$ divider. Each SPE contains the following blocks

- A Nwell–Psubstrate photodiode that transforms incident light into a photogenerated current.
- A re-configurable current to voltage conversion unit, which transforms this photogenerated current into a voltage level.
- An analog buffer which transmits this voltage to a bank of five 3-bit programmable current sources.
- Five 3-bit programmable current sources which receive an input voltage from the analog buffer and transform it into five independent output currents.
- A 15-bit shift-register, which stores the values of each of the 3-bit numbers that define the scaling factor that the SPE will apply on each of its five output currents.

Basically – leaving aside the optional features in some operations – the SPE operates as follows:

1. During the programming phase, shift registers in all SPEs are connected in series (receiving data from the left neighbor and providing data to the right neighbor) to form a 3000-bit shift register. Once the programming stream has been loaded into the array (through 3,000 clock cycles), each SPE register contains five 3-bit numbers, which define its scaling coefficients in (10).
2. The photodiode creates a photogenerated current which is – approximately – proportional to the power of the incident light.
3. The information stored in the shift register is automatically driven – its is wired[9] – to the programmable current sources.
4. This photogenerated current is transformed into a voltage by the input stage of a cascode PMOS current mirror, and copied, by the analog buffer, to the input node of five identical programmable cascode current sources. These 3-bit programmable current sources are designed as seven unitary elements with *common centroid* layout configuration – also including dummy elements to improve the matching –, in such a way that the disposition (from left-to-right) is Dummy-$b_2$-$b_1$-$b_2$-$b_0$-$b_2$-$b_1$-$b_2$-Dummy. Since we are using current mode outputs, we get the summation in (10) simply by connecting outputs in different SPEs to the same low-impedance node. This node, which is indeed the input stage of the CM-PGA in each correlation output channel of the chip, is described in Sect. 5.3.

The following subsections describe the main subsystems in the SPE in more detail.

[9] Due to this direct connection, we could get big current peaks during programming since we are moving all bits in the shift register every clock cycle. To avoid this, the analogue part of the chip can be switched off during the programming phase by asserting a particular bit in CRF20. Indeed, by default – i.e., after power-on or reset –, this bit is set to 0, to avoid any kind of trouble with this issue, and the user is always requested to activate the analogue part of the chip to get some current through its outputs. This option does not switch off the three rows of APS pixels, and nor their output amplifiers, thus allowing to get information about correct positioning of the chip during the mounting of the head without needing to activate the five correlation outputs.

### 5.1.1 The Sensory Block

The sensory block is, obviously, one of the most important elements in the SPE. This block is formed by the photodiode that senses the light and the analogue circuitry, which transforms this current into voltage, which is suitable to be transmitted to the programmable current sources.

The sensory block, illustrated in Fig. 9, consists of:

- A $2{,}000 \times 10.9\,\mu m^2$ Nwell–Psubstrate photodiode which provides the photogenerated current. According to the sensitivity value provided by the factory, and the expected incident power at the focal plane (see Sect. 2.1), expected photogenerated current will be in the range of [35–350] nA. The layout of this large photodiode includes contacts to the Nwell every $10\,\mu m$ to reduce transit time of photogenerated carriers from the place where they are created to the place where they are collected. In addition to that, left and right (long) sides of the diode incorporate substrate contacts every $20\,\mu m$ as well, placed in such a way that substrate contact on one side, Nwell contact, and substrate contact on the other side are in a zig-zag disposition.
- An NMOS transistor, which keeps the reverse biasing of the photodiode to an almost constant voltage independently of the amount of photogenerated current. This transistor also serves to speed improvement purposes since the effect of the big parasitic capacitor of the photodiode over the frequency response of the system is largely attenuated by the cascading effect.
- The input state of a cascaded PMOS current mirror, which transforms the photogenerated current into a voltage. Unfortunately, the need for a continuous time operation avoids any possibility of using offset correction during photodiode's reset phases, then, one has to meet accuracy constraints by using large devices. However, making so big the transistor performing the current to voltage ($I$–$V$) conversion has a direct impact on its gate capacitance and therefore on the

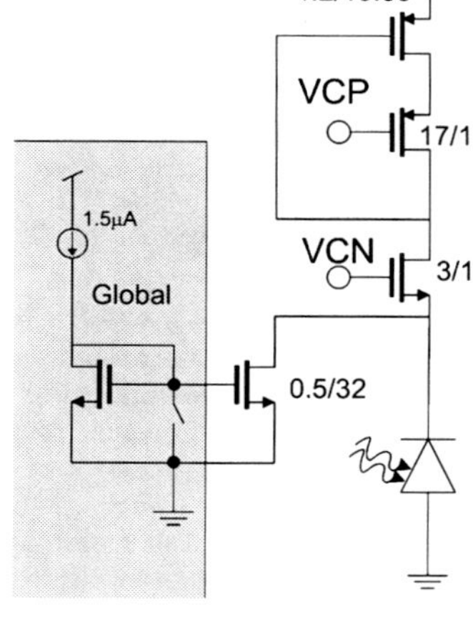


**Fig. 9** Schematic of the sensory block

frequency response of the system. To overcome this limitation we introduced the next option in this block.

- An NMOS optional current source which can be added to the photogenerated current to improve the frequency response of the I–V block. This additional current will produce a shift in the location of the first pole of the system, which instead of being proportional to $\sqrt{(I_{\text{photo}})}$ becomes proportional to $\sqrt{I_{\text{photo}}+I_{\text{BIAS}}}$. Obviously, since incident light power may vary within an order of magnitude, we may not require this block in the case of maximum illumination. Besides, adding this current degrades accuracy. First, it is evident that we must subtract the added offset current in a later stage, and, of course, this subtraction is not error free. Second, and not so evident, we must also consider mismatching dependence on the absolute current circulating through the mirror. We know that, neglecting output resistance effects, the mismatch in a simple mirror can be approximately expressed as[10]:

$$\frac{I_{\text{out}}}{I_{\text{in}}}=\frac{(\beta+0.5\Delta\beta)(V_{\text{GE}}+0.5\Delta V_{\text{TH}})^2}{(\beta-0.5\Delta\beta)(V_{\text{GE}}-0.5\Delta V_{\text{TH}})^2}\approx 1+\frac{\Delta\beta}{\beta}-\frac{2\Delta V_{\text{TH}}}{V_{\text{GE}}}, \tag{12}$$

where $V_{\text{GE}}$ is the *well-known* effective gate to source voltage.[11] As we see, there is a term, which does not depend on the current through the mirror whereas there is another term which, via $V_{\text{GE}}=\sqrt{I/\beta}$, does depend on the current. Therefore, one can simply state that, for a given current mirror, while in saturation, matching improves as the current improves. However, this is true for the total current through the mirror, and in our case, the signal is only a part of it. Therefore, we can find that the relative errors, defined as $\varepsilon=(I_{\text{OUT}}-I_{\text{IN}})/(I_{\text{photo}})$, are expressed as:

$$\begin{aligned}\varepsilon_{\text{no\_IBIAS}}&=\frac{\Delta\beta}{\beta}-2\Delta V_{\text{TH}}\sqrt{\frac{\beta}{I_{\text{photo}}}}\\ \varepsilon_{\text{IBIAS}}&=\frac{I_{\text{BIAS}}+I_{\text{photo}}}{I_{\text{photo}}}\frac{\Delta\beta}{\beta}-2\Delta V_{\text{TH}}\sqrt{\frac{\beta(I_{\text{BIAS}}+I_{\text{photo}})}{(I_{\text{photo}})^2}}\end{aligned} \tag{13}$$

hence, the error when adding $I_{\text{BIAS}}$ to the photogenerated current is always larger and, therefore, one should only employ this extra current in those cases where the head is moving at top speed. Regarding the selection of a proper value for this offset current, we did a parametric analysis in which this current was varied between [0, 3] μA to find an optimum value. The result of this parametric analysis is shown in Fig. 10, where $x$-axis is the bias current and $y$-axis shows is

[10] We use the NMOS version for simplicity – i.e., not including $V_{\text{DD}}$ in the equations.

[11] $V_{\text{GE}}=V_{\text{GS}}-V_{\text{TH}}$.

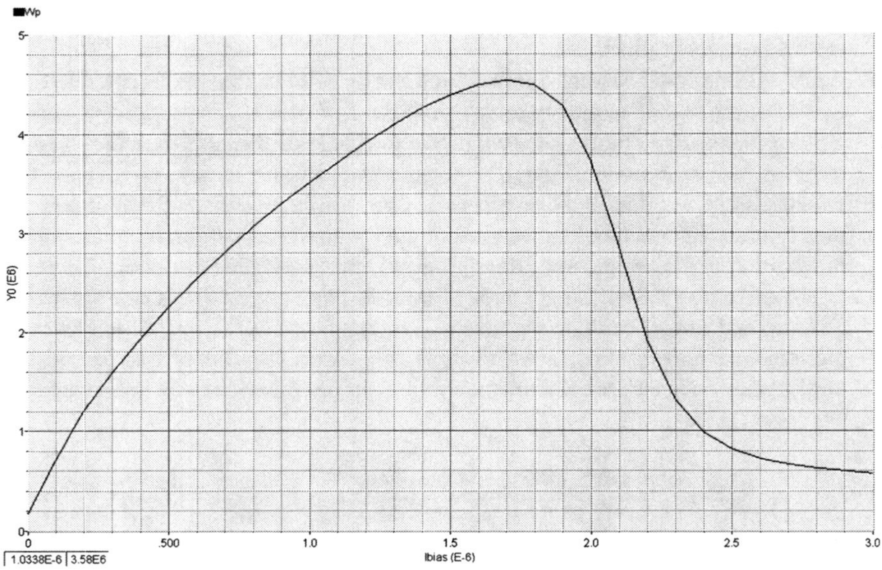


**Fig. 10** Effect of $I_{\mathrm{BIAS}}$ level over the location of the first pole of the I–V block

the position of the first pole. According to this result, we have selected a *near*[12] to the peak value of 1.5 μA, which moves the first pole of the system to about 4.4 MHz.

As shown in Fig. 8, we have also added a buffer inserted between the gate of the input transistor in the current mirror and the input of the 35 (7×5) programmable current sources. This buffer, which can be switched off and bypassed when not required, has been added to reduce the capacitive load at the input node of the current mirror. Thus, instead of having 36 (35+1) equal transistors connected to this node, we only have 2 (the input transistor of the current mirror and the transistor at the positive input of the buffer). This buffer is a PMOS-input standard 5T operational transconductance amplifier (OTA), which employs a 2.5 μA bias current. Obviously, and similarly to the case of adding $I_{\mathrm{BIAS}}$, the use of the buffer degrades accuracy performance due to the effect of its offset voltage.[13]

[12] Since degradation of performance beyond the optimum is quite abrupt – the PMOS input transistor of the mirror leaves the saturation region –, we preferred to move a little bit from the optimum value.

[13] Indeed, this offset voltage plays the same role as a variation in the threshold voltage of input transistors in the programmable current sources.

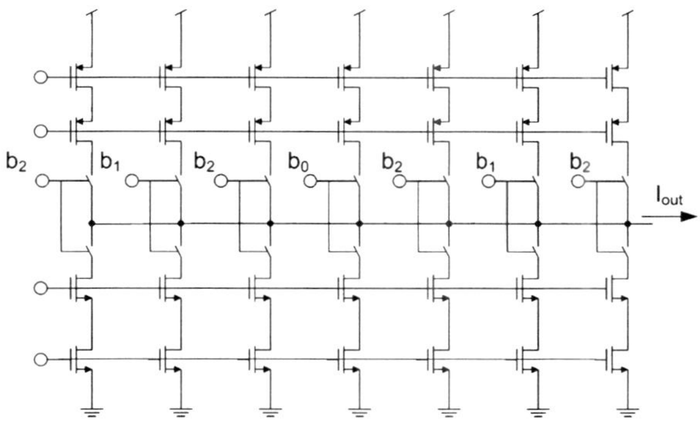


**Fig. 11** Schematic of one of the programmable current sources in the SPE (transistor sizes as shown in Fig. 9)

### 5.1.2 The Current Scaling Block

The current scaling block provides the output current contribution of each SPE to (10). It consists of 35 identical cascode current source units (seven units per output), which also include the $I_{BIAS}$ suppression circuitry. As described above, current sources in each output are laid out in a common centroid configuration (with dummies at both ends) to improve matching. Figure 11 shows the schematic of one of this 3-bit programmable current sources.

### 5.1.3 The Memory Unit

The memory unit within the SPE is a simple 15-bit shift register which uses flip-flops from the available standard cells library. Its 15-bit parallel output (in parallel) drives the corresponding switches in the current scaling block. This memory unit also contains some clock buffering circuitry – end branch of the clock tree which is created for the whole array – to avoid any data corruption during shifting due to the use of very long clock wires with such huge (3,000 registers) capacitive load.

## *5.2 Physical Details*

As detailed in Sect. 5.1.1, the photodiode capturing fringes information has a pitch of 10.9 μm (7.9 (active area) + 3 (separation between hot-wells)), and therefore, the pitch of the SPE should match this value. However, since standard cells have a height of 12 μm in this technology, we opted for using a double-pitch layout for the SPE and locate SPEs both at the top and at the bottom of the photodiodes. Thus, every module within the SPE has been designed to match a pitch of 21.8 μm. Obviously, with this configuration, odd SPEs have processing circuitry at one side of the

**Table 3** Area occupation per block within the SPE

| Block | Height (μm) | (%) |
|---|---|---|
| Photodiode | 2,000 | 66.48 |
| I–V | 42 | 1.40 |
| Current scaling blocks | 5 × 102 | 16.95 |
| Buffer | 24 | 0.80 |
| Test[a] | 67.5 | 2.24 |
| Registers | 365 | 12.13 |

[a]The test circuitry consists of a switch, which allows transmitting photodiode's current to a test-purpose output of the chip instead of to the input of the current mirror. Besides, the test block also contains a digital circuitry, which acts as a pointer. This pointer selects the photodiode, whose output is to be connected to the test pad. A reset pulse points to photodiode #0 (the leftmost device). Afterwards, consecutive clock pulses move this position to the right. In addition, another signal (a bit in CRF20) selects all diodes, only available in test mode, simultaneously, providing a fast mechanism to get total photogenerated current

array (bottom) whereas even SPEs have it on the other side (top). Consequently, the *bitstream*, which is loaded into the chip to configure the different scaling coefficients must take this into account.

The processing part of each SPE is 1008.5 μm height, with the occupation ratio detailed in Table 3.

## *5.3 The Current-Mode Programmable Gain Amplifier*

The chip provides its correlation outputs in current form through five Current-Mode Programmable Gain amplifiers (CMPGA). Each CMPGA must perform two important functions. First, it must accumulate current contributions from individual SPEs – implement the summation operation in (10). Second, it must scale this current up or down according to the gain programmed in the corresponding CRF register (CRF21–CRF25). Each function is implemented by a different subsystem, in the first case, accurate accumulation of SPEs current contributions is accomplished by a class-II current conveyor whereas its output – accumulated current – is scaled by a programmable current mirror, both subsystems are described in what that follows.

### 5.3.1 Accumulating the SPEs Contribution

The accumulation of the contribution of SPEs to the correlation output is accomplished by means of the virtual ground provided by a class-II current conveyor as shown in Fig. 12a. The PMOS transistor and the amplifier are connected in a negative feedback look that maintains the voltage level at the input node independently of the input current flowing through the transistor. Obviously, this simple

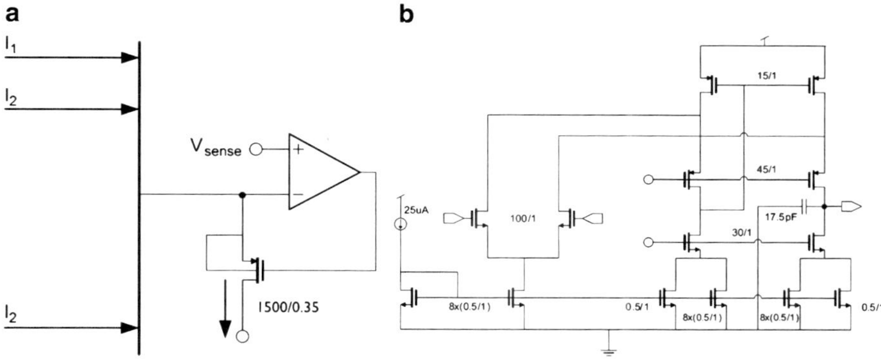


**Fig. 12** Channel accumulation circuitry: (**a**) Current summation block. (**b**) Schematic of the amplifier

description is far from what happens in practice, where one needs to consider the real input impedance at this virtual ground, and the output impedance of all current sources connected to it in order to extract useful design equations. Let us first consider the total output impedance of all (200) SPEs connected in a channel. Since we are using[14] cascaded current sources, the output conductance of the $k$th SPE is:

$$G_{\mathrm{o}}^{k} = \frac{g_{DS_p} \cdot g_{DS\mathrm{casc}_p}}{g_{M\mathrm{casc}_p}} \times m_k, \tag{14}$$

where $m_k$ is the scaling coefficient implemented by this SPE,[15] and all other symbols are common in CMOS literature. Therefore, the total output impedance of all current sources in a correlation channel is simply:

$$G_{\mathrm{o}} = \sum_{k=1}^{k=200} G_{\mathrm{o}}^{k} = \frac{g_{DS_p} \cdot g_{DS\mathrm{casc}_p}}{g_{M\mathrm{casc}_p}} \times \sum_{k=1}^{k=200} m_k \tag{15}$$

As we are using a PMOS transistor in a negative feedback loop to collect current contributions from SPEs, we can define the error in current transmission as the difference between the current that is ideally provided by SPEs – let us denote it by $I_{\mathrm{in}}$ and the current flowing through the transistor in the feedback loop – $I_{\mathrm{out}}$. After simple calculations, one finds that:

$$I_{\mathrm{out}} \approx I_{\mathrm{in}} \times (1 - \varepsilon) \quad \text{with} \quad \varepsilon = \frac{G_{\mathrm{o}}}{g_{DS_{\mathrm{feedback}}} + (A+1) g_{M_{\mathrm{feedback}}}}, \tag{16}$$

[14] Assuming, for simplicity, that we are not using $I_{\mathrm{BIAS}}$.

[15] Or, equivalently, the number of unitary current sources connected in parallel in this SPE to this accumulation node.

where the *feedback* subindex refers to parameters of the PMOS transistor in the feedback loop, and we have assumed – which is indeed a design constraint – that:

$$g_{M_{\text{feedback}}} \times g_{M\text{casc}_p} \times (A+1) \gg g_{DS_p} \times g_{DS\text{casc}_p} \times \sum_{k=1}^{k=200} m_k \tag{17}$$

### 5.3.2 Scaling the Accumulated Current

The current flowing through the transistor in the feedback loop enters the input node of the Current-Mode Programmable Gain Stage. This unit – see Fig. 13 – is simply an all-NMOS current mirror with 16 identical input branches and 16 identical output branches. Clearly, the output current provided by this block is simply given by:

$$I_{\text{Channel}} = \frac{N}{D} \times I_{\text{in}}, \tag{18}$$

where $N$ is the number of active output units and $D$ is the number of active input units – diode-configured transistors. $N$ and $D$ are configured by the user in CRF21–CRF25. There, each byte is divided into two octets (4-bit number) as $\{N_3N_2N_1N_0D_3D_2D_1D_0\}$, with two important considerations:

- This current gain stage has been included to guarantee that the chip will provide a sufficient amount of current in cases of very poor illumination conditions. By design, the maximum current through each bit-element in the input stage is $10\,\mu\text{A}$. Currents beyond this limit will produce a saturation in the output channel.[16] Thus, for instance, if a channel is producing a maximum current of $100\,\mu\text{A}$

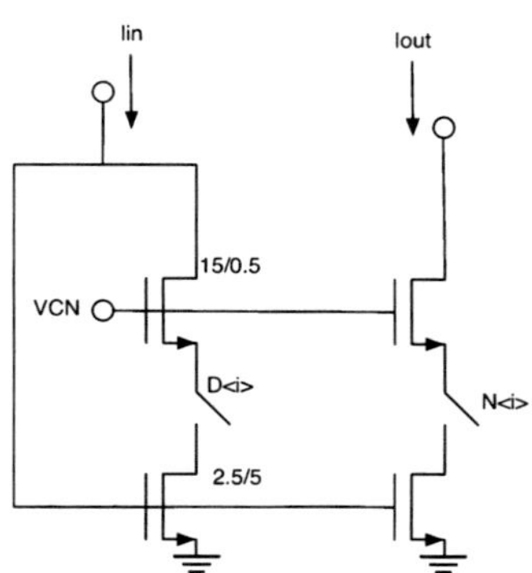


**Fig. 13** Schematic of one-bit in the programmable gain current mirror. The programmable gain amplifier contains 16 of identical items in both input and output branches

[16] It will make voltage at the input node to go above the limit imposed by the amplifier ($V_{\text{sense}}$), producing an instability in the circuitry due to the continuous transition from cut-off to conduction of the transistor in the feedback loop.

the user must program **D** to be 10 or greater. It is clear that this limitation imposes a maximum output current per channel of 150 μA, which is a design specification fixed since the beginning of the project.

- One can wonder what happens if the user programs **D** to 0. In this case, there is no input stage receiving the current from the current conveyor. Hence, the input node would increase its voltage until producing the same instability.[17] To avoid this, and to provide an additional feature, the controller checks whether D=0 for any of the output channel gains, and, if true, bypasses the current scaling stage providing the output current as it is collected by the current conveyor – including sign inversion. This option allows us to evaluate the operation of the current scaling block and to read correlation output currents larger than 150 μA.[18]

## 6 Chip Layout

Figure 14 shows the layout of the chip. It occupies $3.4 \times 4.9\,\text{mm}^2$ and has been fabricated in a 0.35 μm 4Metal-2Poly technology. The fringe sampling diodes are the vertical structures in the middle of the plot, being also easily visible the central row of APS pixels and the digital subsystem at the left side of the chip. To ease chip mounting on the encoder's head, the chip only contains pads in left and right sides. The CM-PGA are the vertical structures on the right side of the chip. Thus, all digital pads are on the left side whereas all analog pads are on the right side of the chip. On the one hand, it reduces noise in the analog lines, and, on the other hand, it allows for a *cleaner* design of the board which will host the chip.

## 7 Experimental Results

This section presents the experimental results obtained with the chip. All modules have been satisfactorily measured with experimental results meeting design specifications.

### 7.1 *Test Setup*

Figure 15a shows a block diagram of our test setup. A control software, designed for MATLAB, communicates with the test board via RS232 protocol. Via this software,

[17] See footnote 16.

[18] This situation is not very likely though. Considering that the average coefficient in each correlation channel is around 3 that the maximum expected photogenerated current is about 300 nA (very optimistic supposition), and that we have 200 SPEs, the maximum expected output channel would be 180 μA.

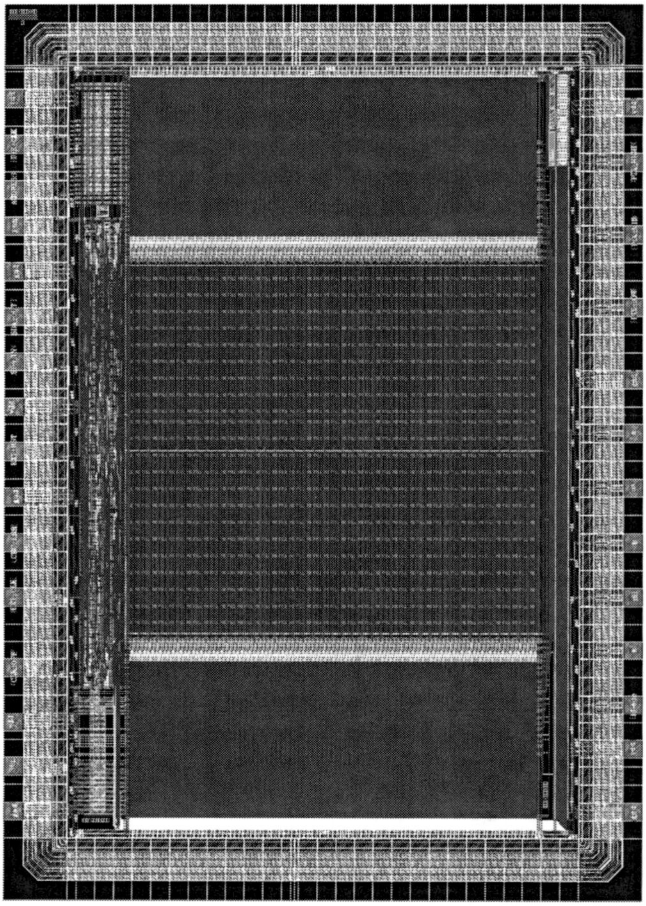

**Fig. 14** Chip layout (rotated 90° ccw)

we can command orders to the board controller – an 18LF8722 PIC, and, through this PIC, interact with the chip. Chip controls, CRF, and 3000-bit FIFO,[19] are loaded into the chip by the PIC through its native $I^2C$ interface. Analogue outputs of the chip can be read in two modes. In the first mode, a bank of switches connects all analogue output to different test points in the board. These test points are then read and digitized with an oscilloscope – which can be connected to MATLAB in the main computer as well. In the second mode, the bank of switches connects the analogue outputs of the chip to different analogue input channels on the PIC.

[19] Although we can transmit byte by byte the information to be written in the FIFO from the PC via RS232 to the PIC and from the PIC to the chip, we have implemented a faster method by defining some preloaded FIFO configurations in the PIC memory so that one can simply select the configuration to with which write into the chip instead of transmitting it from the computer.

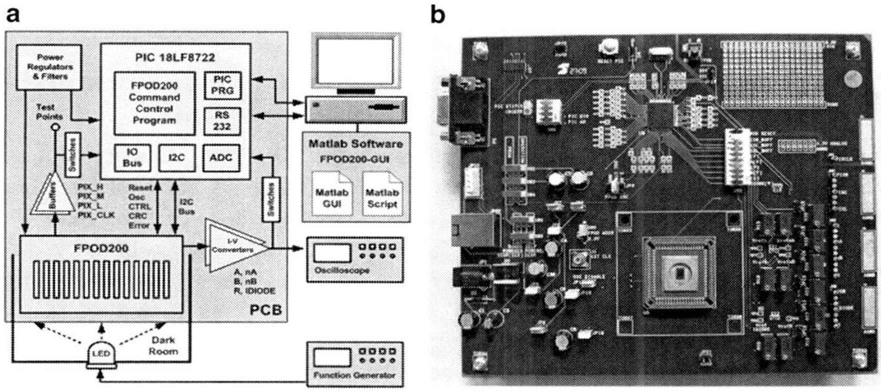


**Fig. 15** Chip test (**a**) block diagram of the test setup (**b**) the test board

Information is digitized by the ADC in the PIC and transmitted to the main computer via RS232. Using this second option is limited to DC – or low frequency – characterization measurements since the PIC only has one ADC, which is time-multiplexed when it is required to convert inputs from more than one channel.

Figure 15b shows the 4-layered test board designed to host the chip and the PIC. The chip is located inside the white square. Holes in the corners of this square are used to insert the screws that fix a black box, which is placed on top of the chip during optical measurements. This black box contains on its top an 880 nm LED whose current is modulated by a programmable function generator – we cannot produce fringes in this setup but modulated intensity patterns.

## 7.2 Scaling Coefficients Test: DC Response

Once we have checked that the digital subsystem operates correctly, we have evaluated the DC performance of the correlation channels. The test runs as follows:

1. We configure the chip in test mode and enable all diodes simultaneously. With this configuration, we read the output current through the test pad, and vary the current through NIR LED until we get an equivalent current of 30 nA[20] from each photodiode.
2. We load an FIFO stream (3,000 bits) that configures coefficients in all SPEs to 1.
3. We measure the output current through all channels in four situations:

   a. In normal mode, i.e., not enabling the buffer neither the $I_{BIAS}$ extra current.
   b. Enabling the buffer.
   c. Enabling the $I_{BIAS}$ extra current.
   d. Enabling both the buffer and the $I_{BIAS}$ extra current.

---

[20] Indeed, we read $200 \times 30$ nA=6 μA.

4. We re-write the FIFO by increasing the equivalent coefficient by 1 (while <7) and repeat measurements.
5. We go back to the first step, we increase the equivalent photogenerated current in 30 nA (while <150 nA), and we repeat all measurements again.

Figure 16 shows the equivalent gains obtained – we are displaying only one channel for visibility purposes – for photogenerated currents of 30 nA and 60 nA – worst cases regarding accuracy – and in the four previously described operation modes.

Let us first comment that these results have been digitized by the DAC on the PIC. This DAC is a voltage mode DAC and, therefore, correlation output current from the chip has been converted to a voltage by means of a bank of programmable gain I–V converters – just a programmable resistor in a negative feedback loop around an operational amplifier. Due to noise in the board, and limitations imposed by the PIC DAC, our current mode LSB is about 150 nA. Besides, in order to obtain the implemented coefficient, we are normalizing correlation output currents to the current obtained when we program all coefficients to 1 (this is why the coefficient 1 seems to be errorless). Results show that coefficients are satisfactorily implemented for a 3-bit representation of the information. Maximum obtained error is below 8% when we compute them by normalizing to the result of the implementation of coefficient one.

We can also compute the errors as the difference between the ideally produced output currents ($I_{\text{photo}} \times N_{\text{pixels}} \times \text{Coeff}$) vs. the measured ones. It is obvious that these errors would be bigger than those obtained when we normalized to the output current produced for coefficient one. However, we include them here to show overall deviation from ideality in the response of the chip. Figure 17 shows these error computations (in %) for two extreme cases, the default configuration, in which we obtain the smallest error, and the result of adding $I_{\text{BIAS}}$, which produces the largest error.[21] Note that we obtain errors that move between 5% and 8%, confirming that the required 3-bit implementation of the scaling coefficients is satisfactorily met.

## 7.3 Scaling Coefficients Test: Frequency Response

Frequency response of the correlation channels has been characterized using a 4-channels@500 MHz Tektronix 3054DPO oscilloscope. The test works as follows:

1. We configure the chip in test mode and enable all diodes simultaneously – by enabling two bits in CRF20. With this configuration, we read the output current

[21] Surprisingly, when we use the buffer and $I_{\text{BIAS}}$ the resulting error is smaller due to their different signs. Indeed, adding $I_{\text{BIAS}}$ introduces a small systematic offset – due to non-total suppression of the $I_{\text{BIAS}}$ in the SPE output current – which somehow compensates the small systematic – obviously not the random, but we observe averaged results since we measure the current from 200 SPEs – component of the offset voltage of the buffer. Besides, this compensation is observed independently of the implemented coefficient since both terms scale as a function of the number of current sources connected to the SPE output node.

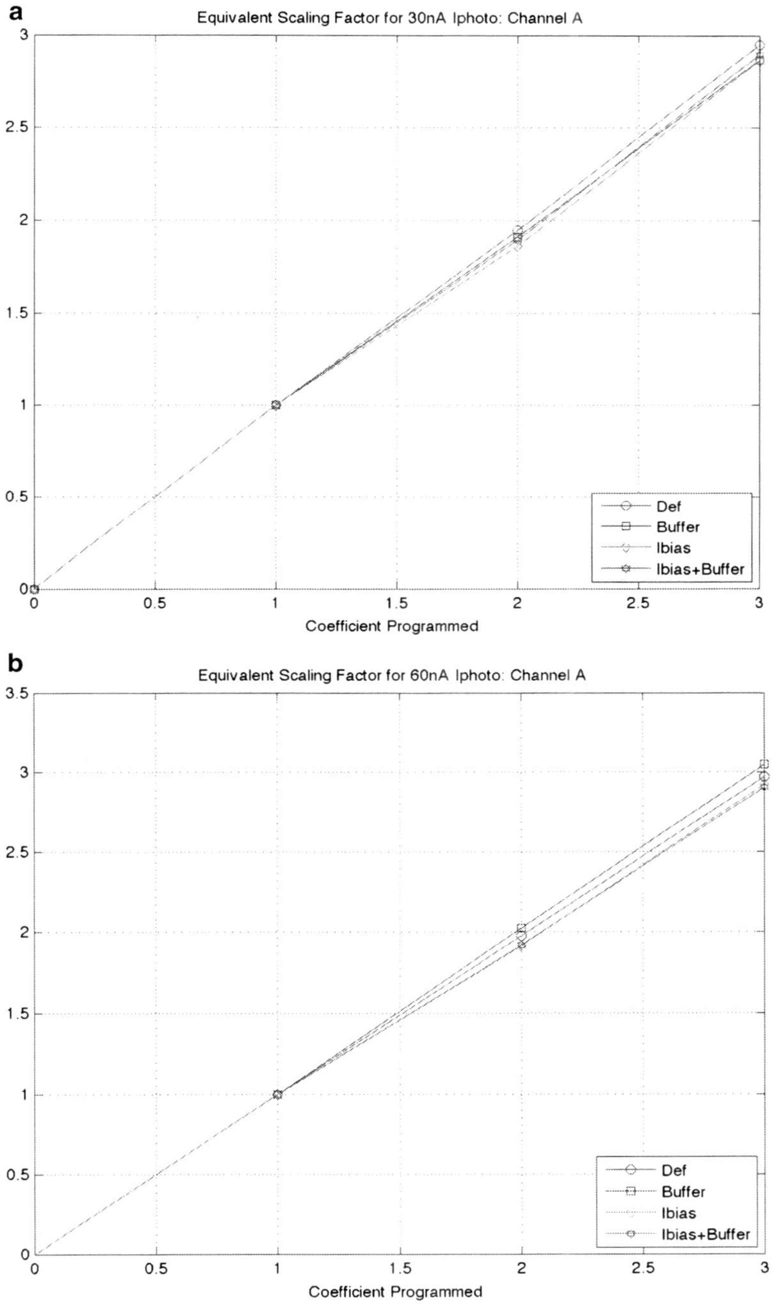


**Fig. 16** Evaluation of equivalent gain for different photogenerated currents using the four modes of operation of the SPE. (1) Default (2) Enabling the buffer (3) Adding $I_{\mathrm{BIAS}}$ (4) Using the buffer and adding $I_{\mathrm{BIAS}}$

a

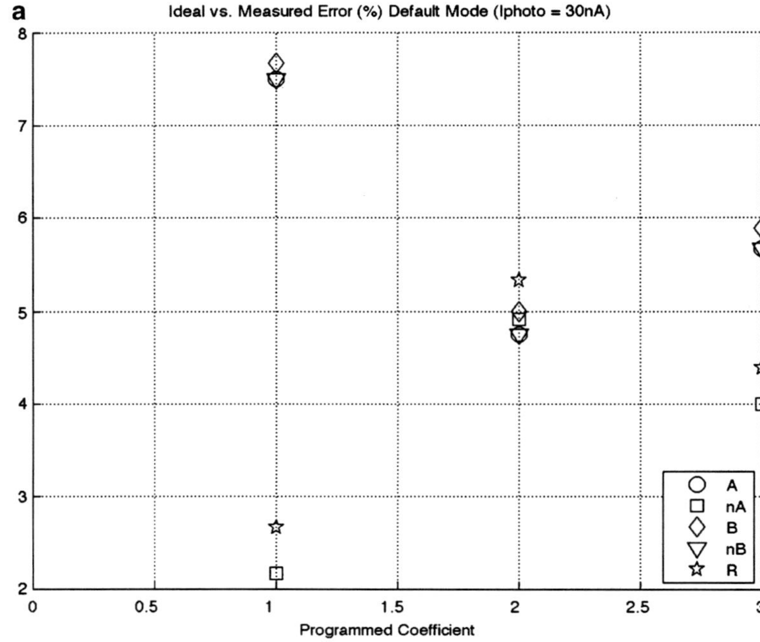


b

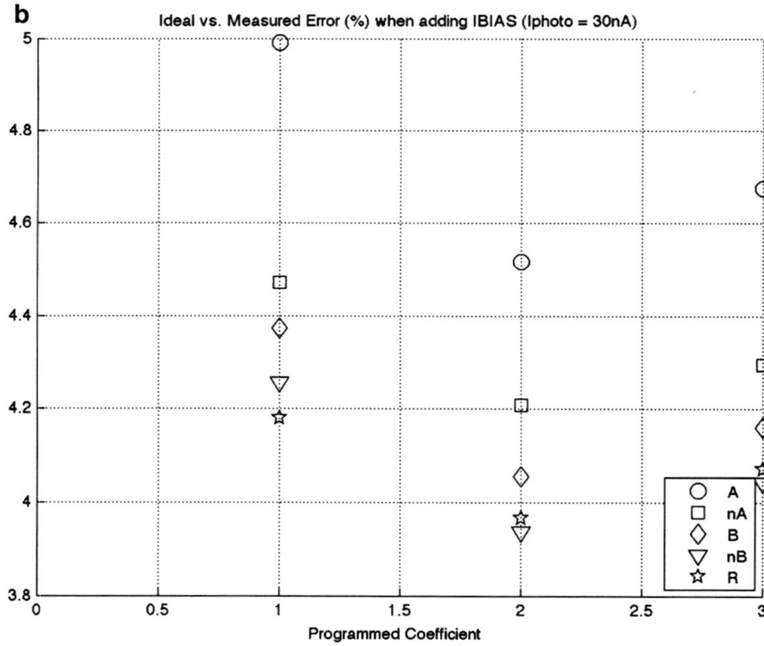


**Fig. 17** Error computations for all output channels. (**a**) Default mode – lowest error – (**b**) Adding $I_{\mathrm{BIAS}}$ – largest error

**Table 4** Frequency response of the correlation channels in the different configuration modes

| $I_{BIAS}$ | Buffer | −3dB frequency (kHz) |
|---|---|---|
| OFF | OFF | 40 |
| ON | OFF | 140 |
| OFF | ON | 130 |
| ON | ON | 2,700 |

through the test pad, and vary the current through the photodiode until we get an equivalent sine current of 30nA with an optical contrast of 12% from each photodiode.
2. We load an FIFO stream (3000-bit) that configures all coefficients in all SPEs to 7.[22]
3. We measure the output current through all channels in four situations at a very low frequency (200 Hz):

   a. In normal mode, i.e., not enabling the buffer neither the $I_{BIAS}$ extra current.
   b. Enabling the buffer.
   c. Enabling the $I_{BIAS}$ extra current.
   d. Enabling both the buffer and the $I_{BIAS}$ extra current.

4. We increase the frequency (while <5 MHz) and repeat the measurements.
5. We find the −3dB frequencies for the different modes.

Table 4 shows averaged results of the cut frequency for different samples of the chip. Results show that, if properly configured, the chip can operate with frequency fringes moving at 500 kHz (20 m s$^{-1}$).

**Acknowledgements** The authors thank Dr. E. Roca from IMSE-CNM for her useful comments during pixel design. This work has been partially funded by CICE/JA, MICINN, and CDTI (Spain) through projects 2006-TIC-2352, TEC2009-11812, and Cenit EeE.

## References

1. Fagor encoders catalog. http://www.fagorautomation.com/pub/doc/File/Catalogos/ingl/cat_captacion_general.pdf
2. D. Crespo, P. Alonso, T. Morlanes, E. Bernabeu, Opt. Eng. **39**, 817 (2000)
3. D. Crespo, Nuevas herramientas aplicadas a la codificación óptica. Ph.D. thesis, Univ. Complutense de Madrid (2001)
4. T. Morlanes, Optical length measuring device with optoelectronic arrangement of photodetectors. European Patent Specification EP1164359B1
5. J. Tu, L. Zhan, Optic. Commun. **82**(3–4), 229 (1991)
6. H.F. Talbot, Philos. Mag. **9**, 401–407 (1836)
7. L. Liu, Appl. Opt. **28**, 4668 (1989)

[22] This is the worst case since we are programming the maximum capacitive load.

# VISCUBE: A Multi-Layer Vision Chip

**Ákos Zarándy, Csaba Rekeczky, Péter Földesy, Ricardo Carmona-Galán, Gustavo Liñán Cembrano, Soós Gergely, Ángel Rodríguez-Vázquez, and Tamás Roska**

**Abstract** Vertically integrated focal-plane sensor-processor chip design, combining image sensor with mixed-signal and digital processor arrays on a four layer structure is introduced. The mixed-signal processor array is designed to perform early image processing, while the role of the digital processor array is to accomplish foveal processing. The architecture supports multiscale, multifovea processing. The chip has been designed on a 0.15 um feature sized 3DM2 SOI technology provided by MIT Lincoln Laboratory.

## 1 Introduction

The main trade-off of the focal-plane sensor-processor design considerations is the silicon resource distribution between the sensors and the processors. The problem already starts at the technology selection. Conservative technologies with thicker depletion layer and fewer metal/dielectric layers have larger light response; hence, they are better suited to be sensor materials. However, processor circuits can benefit from modern deep sub-micron technologies because both the large number of metals and the small minimal size transistors needed to build complex circuits, especially in the digital domain. After the technology is selected, the next hard trade-off is the relation between the sensor and processor areas. The minimal sensor area is determined by the sensitivity requirements, while the fill factor and the maximum chip area limit the processor size. If we cannot squeeze the processor circuits into the required area, we need to choose a finer silicon technology, and the iteration starts again.

Vertical integration technology [1] makes possible the breakout form this circle. It enables the integration of multiple silicon and/or other semiconductor layers above each other by interconnecting them with *thru silicon vias* (TSVs) [1, 11].

Á. Zarándy (✉)
Computer and Automation Research Institute of the Hungarian Academy of Sciences, (MTA-SZTAKI), Budapest, Hungary
e-mail: zarandy@sztaki.hu

Á. Zarándy (ed.), *Focal-Plane Sensor-Processor Chips*,
DOI 10.1007/978-1-4419-6475-5_8, 

In this way, the sensors may occupy the entire top layer, leading to 100% fill factor [2]. Moreover, besides the visual domain, infrared (IR) image ecensor can be implemented also by applying InGaAs [3] or other materials. Analog or mixed-signal layer for signal conditioning, early image processing and analog-digital (AD) conversion can be implemented on a different silicon layer below, and digital post processor layer(s) can be built even below. In ideal case, we can even choose different silicon technology for the different layers, which are well suited to the requirements of the implementation of the optimal sensor, analog/mixed-signal cells, and digital processors.

This chapter introduces a focal-plane sensor-processor circuit design that uses vertical integration technology. The work was done in the framework of the VISCUBE project [4], which is led by Eutecus Inc. Berkeley, California, and financed by the Office of Naval Research (N00173–08C-4005). Two European design groups are involved in the VLSI design, one from Spain (AnaFocus LLC), and the other from Hungary (MTA-SZTAKI). The goal of the project is to build a visual surveillance, reconnaissance, and navigation device, which can be carried by miniature *unmanned aerial vehicle* (UAV). Because of the limited payload and power supply on board, a single chip vision system under 1 W was the goal.

The algorithmic requirements are to perform moving platform video analytics, including feature point extraction, displacement calculation (optical flow), and feature extraction in multiple windows [5]. The image capturing and processing speed should reach 1,000 FPS. To be able to fulfill these goals, the design supports multi-scale, multifovea processing.

This chapter is organized as follows. After the introduction, the architecture and the technology are introduced. It is followed by the implementation details. Then, the simulated operational examples, and finally, the conclusions are shown.

## 2 Architecture

In general, the architecture is derived according to the required functionalities and the available technologies. Here, the required functionalities are

- Image sensing.
- Moving platform image analysis:
  - Feature point extraction;
  - Displacement calculation;
  - Feature extraction;
- Analog to digital conversion.

To fulfill these computational requirements in the given strict power budget, we have to apply advanced algorithmic approaches such as multiscale and multifovea techniques. The common feature of these techniques is that they significantly reduce the amount of data to be processed. The multiscale algorithms based on repeated subsampling, hence, decreasing amount of information is processed.

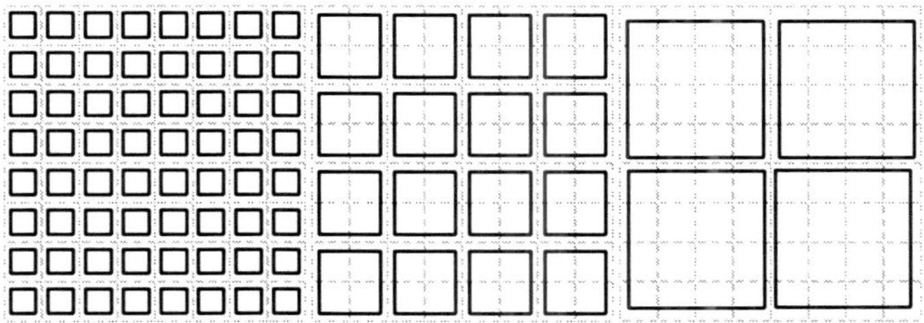


**Fig. 1** The three different scales used in the VISCUBE design

Fovea processing techniques use detailed image analysis on small windows of the image only. When these two methods are combined, a low resolution version of the image is fully preprocessed, and as a result of this, the areas of interests are detected. These areas of interests (windows) require further detailed processing. The size and the resolution (scale) of these windows depend on the type of analysis applied.

When selecting image processor devices, we need to study the efficiency of different architectures. For early image processing, the mixed-signal locally interconnected processor arrays and the digital pass-through (pipeline) pixel flow processor arrays can be efficiently applied [6]. However, above 1,000 FPS under low latency requirements in near-sensor-processing, the fine-grain mixed-signal processor array is the best choice. For fovea-type postprocessing, coarse-grain digital processor arrays can provide efficient solution [6] .

To connect and efficiently use the fine-grain mixed-signal processor array and the digital foveal processor array, we need (1) analog-to-digital converters; (2) a digital frame buffer, which can store the entire frame; and (3) a switch matrix, which makes possible glue-less scaling and windowing, to prepare data for the multiscale, foveal processor array.

In this design, we consider three scales (Fig. 1):

Scale 0: Sensor resolution, $320 \times 240$ grayscale, 8-bit representation
Scale 1: Mixed-signal processor resolution, $160 \times 120$ grayscale or binary
Scale 2: Downscaled Scale 1 resolution, $80 \times 60$ grayscale, 8-bit

## 2.1 Technology

The targeted technology contains three *silicon on insulator* (SOI) layers, 0.15 μm with three metal layers per tier (plus additional metal layer, so called backmetal, on tier 2 and tier 3), provided by MIT Lincoln Laboratory (Fig. 2) [7]. The layers are interconnected with 5-μm pitch Tungsten TSV. The back-illuminated sensor array is implemented on an additional bump-bonded semiconductor layer on top.

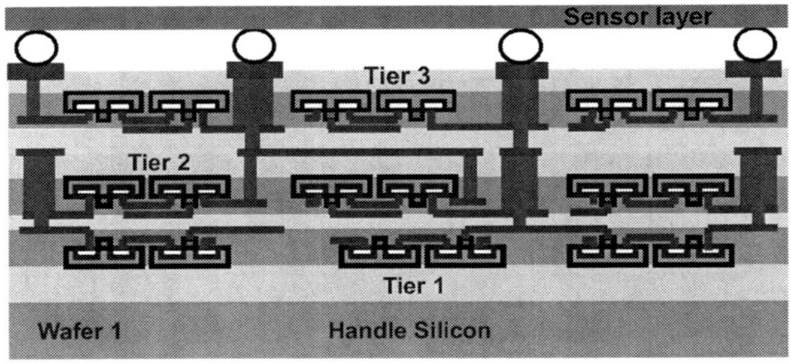


**Fig. 2** Targeted vertical integration technology with three SOI layers and a bump-bonded sensor layer (MITLL 3DM2 0.15) [7]

**Fig. 3** Architecture mapping

| Sensor array |
|---|
| Mixed-signal processor array + AD converter |
| Frame buffer + switch matrix |
| Foveal processor array |

## 2.2 *Architecture Mapping*

After defining the necessary architecture components, the architecture mapping is straightforward to the three functional and the sensor layer (Fig. 3). The mixed-signal layer needs to be next to the sensors, because it processes the raw analog sensor data, and takes part in the AD conversion. Between the mixed-signal and the foveal processing layer, the frame buffer and the switch matrix take place, to provide efficient data communication.

After the architecture mapping is described, we introduce the architecture of the individual functional blocks and layers. The implementation details are introduced in Sect. 3.

## 2.3 *Sensor Layer*

As it was said already, the back-illuminated sensor array is built on an extra semiconductor layer, connected to the processor array using bump-bonding interface. The resolution of the sensor array is $320 \times 240$. The choice of the sensor material defines the wavelength. This can range from visual range to near infrared (NIR), or even short-wave infra red (SWIR). The sensor layer contains the photodiodes only. The integration-type sensor interface, implemented on the mixed-signal layer, keeps a constant bias voltage on the diodes. This supports the usage of a wide range of sensor material, independently of their sensitivity characteristics.

## 2.4 *Mixed-Signal Processor Array*

The fine-grain mixed-signal processor array occupies the top silicon layer. Its resolution is 160 × 120 (Scale 1). The *mixed-signal processing elements* (MSPEs) serve the image acquisition; they can calculate diffusion, subtraction, and identify the local maxima. *Difference of Gaussians* (DOG) filter can be calculated by subtracting different linear diffusion results of the same image.

Each MSPE (Fig. 4) contains (1) electrical interface for four photodiodes (4 pixels), (2) four *local analog memories* (LAMs) for storing an entire Scale 0 image; (3) an analog diffusion unit, and (4) a comparator, which is used either as the local extremum detector, or as a component of a single-slop AD converter. Each MSPE is interconnected with its eight neighboring MSPEs, allowing for programmable real-time spatial processing operations.

The mixed-signal processor array is designed to be able to handle both 160 × 120 and 320 × 240-sized images. It can process 160 × 120-sized images. The spatial operators are designed to handle full Scale 1 images. The 160 × 120-sized image is generated by subsampling or binning the 320 × 240 image. The AD converter can convert either Scale 0, or Scale 1, or Scale 2 images.

### 2.4.1 Image Acquisition

Image acquisition is performed together by the sensor layer and the mixed-signal layer and the mixed-signal layer (see also [12]). The photodiodes are located on the sensor layer. Their photocurrent is integrated by the mixed-signal layer. The integration time is controllable in a wide range from sub-microsecond to hundred milliseconds. The sensor interface applies a transconductance amplifier. The photodiodes are kept on a constant bias voltage during the integration. This bias voltage can be set externally to support various sensor materials.

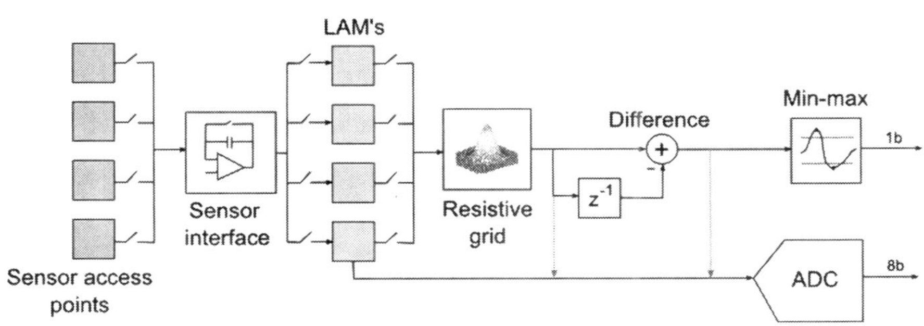


**Fig. 4** MSPE architecture of the mixed-signal processor array layer

### 2.4.2 Diffusion Operator and DOG Filter

The diffusion operator processes Scale 1 resolution (160 × 120) images. This operator has been implemented by *switched capacitor* (SC) circuit, which provides fine control on the diffusion scale. The circuit connects four adjacent neighbors and operates in discrete-time. The diffusion process can be sampled and then continued to be able to generate different images with different smoothing operators. By subtracting these images from each other, we get DOG filter.

### 2.4.3 Extremum Location Detection

The extremum location detection is also defined on Scale 1 grid. The output of the operator is a binary image, which is black in those pixels, which contains local maximum/minimum in a 3 × 3 neighborhood. In order to suppress irrelevant local extrema, a maximum/minimum pixel is accepted if its value exceeds a programmable global threshold limit.

### 2.4.4 AD Conversion

Each MSPE contains a single slope AD converter (Fig. 5). It consists of four components (1) a digital-to-analog (DA) converter, (2) comparator, (3) an 8-bit counter, and (4) a digital memory buffer. When an array of single slope AD converters are implemented, the DA converter and the counter can be shared, and only the comparator and the digital memory are needed to be built in each location. The comparator of the AD is located on Tier 3, in the mixed-signal layer, while the digital memory is located in the frame buffer layer on Tier 2.

This requires a pitch-matched mixed-signal and frame buffer layer design. Each MSPE and its corresponding frame buffer cell, which contains the digital memory, is connected with one wire only (called trigger in Fig. 5).

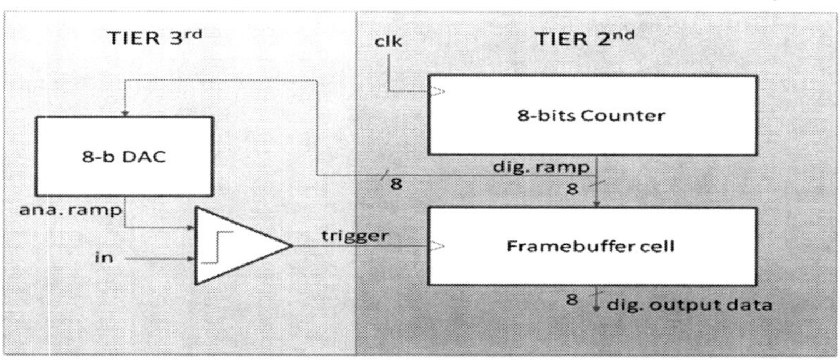


**Fig. 5** Block diagram of the single slope AD. The comparator (*left hand side*) is implemented on the third tier in every MSPE, while the digital register is located on Tier 2

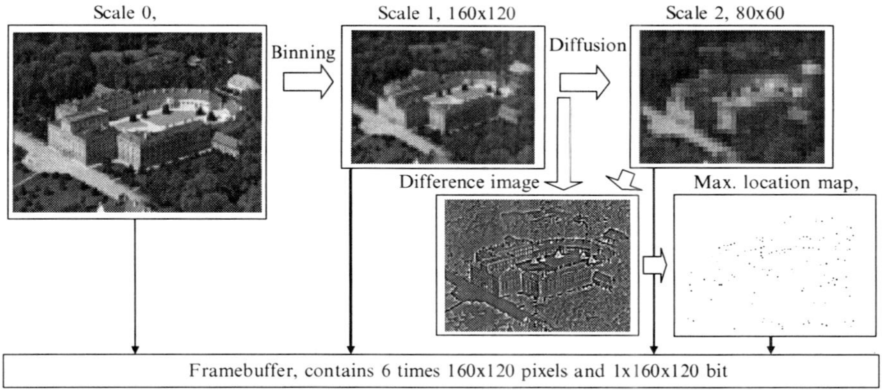


**Fig. 6** Sample images of the algorithm executed on the mixed-signal processor layer (simulation)

Since there is one AD converter in each MSPE, the conversion of a Scale 0 image is done in four conversion cycles, while a Scale 1 image is converted in a single cycle. When a Scale 2 image is converted, every fourth AD converter is used.

### 2.4.5 Operation Example of the Mixed-Signal Layer

Figure 6 illustrates the operation of mixed-signal layer. As it is shown, it can capture a Scale 0 image, convert it into Scale 1 and Scale 2 images by applying either binning, subsampling or diffusion. It can also calculate DOG, and can identify the local extreme points.

## *2.5 Frame Buffer Layer*

The frame buffer layer has three roles. It is used as the (1) digital registers of the single slope AD converters. It is also used as a (2) storage and communication interface between the mixed-signal and the digital layer. Moreover, (3) it supports random access to scaled or not scaled windows of the captured and preprocessed images, which is critical in the multiscale and fovea processing approaches.

The frame buffer layer is constructed of an array of $160 \times 120$ memory units (Scale 1). Each memory unit is corresponding to the mixed-signal processor geometrically located exactly above it.

Each memory unit contains 6 bytes of memory and 2 bits (Fig. 7). Four bytes are needed to store a Scale 0 image, because each unit/cell handles 4 pixels. The remaining 2 bytes can be used to store multiple downscaled images. The single bits can store the outputs of the extremum filters.

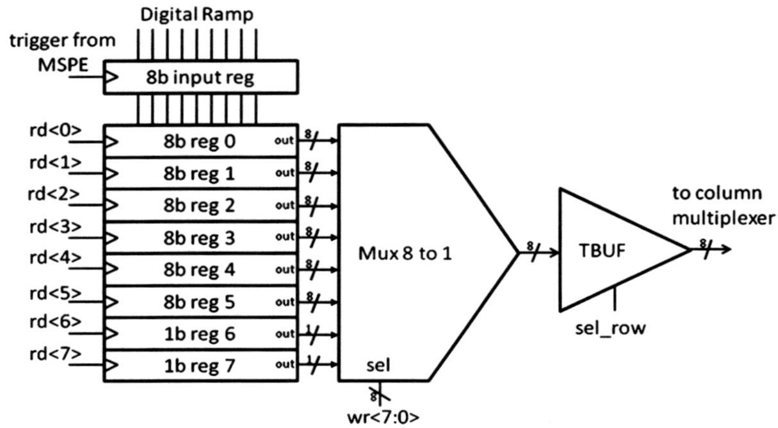


**Fig. 7** Memory unit of the frame buffer layer

The registers of the frame buffer are constructed of two port memories. They are written when an AD conversion or extremum operation is executed. They are read out by the digital processor array layer through a multiplexer. The multiplexer supports automatic windowing and downscaling functions to minimize input and output (I/O) time.

## 2.6 *Foveal Processor Array*

The foveal processor array is intended to be used for both area of interest (window) and full frame processing. This 8 × 8 digital processor array (Fig. 8) is an advanced version of our previous deign [8, 14]. The distinguishing feature of this new implementation is the increased memory size and higher flexibility in the processed window size and distribution.

The processors can work in two different modes. In the first mode, the 64 processors are joining forces, and process a large or medium-sized image (160 × 120, 80 × 60, or 64 × 64) by topographically distributing the image data among the processors. The second mode is used when we have to execute the same operation (e.g., displacement calculation of a feature point) on a large number of smaller windows (24 × 24 or 16 × 16). In this case, each processor processes one window individually. In the formal mode, the neighboring processors are exchanging data intensively, while in the latter mode, the processors are uncoupled. In both modes, the processor array operates in *single instruction multiple data* (SIMD) mode.

The basic constructing element of the foveal processor array is the processor kernel. The kernels are locally interconnected. The processors in each kernel can read the memory of their direct neighbors. There are boundary cells, which are relying data to handle different boundary conditions.

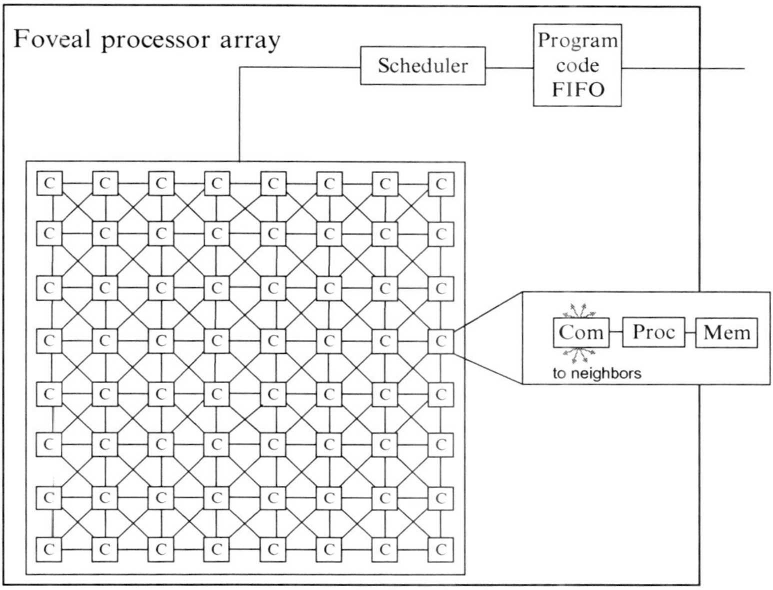


**Fig. 8** The architecture of the foveal processor array

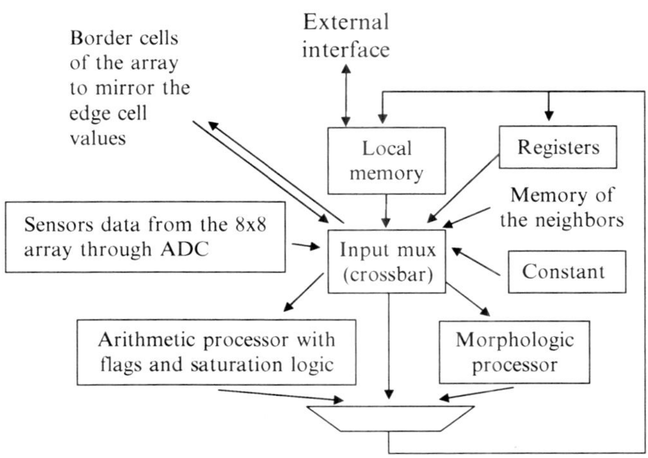


**Fig. 9** The block diagram of the processor kernel

The block diagram of the processor kernel is shown in Fig. 9. Each cell contains an arithmetic processor unit, a morphologic processor unit, data memory, and internal and external communication unit. The arithmetic unit contains an 8-bit multiple-add core with a 24-bit accumulator, and eight pieces of 8-bit registers. This makes possible to perform either 8-, 16-, or 24-bit precision calculations. The arithmetic unit can calculate multiplication, multiple-add, addition, subtraction, and comparison operation. Image processing primitives, such as block matching, convolution, look-up table, diffusion, thresholding, rank-order filters, contour detection, Sobel operator, median, can be efficiently implemented on the arithmetic processor.

The morphology unit supports the processing of black-and-white images. It contains eight pieces of binary morphology processors, for parallel calculation of local or spatial logic operations, pixel-by-pixel binary logic, such as erosion, dilation, opening, closing, hit, and miss operations.

## 2.7 Control and Communication

The VISCUBE design does not include internal control processor. This role can be played by an external control processor. The control processor can execute the main program of the system. It is responsible to initialize subroutines on the individual processor array layers, and image data communication among the three processing units. The control processor continuously evaluates the captured and preprocessed image flow arriving from the mixed-signal layer, and decides which parts (windows) of the input image requires more detailed analysis, and orders the digital processor array to accomplish the necessary routines. It is also responsible to switch between distributed or window processing modes, or modify parameterization of the routines according to the input image content.

The interconnection scheme of the main blocks is shown in Fig. 10. The sensor array, the mixed-signal processor array, and the memory units on the frame buffer layer are pitch matched; hence, direct parallel interconnections are used among them, utilizing through silicon vias. In this way, four sensor units are connected to one *mixed-signal processing element* (MSPE) cell. The interconnection between the MSPEs and the memory units is one to one. There is no parallel interconnection

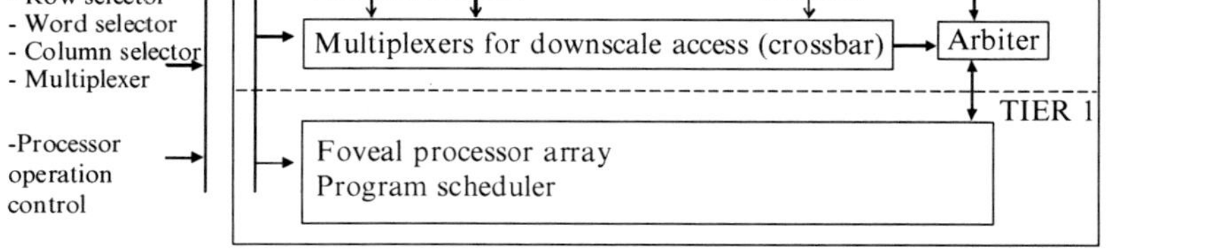


**Fig. 10** The interconnection and the control scheme of the VISCUBE architecture

between the frame buffer and the foveal processor array. These layers are connected only through a multiplexer and a fast digital bus, because the windowing technique does not need full parallel interconnection topology.

# 3 Implementation

After the description of the architecture, the implementation is detailed in this section. In the following, the circuits on the three active tiers are described.

## *3.1 Tier-3: Mixed-Signal Processor Layer*

The entire layout of Tier-3 is shown in Fig. 11. As it is shown there, the mixed-signal processor layer has a topographic structure. Indeed, it is constructed of an array of $160 \times 120$ MSPE (Fig. 12). The size of an MSPE is $50 \times 50\,\mu\text{m}$.

Each MSPE (Fig. 4) includes four diodes/sensors interface circuits, a transimpedance amplifier to adapt the sensor signal, LAMs to store the intermediate data, a programmable diffusion network to perform Gaussian filtering, a susbtractor to obtain DOG filters, a local minimum and maximum locator and a single-ramp Analog Digital Converter.

### 3.1.1 Sensor Interface

Each of the processing elements of the mixed-signal array ($160 \times 120$) provides access points to four diodes/sensors. The sensor interface has two stages, namely: a front-end composed of a *capacitive transimpedance amplifier* (CTIA) with offset sampling and a *switched capacitor* (SC) circuit, where the input data is stored in voltage form. These analog storage places (the LAMs) are also used to realize some arithmetic operations on the pixels data, such as averaging, diffusion, and subtraction.

The CTIA architecture consists of a discrete time integrator (Fig. 13). The CTIA integrates the $I_{\text{ph}}$ photocurrent during the integration period ($\Phi_{\text{int}}$ is on) and provides the following output:

$$V_{\text{o}} = V_{\text{init}} + \frac{I_{\text{ph}} * T_{\text{int}}}{C_{\text{int}}}, \tag{1}$$

which can be tuned by changing the integration time.

LAMs are designed by using a sample and hold structure (Fig. 14), which contains some extra switches to perform arithmetic operations. $\Phi_{\text{av01}}$, $\Phi_{\text{av02}}$, and $\Phi_{\text{av23}}$ switches are used to perform the averaging of the four memories containing the four sensor readings. $\Phi_{\text{dif1}}$ and $\Phi_{\text{dif2}}$ are used to connect independently the different

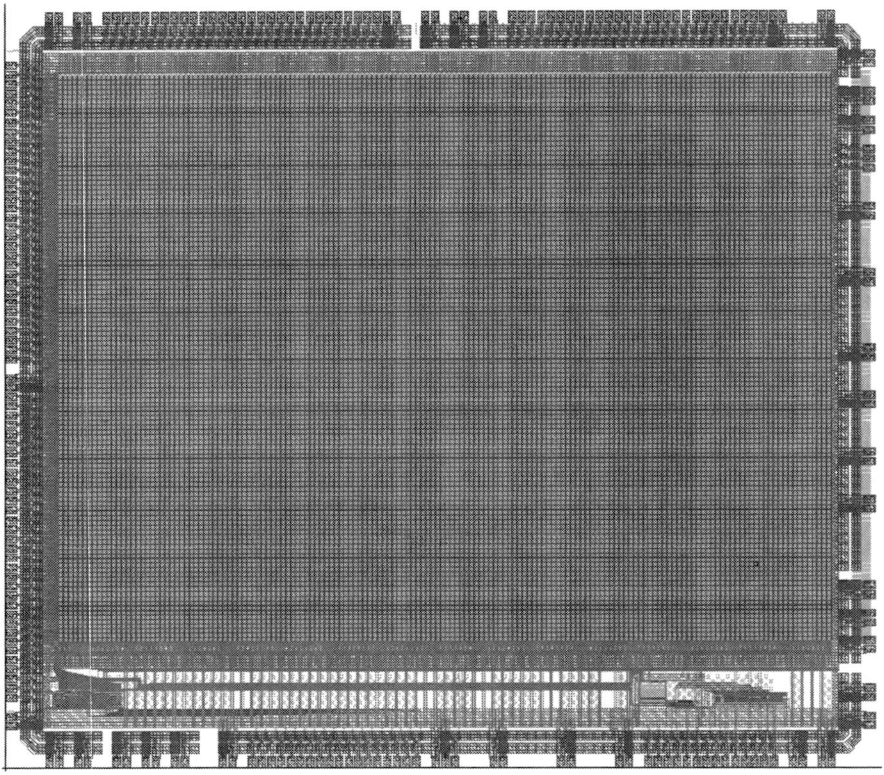

**Fig. 11** Layout of Tier-3. It is constructed of a 120 × 160-sized, topographically arranged mixed-signal cell array (*middle*), and buffers, analog reference generators, a ramp generator, a programmable control unit, and IO circuits on the periphery

capacitances with the diffusive network. Finally, the switches $\Phi_{MT}$, $\Phi_{MA0}$ $\Phi_{MA1}$, $\Phi_{MA2}$, and $\Phi_{MB3}$ reconfigure the capacitors to calculate the subtraction of the stored data.

### 3.1.2 Diffusion Network for Gaussian Pyramid Generator

Gaussian filtering of the quarter image (160 × 120) is realized by a SC circuit. Figure 15 shows the block diagram of the diffusion network unit. This circuit allows the transference of charge packages among nearest neighbors, thus emulating the behavior of a Gaussian filtering.

The image on the nodes is sampled at different point of the discrete-time evolution. These sampling points correspond to particular scale representations. The accuracy of this correspondence is given by the performance of the SC emulator. The samples of the diffused image is either processed to calculate significant points on

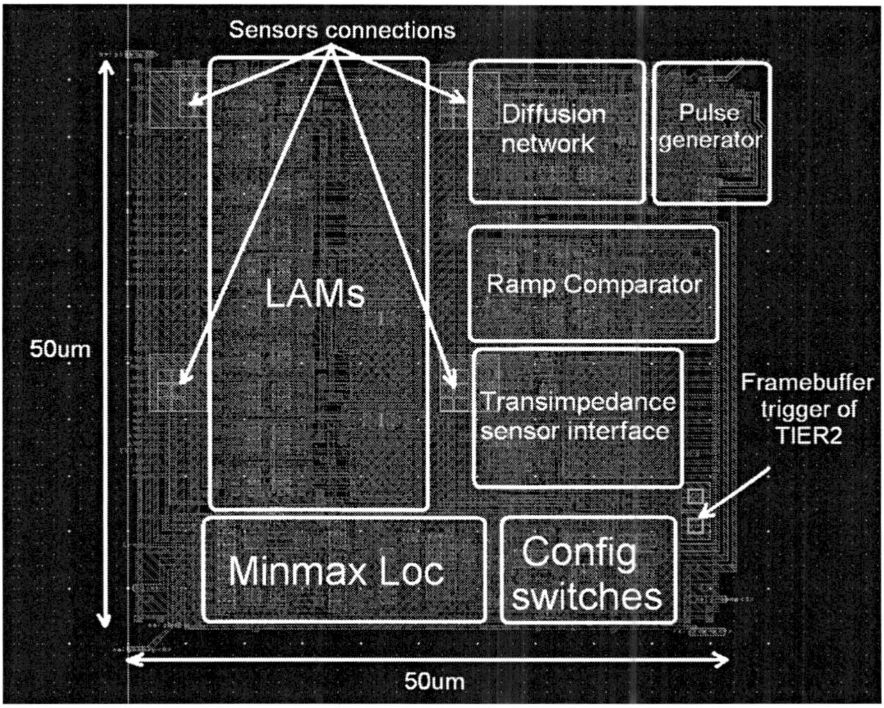


**Fig. 12** Layout of the mixed-signal processor cell

**Fig. 13** Transimpedance sensor interface

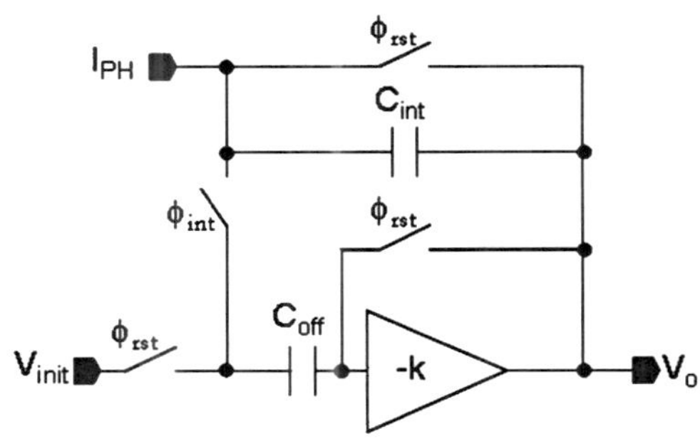


the Gaussian pyramid, or combined to generate the Laplacian pyramid, or converted to be stored in the frame buffer layer because of the needs of a particular algorithm.

### 3.1.3 Local Extrema Values Detector

The local extrema value detector can identify the position of the local maxima or minima in $3 \times 3$ each neighborhood. In order to avoid the implementation of eight

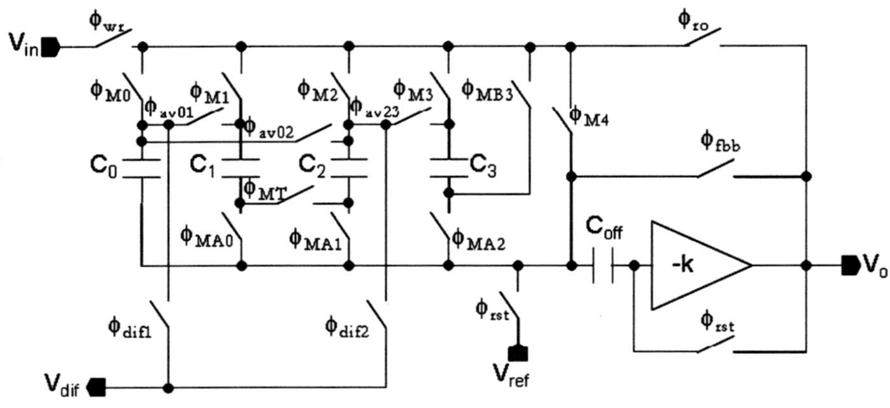


**Fig. 14** Discrete-time sample and hold amplifier

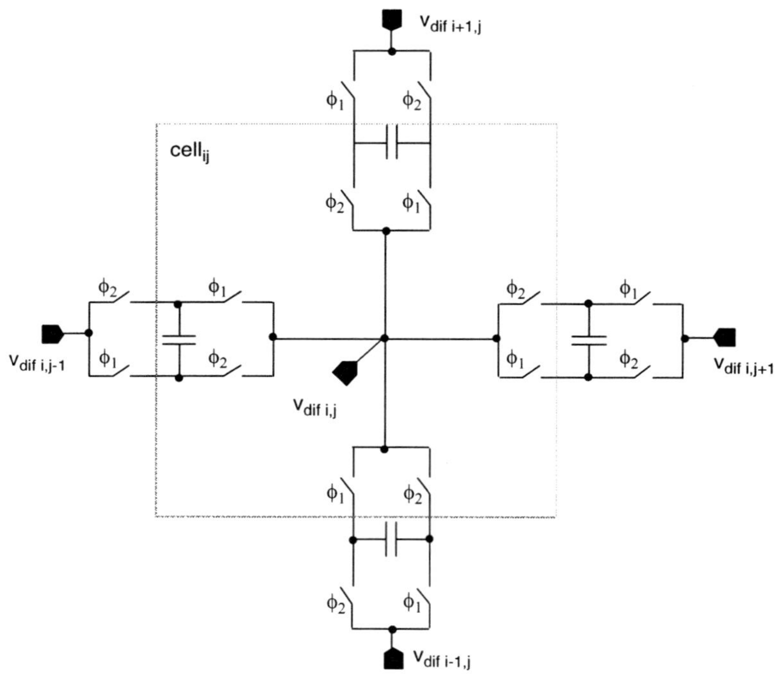


**Fig. 15** SC Gaussian diffusion unit

comparators at each pixel, the calculation is done in time, using the comparator and the ramp generator (DA converter) of the AD converter, a flip-flop and some logic circuit. The local minimum is calculated in a way that a rising ramp starts, and when it becomes larger than the value in a certain pixel position, it tries to set the flip-flop.

However, the flip-flop can be set, if no other flip-flops in the 3 × 3 neighborhood is set already. In this way, maximum one flip-flop is set in each 3 × 3 neighborhood, and this flip-flop belongs to the smaller pixel value in that neighborhood. Figure 16 shows the block diagram of the circuits that are used to calculate local maximum and minimum, respectively.

### 3.1.4 AD Converter and the Comparator

The ADC is a per cell single slope converter. The comparator of the converter is in the third tier, while the latch is implemented on the second tier (Fig. 5). The critical part of the design is the minimization of the offset variance of the large number of the comparators, because this directly leads to fix pattern noise. To achieve this, we used an offset compensation circuit in the comparator design (Fig. 17).

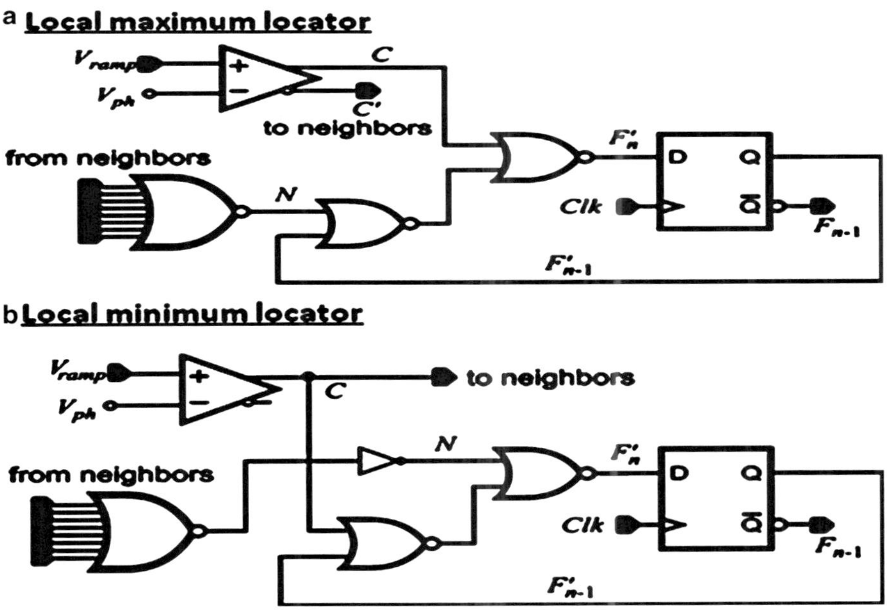


**Fig. 16** (**a**) local maximum locator (**b**) local minimum locator

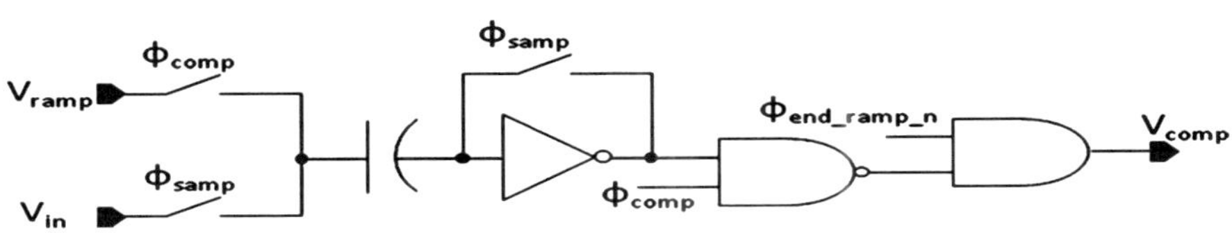


**Fig. 17** Circuit diagram comparator

## 3.2 Tier-2: Frame Buffer Layer

The layout of the frame buffer layer is shown in Fig. 18. Its main function is to provide connection between the mixed-signal layer and the foveal processor layer. It is constructed of an array of 160 × 120 memory units. The array of the memory units are designed on a pitch-matched way with the mixed-signal processor array, and the corresponding units are above each other, to be able to interconnect the two units with a TSV. This condition determines the size of the memory units to be 50 × 50 μm.

The internal structure of the memory units in the frame buffer is derived from the storage and special accessing needs of the memory content. Each memory unit contains 6 bytes and 2 bits, plus the input register of the AD converter. This is 117 kbytes of memory, which can store one piece of Scale 0 image, two pieces of Scale 1 grayscale, and two pieces of Scale 1 binary images. Since the images do not have a predefined position, the images can be interchanged. In this way, one Scale 0 image

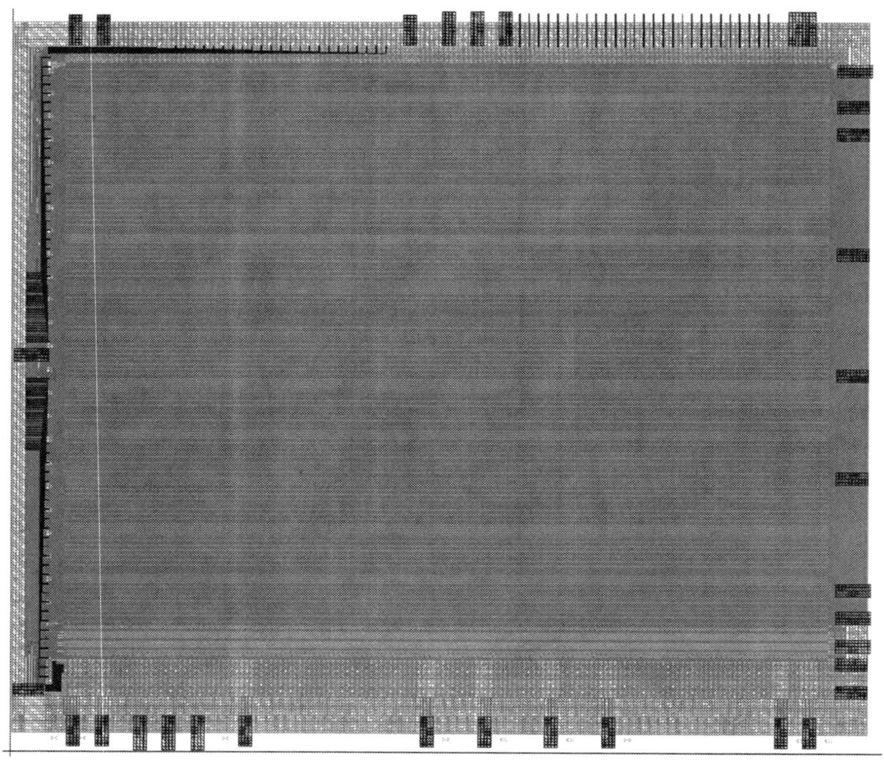

**Fig. 18** Layout of Tier-2

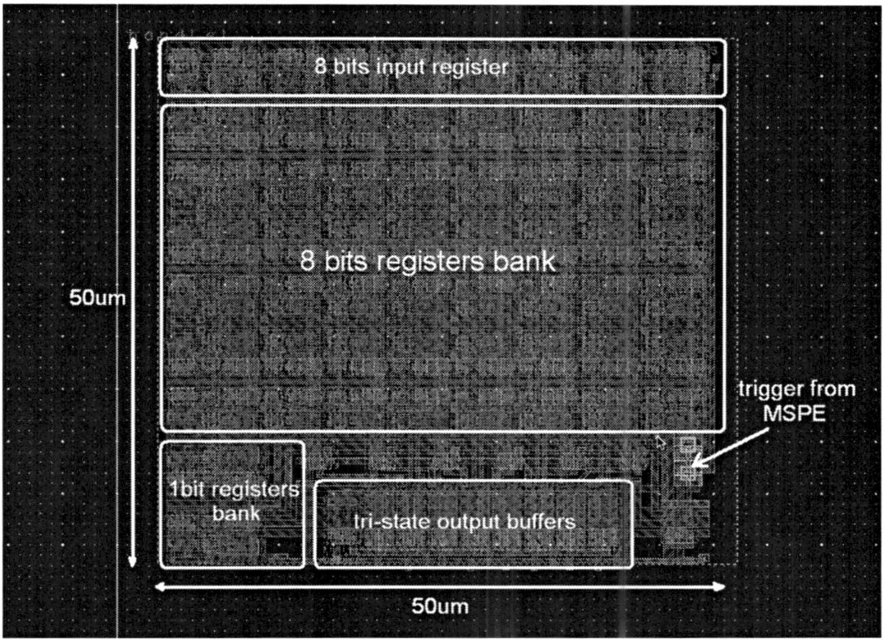


**Fig. 19** The layout of a memory unit of the frame buffer

can be replaced with four pieces of Scale 1 image, and in the same way, one Scale 1 image can be replaced with four pieces of Scale 2 images. The layout of a memory unit is shown in Fig. 19.

The output multiplexer of the memory units and the entire frame buffer is designed to efficiently serve the arbitrary window access demands of the foveal processor array. The frame buffer read-out is random access, the output data are packaged in 32-bit word. During the read-out, the data of a whole row are sent to the output column selector crossbar. The column selector crossbar connects the four selected columns to the 32-bit output bus. This bus delivers data to the foveal processor array, and provides external access to the frame buffer (Fig. 10). The block diagram of a memory unit is shown in Fig. 7, while the structure of the frame buffer, including the column multiplexer, is shown in Fig. 20.

## 3.3 *Tier-1: Digital Processor Layer*

The primary role of the foveal processor array is to perform digital processing on either Scale 0, or Scale 1, or Scale 2 images or image parts. It is constructed of an $8 \times 8$ SIMD processor cell array (Fig. 8). The cells are locally interconnected.

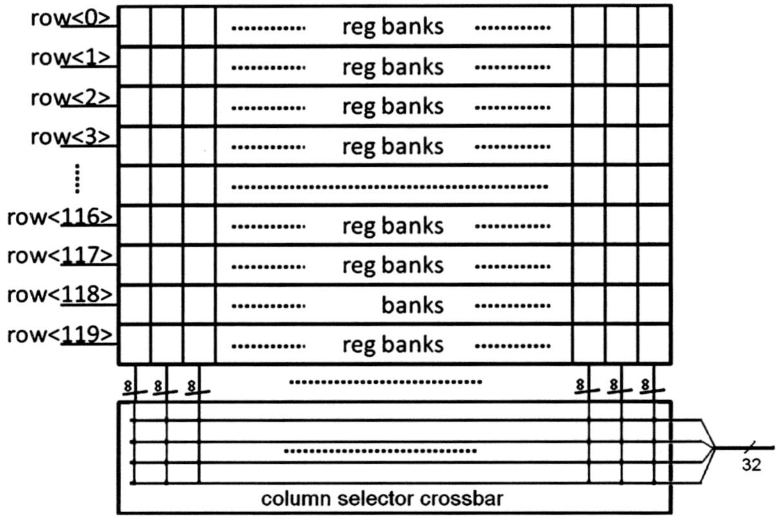


**Fig. 20** Block diagram of the frame buffer and the columns selector crossbar

Figure 21 shows the layout of the foveal processing array. The processor circuits are synthesized, while the memory blocks (the colored squares) are custom design. Neighboring memory blocks denoted with the same color belong to the same processor. The size of the memory blocks is 512 bytes. As it can be seen in Fig. 21, the upper two rows contain four memory blocks for each cell, while the lower six has two only.

### 3.3.1 Memory Access

The communication between the processors is solved in a way that the processors can read the memory of their direct neighbors. This has been obtained by inserting an input multiplexer (crossbar) between the processor cores and their main memory (Fig. 9). The neighborhood memory access works concurrently all over the array, providing the avoidance of data congestion. The multiplexer is capable to access the neighboring memories and provide the required data. In this way, the processors need no distinguished instructions to operate on pixels that are actually stored in a neighbor processor. Furthermore, there are special boundary modules that are straightforward extensions of the arbiter mechanism. These modules provide the boundary condition of the array processor.

The upper two rows of processors (16 pieces) have 2 kbytes, while the lower six rows (48 pieces) have only 1 kbyte of memory. There is no prewired special purpose usage of the memory from hardware point of view, which means that user may store any pixel of the images in any memory locations.

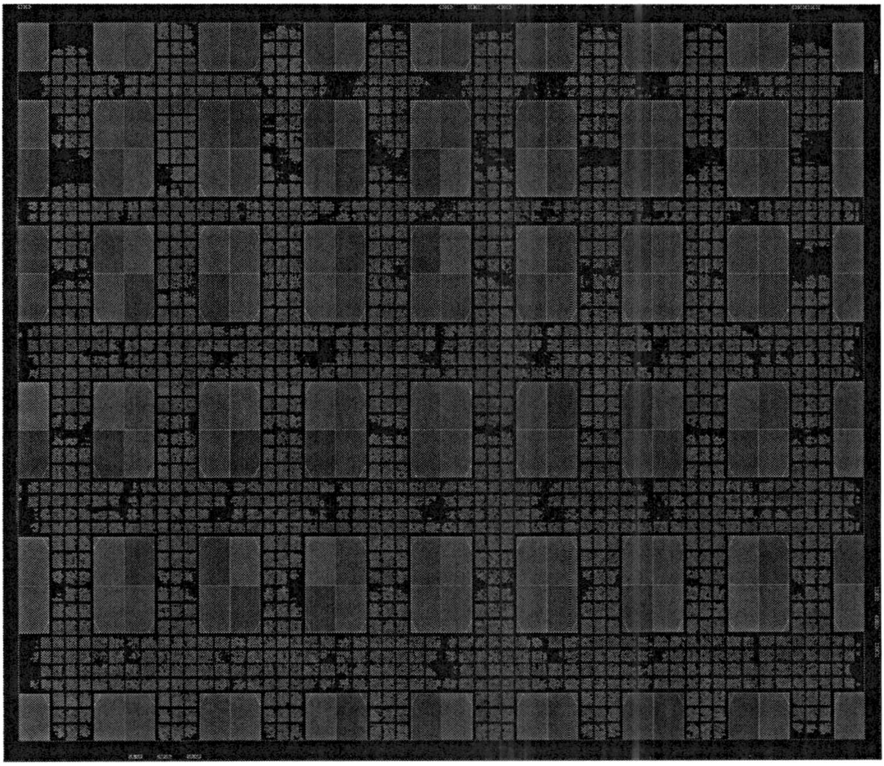

**Fig. 21** Layout of Tier-1

Two memory addressing types are supported. The first is the direct addressing mode, when all the processors access the same memory location in the same step. In this mode, the processors may access to their direct neighbors memory also. The second is the offset addressing method. In this mode, the content of one of the offset register is added to the global address coming from the scheduler. In offset addressing mode, the memory of the neighboring processors cannot be accessed.

### 3.3.2 Image Data Distribution

There are two different operation modes in the processor array. The first is the topographic processing mode, when at least $32 \times 32$-sized images are cut to segments, and these segments are topographically mapped to the array. In this case, the neighboring processors read each others memories to acquire pixel data for the neighborhood processing. The processor cell level memory requirements of the images are summarized in Table 1.

Table 1 Processor level memory requirements of an image in the digital processor array

| Image resolution | Size of a subimage handled by a cell | Memory requirements for grayscale images bytes/cell |
|---|---|---|
| 160 × 120 (Scale 1) | 20 × 15 | 300 |
| 80 × 60 (Scale 2) | 10 × 8 | 80 |
| 64 × 64 | 8 × 8 | 64 |
| 32 × 32 | 4 × 4 | 16 |

The second operation mode is the nontopographic processing mode, when multiple (max 64) windows are cut out from a large image (typically Scale 0), and each of these windows are assigned to one processing cell. Since there are no independent program memories of the cells, the task to be completed should be the same, and cannot contain data-dependent branches at the instruction level. However, thanks to the offset addressing mode, the algorithm may implement the same processing steps on different locations of the windows at the same time. This can be considered data-dependent branch at the data level. This enables, for example, gradient-based searches tasks, such as diamond search.

### 3.3.3 Arithmetic Unit

The arithmetic unit contains an 8-bit multiply-add data path logic with a 24-bit accumulator. The data path enables either 8- or 16-bit precision calculations. The arithmetic unit can calculate multiplication, multiple-add, addition, subtraction, and saturation operations (Fig. 22). Adopted from the common practice of handling both signed and unsigned data by the same unit, the hardware multiplier is of signed 9-by 9-bit precision, and the barrel shifter and the accumulator logic supports sign extension as well. The saturation mechanism also has a great importance in image processing, allowing the user to avoid the time-consuming overflow and underflow management. Besides the arithmetical operations, this unit encompasses bit-field access as well.

The comparator unit is capable to evaluate the relation between signed and unsigned data of any modules. Depending on the outcome of the relation, it sets several flags that are used in later operations.

### 3.3.4 Morphology Unit

The morphology unit supports the processing of black-and-white images (i.e., the pixel representation is one bit per pixel). It contains eight pieces of identical single bit morphology processors (Fig. 23). Hence, it accelerates greatly the parallel calculation of local or spatial logic operations, such as erosion, dilation, opening, closing, hit and miss operations. It can be efficiently used when each processor handles an 8-pixel-wide image segment.

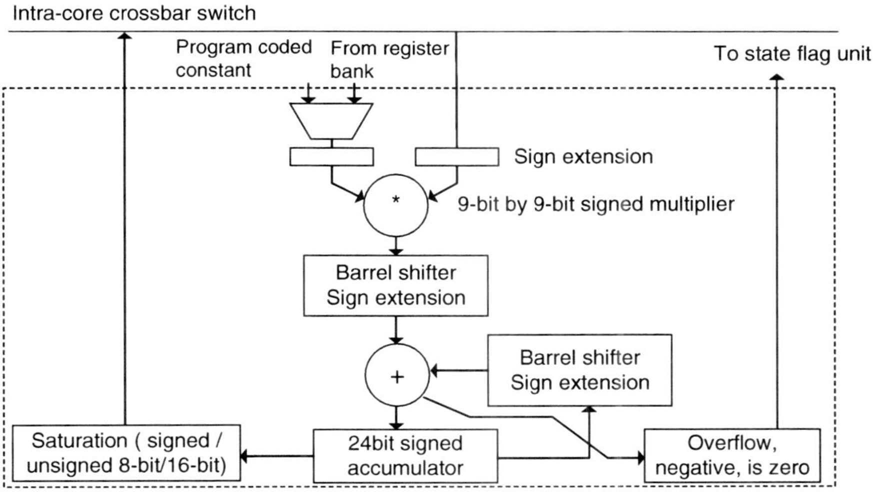


**Fig. 22** The architecture of the arithmetical units

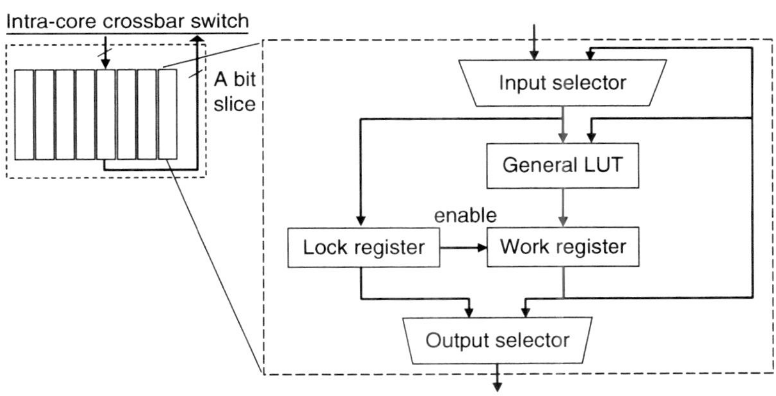


**Fig. 23** The architecture of the binary morphology units

### 3.3.5 Scheduler

The cells do not have local program memory. The program is coming from a global off-array scheduler. Each processor receives identical command, parameters and attributes in each instruction cycle, which makes it an SIMD processor array architecture. The individual processors are maskable. This means that content-dependent masks may enable or disable the execution of a certain image processing operation in a few pixel locations. This masking can make the process locally adaptive.

The program scheduler receives the binary executable code from an on-chip first-in first-out (FIFO). The depth of the FIFO is 16 words. When the program execution is enabled, the scheduler takes an instruction out from the FIFO, decodes it, and sends it to the digital processor array. This process will be going on as long as the execution is enabled and the FIFO contains code. This FIFO is fed from an external source. It uses the digital I/O bus of the system.

### 3.3.6 Instructions of the Processor Array

The instruction set of the digital processor array consists of five groups:

- Initialization instructions;
- Data transfer instructions
- Arithmetic instructions;
- Logic instructions;
- Comparison instructions.

The Initialization instructions are needed to clear or set the accumulator, the boundary condition registers, the masks, and other registers of the cells.

The arithmetic instruction set contains addition, subtraction, multiplication, multiple-add operation, and shift. These operators certainly set the flags of the arithmetic units. These flags can be used in the next instruction as conditions.

The logic operations strongly support the execution of the binary mathematical morphology operations, such as erosion and dilation. Since the internal operand width is 8 bit, each processor can handle 8 pixels in one clock cycle when executing logic operation.

The comparison instructions are introduced to calculate the relation between two scalars. These operators can be used for statistical filter implementations.

Using these instructions, we can efficiently implement the basic image processing functions (convolution, statistical filters, gradient, grayscale and binary mathematical morphology, etc.) on the processor array.

## 4 Operational Example: Displacement Calculation

An operation example is shown in this section to demonstrate the operation of this VISCUBE architecture. In this example, we show how to calculate displacement for multiple points. The goal of the displacement calculation is to support image registration and optical-flow estimation also. In our UAV reconnaissance project, we assume high image sampling rate (500–1,000 FPS); hence, the displacement and the rotation, introduced by the ego-motion of the on-board camera, is supposed to be small.

Our strategy is to calculate the displacement of two consecutive images in 64 feature points. The displacements in the different positions may be different, due to (1) 3D structures in the ground, (2) the rotation of the camera, and (3) false measurement results. The displacement calculation starts with the identification of those points, where the displacement is calculated. These points are called feature points. Here we show two possibilities for the identification of the feature points, and the displacement calculation.

## 4.1 Identification of the Feature Points

We can define feature point as a point of low self-similarity [13]. A point is considered to be feature point, if the contrast changes are high in different orientations, and its local interconnection topology is unique in small neighborhood. This uniqueness is important, because during the displacement calculation, we calculate matches around these feature points, and if there are points with similar topologies in the neighborhood, the displacement calculation will not be unambiguous. Typical situation is, when we are looking for matching on a straight line. In this situation, multiple matching positions can be found. This is called aperture problem (Fig. 24).

However, calculation of the uniqueness, which is practically an autocorrelation, is very expensive, hence it is not used. Rather than that special locations (such as local extrema or corners) are selected as feature points, with the assumption that the image is not periodical; therefore there are no similar locations in the neighborhood.

The mixed-signal layer of the VISCUBE chips is designed to identify the local extrema in different scales, which is considered to be the primary method of selecting feature points. However, we show a feature point selection method implemented on the digital processor array as an alternative option as well. In both cases, the feature point identification is done in $160 \times 120$ (Scale 1) resolution.

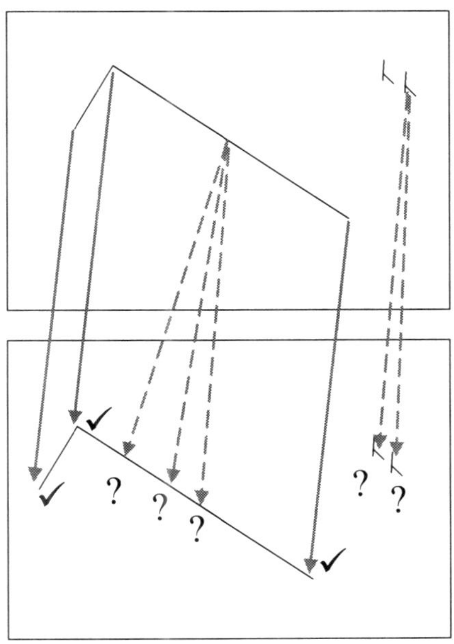


**Fig. 24** Aperture problem in the displacement calculation. Two consecutive images with shifted pattern. The *green solid arrow* shows those locations, where the displacement can be calculated unambiguously, while the *dotted red arrow* shows a location, where the displacement is ambiguous

### 4.1.1 Local Extrema Identification Using the Mixed-Signal Processor Layer

As we have seen, the mixed-signal layer is prepared to calculate DOG operators, and identify the local extrema positions on them. The variance of the Gaussians can be tuned; hence local extrema can be sought in different frequency components. To be able to reduce the number of the local extrema points, the minimal value of the maximum locations, and the maximal value of the minimum locations can be set. This avoids picking up the extrema coming from local noises in flat locations, which cannot be considered as feature points. The extrema calculation is done in $3 \times 3$ neighborhood. The DOG operation takes about 200 μs (sigma dependent), while the local minima and maxima calculation takes 25–25 μs

### 4.1.2 Harris Corner Detection on the Digital Processor Array

One of the most frequently used standard methods for identifying the feature points is the Harris corner detection [9]. Harris corner detector calculates the vertical and horizontal spatial derivatives of the image and puts them to the socalled Harris matrix (2).

$$\mathbf{A} = \begin{bmatrix} \langle I_x^2 \rangle & \langle I_x I_y \rangle \\ \langle I_x I_y \rangle & \langle I_y^2 \rangle \end{bmatrix}, \tag{2}$$

where $I_x$ and $I_y$ are the spatial derivatives and the angle brackets denote averaging around the certain point (e.g., Gaussian averaging).

The special features of this matrix [9] are that

- If the eigenvalues are small, the area is flat;
- One large eigenvalue indicates an edge;
- Two large eigenvalues indicate intersection of edges (corners, feature points).

However, the calculation of the eigenvalues is computationally expensive due to the square root. To be able to reduce the computational burden, it is enough to calculate (3) [10]

$$M_c = \det(\mathbf{A}) - \kappa \, trace^2(\mathbf{A}), \tag{3}$$

where $\kappa$ is the sensitivity parameter. The value of $\kappa$ has to be determined empirically, and in the literature values in the range 0.04–0.15 have been reported as feasible. The good feature points are indicated by the large values of $M_c$ [10].

In the digital processor array, the $160 \times 120$ image is distributed among the processors, which means that each receives a $20 \times 15$ sized image. The image is smoothened first with an averaging filter and then, the spatial derivatives are calculated applying Sobel operators. After that, $M_c$ is calculated, and in each $20 \times 15$ segment, the location of the largest $M_c$ is selected to be the feature point. In this way the feature points are distributed over the image, and are not concentrated to the vertexes of a few high contrast objects.

The feature point search on an entire Scale 1 image takes 30,005 clock cycles, which is about 300 μs assuming 100 MHz clock frequency.

### 4.2 Displacement Calculation

The displacement calculation is done in a way that a certain location is cut out from the currently captured image around the feature point, and the best matching position is sought on the previous frame.

In our case an $11 \times 11$ window around the feature point is cut out from the original $320 \times 240$ image. In the first method, we assume that the displacement is not larger than 4 pixels, hence we are seeking the best matching position in a $9 \times 9$ window.

The matching is calculated by using $L_2$ *norm* (4)

$$L_2 norm = \sqrt{\sum_{x \in N} [I_n(x+h) - I_{n-1}(x)]^2}, \tag{4}$$

where $I_n$ are the intensity values in the current frame and $I_{n-1}$ are the intensity values of the previous frame.

The best matching position is in that location, where the $L_2$ *norm* is minimal. (Certainly, the square root is not calculated.) To further reduce the computational needs, we do not calculate the $L_2$ *norm* in each location. There are several methods to reduce the number of the points, where the $L_2$ *norm* is actually calculated. These strategies assume that there is a monotonic decrease in the $L_2$ *norm* as we are proceeding toward the matching point.

In our case, first we calculate the $L_2$ norm in nine positions as it is shown in Fig. 25. Then, we perform eight new calculations around the minimal value, and finally select the best one. This means that we perform 17 $L_2$ *norm* calculation altogether. This calculation takes 102,791 clock cycles, which is roughly 1 ms.

In those cases, when the feature point is not intense (e.g. the entire $20 \times 15$ window was from a flat area) we reject the calculated displacement vector. These vectors tend to be false and misleading with a high probability. Figure 26 shows an example of the identified displacement vectors.

If the displacement is larger than 4 pixels we have to use multi-scale (pyramid) method (Fig. 27). This means that we make the displacement calculation on different scales. It is calculated first on Scale 2. When the minimum of the $L_2$ *norm* is found, we start out the displacement calculation on Scale 1 from that point, which was the calculated displacement (result) on Scale 2. After that, it is repeated for Scale 0 also in the same way. To calculate proper downscaling the diffusion operator of the mixed-signal layer is used. In this way the largest identifiable displacement goes up to 16 pixels. Figure 28 shows a simulation example for the displacement calculation, where the pyramid method was used. As it can be seen both small and large displacements were identified.

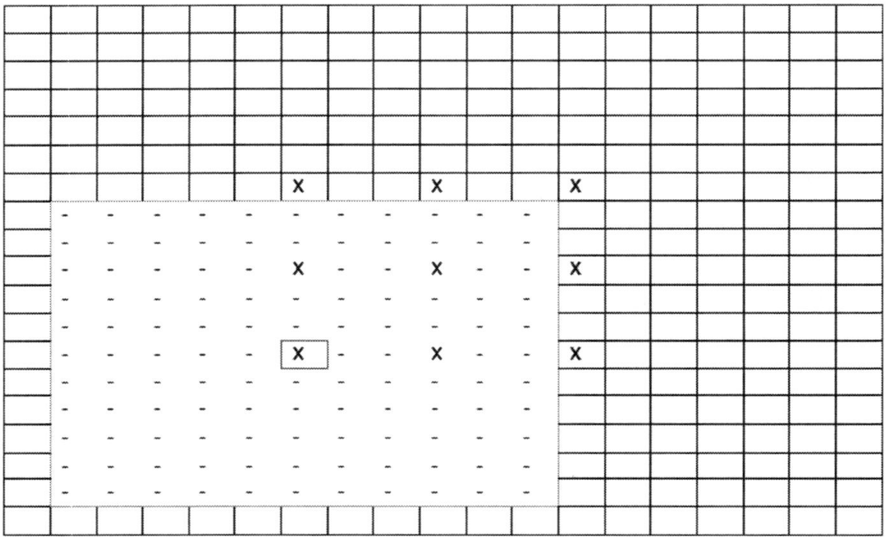

**Fig. 25** Matching position searching strategy. The $11 \times 11$-sized window (indicated with *dots*) is cut out from the current image around the feature point. The large $19 \times 19$ window is cut out from the previous frame from the same location. The *x* shows those nine positions, where $L_2$ norm is calculated in the first step

**Fig. 26** Result of the displacement calculation

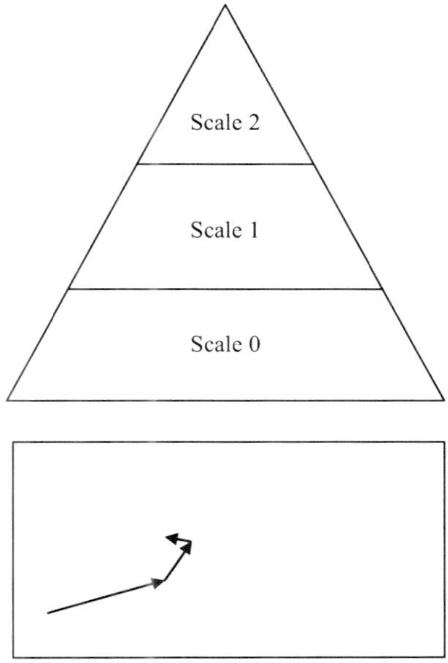


**Fig. 27** Pyramid layers, and a possible sequential displacement calculation result

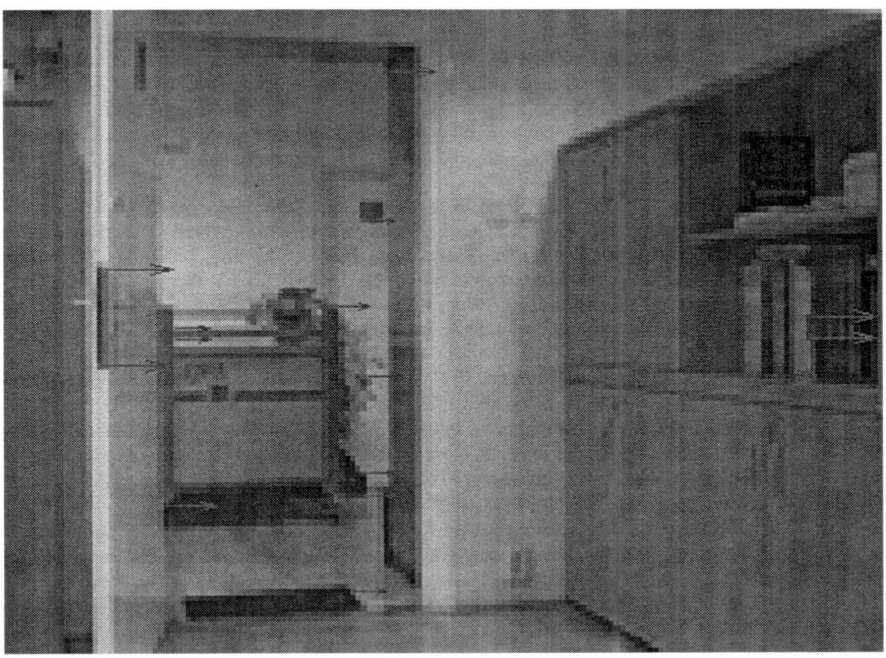

**Fig. 28** Result of the displacement calculation on an image, where the displacements are larger than 4 pixels, hence pyramid methods applied

## 5 Conclusion

Advanced focal-plane sensor-processor design, based on vertical integration is introduced. The unique features of the design are that it incorporates heterogeneous processor arrays, one high-resolution fine-grain processor array operating in the analog/mixed-signal domain, and one low-resolution digital processor array. Algorithmically, the senor-processor architecture is designed to perform multi-scale, multi-fovea processing. The targeted application is UAV visual navigation and reconnaissance.

## References

1. P. Garrou, C. Bower, P. Ramm, Handbook of 3D Integration: Technology and Applications of 3D Integrated Circuits, Wiley, ISBN: 978–3–527–32034–9, 2008
2. G. Deptuch, Vertical Integration of Integrated Circuits and Pixel Detectors, Vertex 2008Workshop, July 28–August 1, 2008, Uto island, Sweden
3. M.D. Enriquez, M.A. Blessinger, J.V. Groppe, T.M. Sudol, J. Battaglia, J. Passe, M. Stern, B.M. Onat, Performance of High Resolution Visible-InGaAs Imager for Day/Night Vision, Proc. of SPIE, Vol 6940, 694000, 2008
4. P. Földesy, R. Carmona-Galan, Á. Zarándy, C. Rekeczky, A. Rodríguez-Vázquez, T. Roska, 3D multi-layer vision architecture for surveillance and reconnaissance applications, ECCTD-2009, Antalya, Turkey
5. A. Zarandy, D. Fekete, P. Foldesy, G. Soos, C. Rekeczky, Displacement calculation algorithm on a heterogeneous multi-layer cellular sensor processor array, Proc. of the CNNA-2010, pp 171–176, Berkeley, CA, USA
6. A. Zarandy, C. Rekeczky, 2D operators on topographic and non-topographic architectures – implementation, efficiency analysis, and architecture selection methodology, Int. J. Circ. Theor. Appl., 2010
7. MITLL Low-Power FDSOI CMOS Process Design Guide, 2006
8. P. Földesy, Á. Zarándy, Cs. Rekeczky, T. Roska, Configurable 3D integrated focal-plane sensor-processor array architecture, Int. J. Circ. Theor. Appl. (CTA), 573–588, 2008
9. C. Harris M. Stephens, A combined corner and edge detector (PDF). Proc. of the 4th Alvey Vision Conference. pp 147–151. 1988
10. C. Harris, Geometry from visual motion In A. Blake A. Yuille, Active Vision, MIT, Cambridge, MA, 1992
11. S. Spiesshoefer, Z. Rahman, G. Vangara, S. Polamreddy, S. Burkett, L. Schaper, Process integration for through-silicon vias, J. Vacuum. Sci. Technol., A 23(4), 824–829, 2005
12. A. Rodríguez-Vázquez, R. Domínguez-Castro, F. Jiménez-Garrido, S. Morillas, A. García, C. Utrera, M. Dolores Pardo, J. Listan, R. Romay, A CMOS vision system on-chip with multi-core, cellular sensory-processing front-end In C. Baatar, W. Porod T. Roska Cellular Nanoscale Sensory Wave Computing, ISBN: 978–1–4419–1010–3, 2009
13. T. Kanade B.D. Lucas, An Iterative Image Registration Technique with an Application to Stereo Vision; Computer Science Department Carnegie-Mellon University Pittsburgh, Pennsylvania 15213; From Proc. of Imaging Understanding Workshop, pp 121–130, 1981
14. P. Foldesy, A. Zarandy, Cs. Rekeczky, T. Roska, Digital implementation of cellular sensor-computers, Int. J. Circ Theor Appl (CTA), 34(4), 409–428, 2006

# The Nonlinear Memristive Grid

**Feijun Jiang and Bertram E. Shi**

**Abstract** Nonlinear resistive grids have been proposed previously for image smoothing that preserves discontinuities. The recent development of nonlinear memristors using nanotechnology has opened the possibility for expanding the capabilities of such circuit networks by replacing the nonlinear resistors with memristors. We demonstrate here that replacing the connections between nodes in a nonlinear resistive grid with memristors yields a network that performs a similar discontinuity-preserving image smoothing, but with significant functional advantages. In particular, edges extracted from the outputs of the nonlinear memristive grid more closely match the results of human segmentations.

## 1 Introduction

Nonlinear resistive grids (RGs) have been proposed to smooth images while preserving edges [1–5]. These grids are attractive in part because they can be implemented in VLSI leading to high-speed image processing circuits [6–8]. The structure of the nonlinear RGs is shown in Fig. 1a. This network consists of a set of nodes arranged in a regular rectangular array. Each node is associated with a pixel in the image that is to be filtered. Each node is connected through a series resistor to a voltage source referenced to ground. By convention, we refer to these resistors as vertical resistors. The voltage across this source is proportional to the intensity of the input image at the corresponding pixel. Each node is connected to its four neighboring nodes through a nonlinear resistor, which we refer to as horizontal resistors. The voltage between each node and ground is proportional to the intensity of the output image at the corresponding pixel.

B.E. Shi (✉)
Department of Electronic and Computer Engineering, Hong Kong University of Science and Technology, Clear Water Bay, Kowloon, Hong Kong Sar, Peoples Republic of China
e-mail: eebert@ust.hk

Á. Zarándy (ed.), *Focal-Plane Sensor-Processor Chips*,
DOI 10.1007/978-1-4419-6475-5_9, 

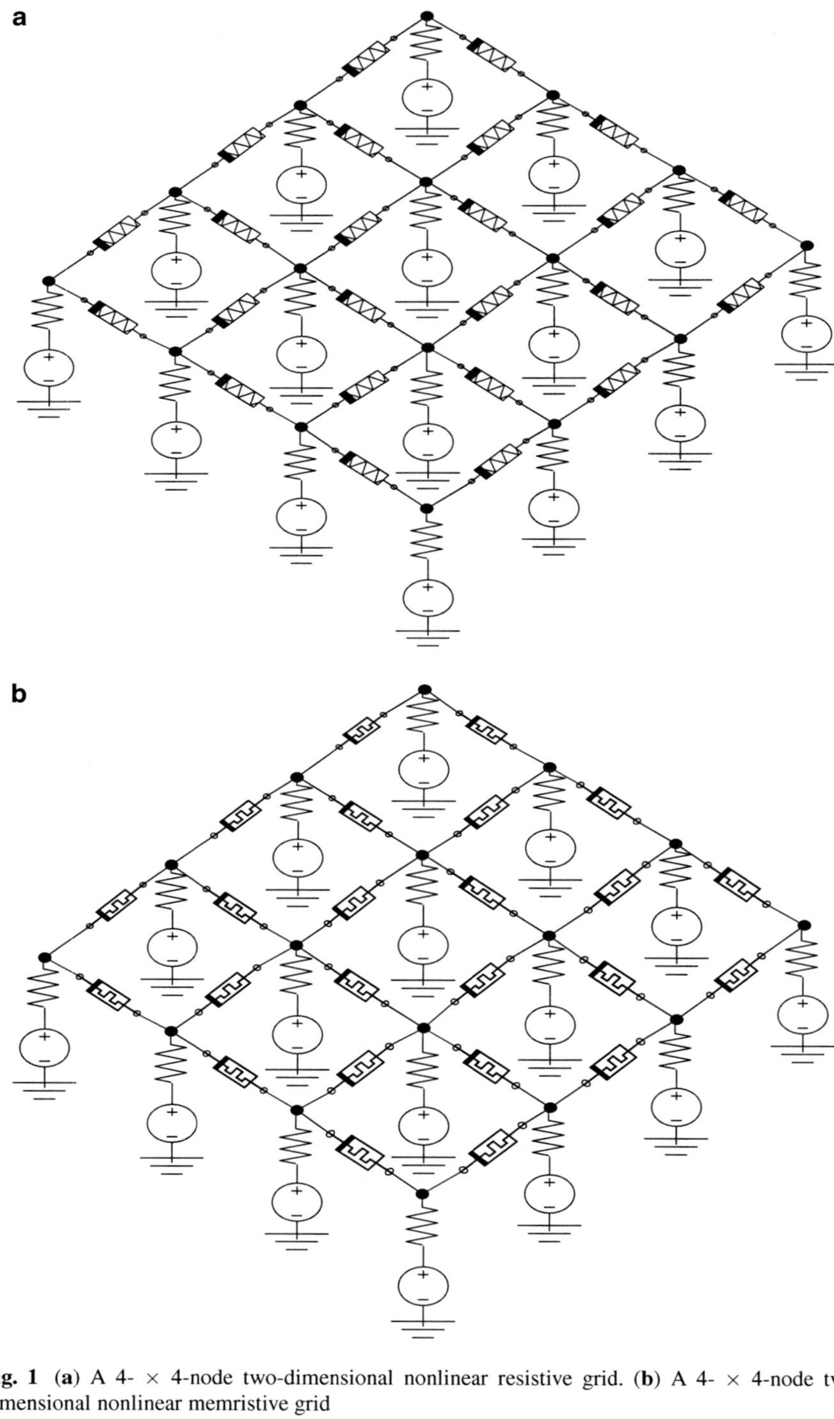

**Fig. 1** (**a**) A 4- × 4-node two-dimensional nonlinear resistive grid. (**b**) A 4- × 4-node two-dimensional nonlinear memristive grid

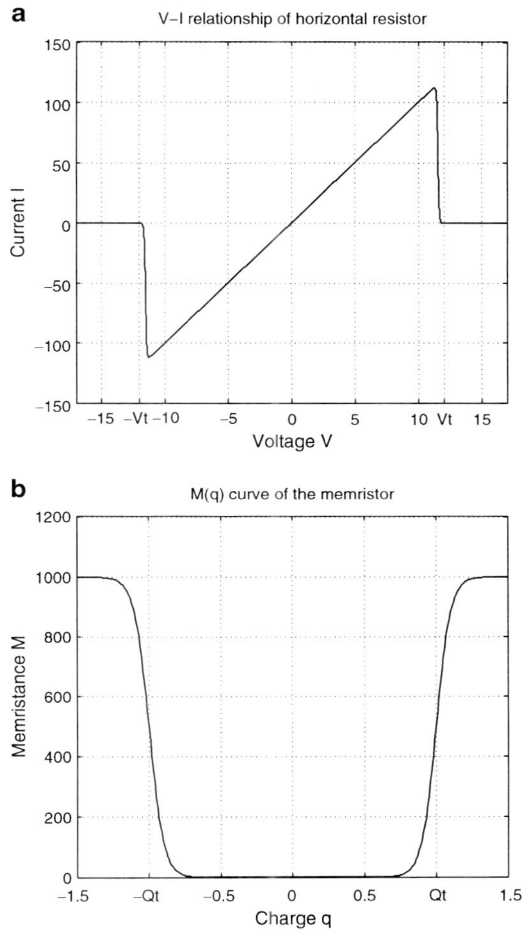


**Fig. 2** (**a**) The V–I relationship of the nonlinear resistor. When the absolute value of $V$ is smaller than $V_t$, the resistor appears to be a linear resistor with resistance $R_{on}$. Otherwise, the resistor conducts the same current as a linear resistor with resistance $R_{off}$. (**b**) The incremental memristance $M(q)$ of the memristor. When the absolute value of the charge q is smaller than $Q_t$, the memristance is $M_{on} = 0.1$. Otherwise, the memristance is $M_{off} = 1{,}000$

The typical V–I relationship of the nonlinear resistors connecting adjacent nodes is shown in Fig. 2a. If the voltage across this nonlinear resistor is larger than $V_t$, the resistor passes the same current as a linear resistor with large resistance, $R_{off}$. For smaller voltages, the resistor passes the same current as a linear resistor with a much smaller resistance, $R_{on}$. The operation of this network has been referred to as biased anisotropic diffusion [5]. Areas where the intensity changes slowly will be smoothed. However, when the voltage difference between two adjacent nodes is large enough (e.g., in areas where the image intensity changes quickly), the nonlinear resistor will pass little current, and the diffusion will be blocked, preserving the discontinuity.

Nonlinear RGs settle to points that are local minima of the global co-content, which can be thought of as a cost function [9, 10]. The global co-content of this network is nonconvex, which means there may be many local minima and makes it difficult to find the global minimum. In addition, since the output voltages of the network are not unique, they depend critically on the initial conditions of the nodes, if the effect of parasitic capacitances is included.

As a way to find a unique solution that is hopefully close to the global minimum, Blake and Zisserman proposed the graduated nonconvexity (GNC) algorithm [3], which finds a solution to a network with a desired threshold voltage $V_t$ by solving a sequence of grids that approach the desired grids. The algorithm starts by solving grids with a value of $V_t$ chosen large enough so that all of the nonlinear resistors appear to be linear resistors with resistance $R_{on}$. In this case, the cost function is convex, and the solution is unique. This solution corresponds to a heavily smoothed version of the input image. The value of $V_t$ is then decreased slowly. As $V_t$ decreases, edges begin to emerge. In the limit where $V_t = 0$, all adjacent nodes are essentially disconnected (assuming that $R_{off}$ is much larger than the resistance connecting the input and output nodes), and the input and output voltages are almost identical.

In 1971, Chua described the memristor, a two-terminal circuit element, which implements a constraint between the magnetic flux across it and the charge that has passed through it [11]. In 2008, scientists at Hewlett Packard successfully fabricated a memristor using nanotechnology [12]. The relationship between voltage and current of a memristor is similar to that of a resistor, except that the constant of proportionality, called the incremental memristance, is a function of charge that has flowed through the memristor

$$v = M(q) \times i \tag{1}$$

Here, we describe a nonlinear circuit where we replace the horizontal nonlinear resistors of the RG with nonlinear memristors. To differentiate it from a purely RG, we refer to the new grids as a memristive grid (MG). MG also contains standard linear resistors (the vertical resistors). Nonetheless, the term memristive grid is still accurate, since linear resistors are special cases of the memristors where the incremental memristance is constant.

If the incremental memristance of the memristors connecting adjacent nodes is chosen appropriately, MG can perform a similar discontinuity preserving image smoothing as the nonlinear RG. We compare the performance of the MG with that of the RG. We find that the segmentation results based on the output of MG are qualitatively better, and quantitatively closer to human level segmentation.

## 2 The Memristive Grid

MG is a modification of the nonlinear RG. We replace the nonlinear horizontal resistors of Fig. 1a with nonlinear memristors, as shown in the Fig. 1b. Similar to the RG, the voltages across the voltage sources are proportional to the gray scale intensities at corresponding pixels in the input image. The voltages between the nodes and ground are proportional to the gray scale intensities of the pixels in the output image.

The memristor is a two-terminal circuit element like the more commonly known resistor, inductor and capacitor [11]. It implements a relation between flux $\varphi$ and charge $q$: $g(\phi, q) = 0$. The memristor is said to be charge controlled if the flux is a single-valued function of charge

$$\varphi = \varphi(q). \tag{2}$$

Differentiating both sides with respect to time, we find

$$v = \frac{\mathrm{d}\varphi}{\mathrm{d}t} = \frac{\mathrm{d}\varphi(q)}{\mathrm{d}q} \times \frac{\mathrm{d}q}{\mathrm{d}t} = M(q) \times i, \tag{3}$$

where

$$M(q) \equiv \frac{\mathrm{d}\varphi(q)}{\mathrm{d}q} \tag{4}$$

is defined as the incremental memristance.

To perform edge preserving image smoothing, we choose the incremental memristance of the memristors connecting adjacent nodes to be

$$M(q) = M_{\mathrm{on}} + M_{\mathrm{off}} \left( \frac{1}{1 + \exp(20(Q_{\mathrm{t}} - q))} + \frac{1}{1 + \exp(20(Q_{\mathrm{t}} + q))} \right) \tag{5}$$

Figure 2b plots the incremental memristance as function of the charge. The incremental memristance makes a smooth transition between two values $M_{\mathrm{on}}$ and $M_{\mathrm{off}}$. When the absolute value of the charge is smaller than a threshold $Q_{\mathrm{t}}$, the memristance is $M_{\mathrm{on}}$. If the absolute value of the charge is larger than a threshold $Q_{\mathrm{t}}$, the memristance is $M_{\mathrm{off}}$, which is large so that the memristor approximates an open circuit. This memristor can be constructed by connecting two memristors similar to those described in [12] in series and with opposite polarity.

We assume that the memristors are initialized with zero charge, so that their memristance is $M_{\mathrm{on}}$. At this point, the memristors appear to be linear resistors with resistance $M_{\mathrm{on}}$, which is chosen to be smaller than the vertical resistance, so that the output node voltages correspond to a smoothed version of the image. This removes high frequency noise in the image.

The current flowing through a memristor determines the rate of change of the charge. The voltage across (and therefore the current through) a horizontal memristor connecting two nodes spanning an intensity discontinuity will be large in comparison with the voltage across (current through) a memristor connecting nodes within a region where the intensity changes smoothly. Thus, the absolute value of charge in horizontal memristors spanning intensity discontinuities will reach the threshold $Q_{\mathrm{t}}$ faster, causing them to shut off first. When this happens, the discontinuities in the input image appear in the output image.

As time goes on, the charges in more and more of the memristors will increase in absolute value to the point that they eventually cross threshold. Eventually, the output image will be very close to the input image, much in the same way that the output of the RG with $V_{\mathrm{t}} = 0$ is close to the input.

Thus, the qualitative operation of the MG over time is somewhat similar to the operation of the RG implementing the GNC algorithm, where the threshold decreases from a large value to a small value. In our simulations, these two grids indeed show similar qualitative behavior. Both initially smooth the image and remove the noise. Later, sharp edges begin to appear.

However, their operating principles are very different, which leads to important functional differences in the image processing performed by the grids. In RG, the characteristics of the nonlinear resistor are controlled uniformly over the entire grid. Thus, at each point in time all of the horizontal resistors have the same $V_{\mathrm{t}}$ value independent of the input image. However, for the MG, the charge, which determines the memristance, is a function of the past history of the grids, which in turn depends upon the input image. As we see below, this qualitative difference in their operation leads to differences in their function.

## 3 Simulating RG and MG

In this section, we will introduce the mathematical models for both the RG and the MG. We also describe how we simulate these grids to generate the experimental results described in the next section.

### 3.1 *The Resistive Grid*

Assuming a unit capacitor attached between each node and ground, each node of the RG evolves according to

$$\begin{aligned}\frac{\mathrm{d}u_{i,j}}{\mathrm{d}t} &= I(u_{i,j-1}-u_{i,j})+I(u_{i,j+1}-u_{i,j})\\ &\quad +I(u_{i-1,j}-u_{i,j})+I(u_{i+1,j}-u_{i,j})\\ &\quad +\frac{d_{i,j}-u_{i,j}}{R_{\mathrm{in}}},\end{aligned} \tag{6}$$

where $u_{i,j}$ represents the voltage at each node, $d_{i,j}$ represents the voltage of the input source, $R_{\mathrm{in}}$ is the resistance of the vertical resistor and the time $t$ is measured in seconds. The $v$-$i$ relationship of the horizontal resistor is given by

$$I(\Delta u) = \frac{\Delta u}{R_{\mathrm{on}}+R_{\mathrm{off}}\left(\frac{1}{1+\exp(20(V_{\mathrm{t}}-\Delta u))}+\frac{1}{1+\exp(20(V_{\mathrm{t}}+\Delta u))}\right)}, \tag{7}$$

where $V_t$ denotes the threshold voltage. Here, we use 8-bit grayscale images, so the input voltages $d_{i,j}$ assume integer values between 0 and 255. For an $M \times N$ pixel image, we obtain a set of $M \times N$ coupled ordinary differential equations (ODEs).

We assume the zero-flux boundary conditions, corresponding to the case where the nodes at the edges of the array are left unconnected, and so are only connected to three horizontal resistors (two for the nodes at the corners).

We are interested in the steady-state voltages at the output notes for different values of the threshold voltage. To find this, we simulate the system of ODEs using the MATLAB function "ode45" with the default parameters. We allow the system of ODEs to evolve for 5 s, which we have found to be long enough so that the system is essentially in steady state, and take the voltages at the last time step of the simulation.

Our simulations use the parameters $R_{in} = 2$, $R_{on} = 0.1$ and $R_{off} = 1000$. If all horizontal resistors have value $R_{on}$, the corresponding space constant measuring the smoothing neighborhood is 4.47 pixels [13]. On the other end, if all horizontal resistors have value $R_{off}$, the corresponding space constant is 0.0447. Very little smoothing is done, so the output image is almost identical to the input image.

For each input image, we perform two sets of simulations. For each set, we find a set of steady-state solutions of the grids for a sequence of threshold voltages, $V_t$. The first value of $V_t$ is chosen so that the steady-state solution is unique. For each subsequent value of $V_t$, we use the steady state solution from the previous value of $V_t$ as the initial condition of the grids. The two sets of simulations differ according to the way in which the sequence of threshold voltages $V_t$ progresses.

In the first set, the sequence of threshold voltages *decreases* from an initial large value (255) to a final small value (0.5). This implements the GNC algorithm. We denote this set by RG–, where the "–" indicates the threshold voltages decrease. For the initial threshold value, all of the horizontal resistors will be operating in the region, where their resistance is well approximated by $R_{on}$. In this case, the global co-content is convex and the solution is unique. The next threshold in the sequence is chosen to be the largest integer smaller than one seventh of the largest absolute intensity difference between neighboring pixels in the input image. For the images we have tested, at this threshold value, all horizontal resistors are still operating in the $R_{on}$ region. Since the output image is a smoothed version of the input intensity, the intensity differences between the neighboring pixels in the output image are smaller than they are in the input image. For the remainder of the sequence, the threshold voltage decreases by 1 until it drops to 1. After that we add a simulation at a threshold of 0.5.

In the second set, which we indicate by RG+, the sequence of threshold voltages *increases* from a small initial value of 0.5 up to a large final value. In the next step, we use a threshold voltage of 1. Afterward, the threshold voltage sequence increases by 10, until it reaches the largest absolute difference between neighboring pixels in the input image. At this point, all of the resistors are operating in the $R_{on}$ region.

## 3.2 The Memristive Grid

MG evolves according to a system of ODEs with constraints, i.e., a system of differential-algebraic equations (DAEs). We denote the charge in the memristor connecting nodes $(i, j)$ and $(i+1, j)$ by $p_{i,j}$ and the charge in the memristor connecting nodes $(i, j)$ and $(i, j+1)$ by $q_{i,j}$. The charges in the memristors are initially zero.

The temporal derivatives of the charges are equal to the instantaneous currents through the memristors,

$$\frac{\mathrm{d}p_{i,j}}{\mathrm{d}t} = \frac{u_{i,j} - u_{i+1,j}}{M(p_{i,j})} \tag{8}$$

$$\frac{\mathrm{d}q_{i,j}}{\mathrm{d}t} = \frac{u_{i,j} - u_{i,j+1}}{M(q_{i,j})}, \tag{9}$$

where $M(\cdot)$ is the incremental memristance given in (5) and the time $t$ is measured in seconds. The currents depend upon the nodal voltages $u_{i,j}$, which are determined by a set of linear equations whose coefficients depend upon the charges in the memristors

$$\frac{d_{i,j} - u_{i,j}}{R_{\text{in}}} + \frac{u_{i-1,j} - u_{i,j}}{M(p_{i-1,j})} + \frac{u_{i+1,j} - u_{i,j}}{M(p_{i,j})} + \frac{u_{i,j-1} - u_{i,j}}{M(q_{i,j-1})} + \frac{u_{i,j+1} - u_{i,j}}{M(q_{i,j})} = 0, \tag{10}$$

where $d_{i,j}$ denotes the voltage across the voltage source and $R_{\text{in}}$ is the vertical resistance. Nodes at the boundaries of the array are left disconnected, corresponding to zero-flux boundary conditions.

To solve this set of DAEs, we use the MATLAB function "ode45" with the default parameters. Given a set of values of $p_{i,j}$ and $q_{i,j}$, we can find the corresponding time derivatives by first solving the set of $M \times N$ linear equations given by (10) for the nodal voltages $u_{i,j}$, and then computing the derivatives according to (8) and (9). This gives a set of time trajectories for $p_{i,j}$ and $q_{i,j}$, which we can convert into a time-varying output image by solving for the values of $u_{i,j}$ at each time instant using (10).

Our simulations use the parameters $R_{\text{in}} = 2$, $M_{on} = 0.1$, $M_{\text{off}} = 1{,}000$ and $Q_t = 1$. These were chosen for consistency with the simulations of the RG. The simulation time is 0.7 s, which we have found is long enough that the final set of output nodal voltages are nearly equal to the input voltage.

## 4 Experimental Results

In this section, we compare the outputs of the RG for the two sequences of threshold voltages with the output of the MG over time. We start with a visual comparison of filtered images produced by the networks to give better intuition about the processing being performed by the networks. We then give a more quantitative comparison by comparing how consistent the edges extracted from these networks are with the results of human segmentation using the methodology proposed by Martin et al. [14, 15].

## 4.1 Edge Preserving Image Filtering

We simulated the three networks for the three images shown in Fig. 3. Two of the images were artificially generated images of a triangle corrupted with zero mean additive Gaussian noise. The original triangle image was the same in both images, but the amount of noise differed (20% and 28.6% of the signal range in the original noise free image). The third image was a natural image taken from the database of Martin et al. [14, 15].

Figures 4–6 show snapshots of the nodal voltage outputs of three grids at different points along the sequence/transient for the three different input images. Both the MG and RG- start from smoothed versions of the image. As time goes on or the threshold voltage decreases, on the one hand, more and more sharp discontinuities appear in the image. Eventually, the output images are nearly identical to the input images. On the other hand, the output images for RG+ start out close to the input images, and sharp discontinuities in the image disappear over time. Eventually, all discontinuities are smoothed, and we obtain the same smoothed image that MG and RG- started with.

We see that noise tends to be preserved in the output RG+, whereas the MG and RG− are less affected by the noise, since they remove it at the initial stage. Comparing MG and RG−, it appears that the locations of the sharp discontinuities in the output image of better match the corresponding locations in the input image for MG than for RG−. To better quantify these observations, we study edge maps extracted from the images in the remainder of this section.

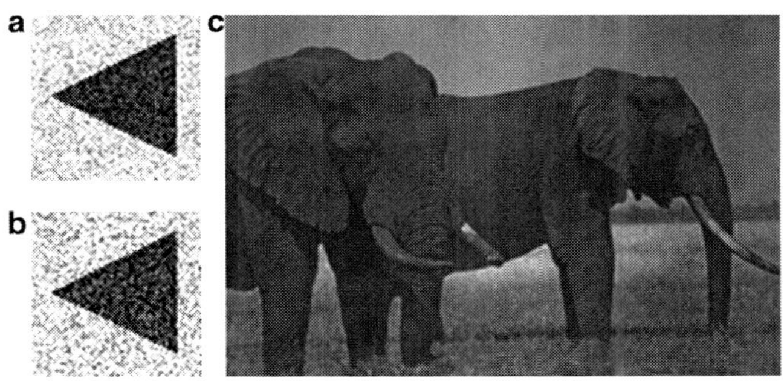


**Fig. 3** (**a**) The input imaage for the simulation of Fig. 4. (**b**) The input image for the simulationof Fig. 5. (**c**) The input image for Fig. 6. The images in (**a**) and (**b**) differ by the amount of noise added to the image. The standard deviation of the noise in (**a**)/(**b**) were 20%/28.6% of the signal range in the original noise-free image

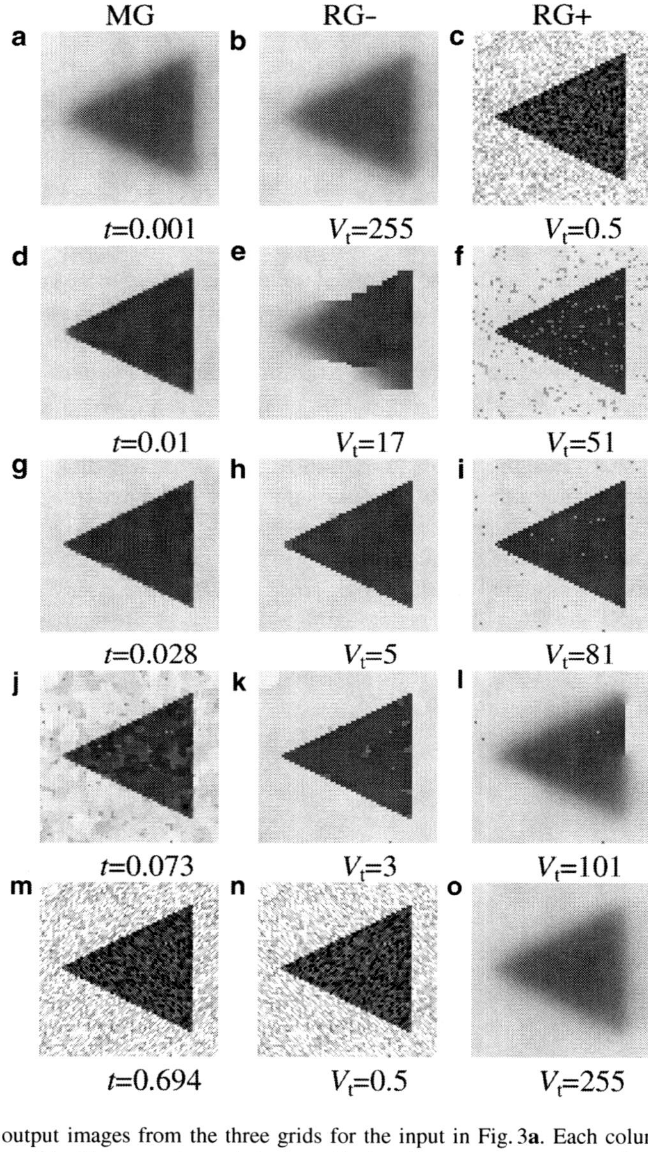


**Fig. 4** The output images from the three grids for the input in Fig. 3**a**. Each column shows the output of one grid with successive points along the transient or sequence organized from *top* to *bottom*. The *first* and *last rows* correspond to the starting and ending points. The *middle row* corresponds to the point where the edge map extracted from the output has the highest edge measure. The *second* and *fourth rows* correspond to the points at which the extracted edge maps have either 60% recall (**d, e** and **l**) or 60% precision (**j, k** and **f**)

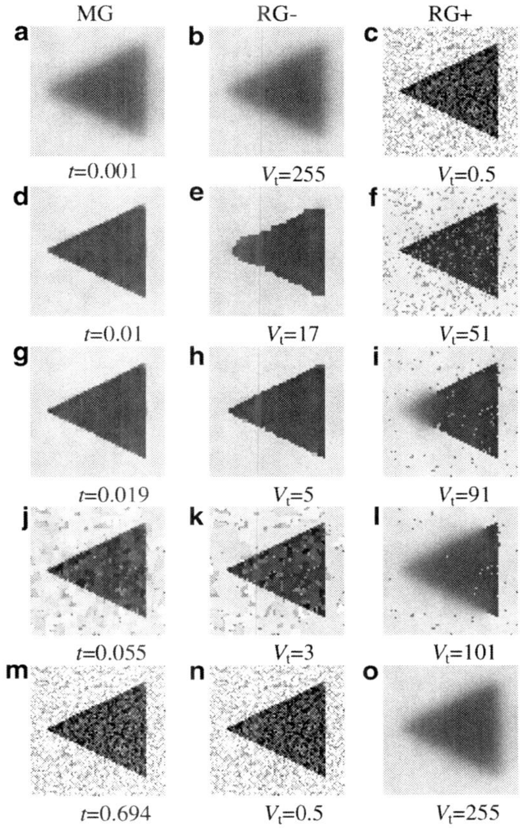


**Fig. 5** The output images from three grids for the input in Fig. 3**b**. The figure is organized similar to Fig. 4

## 4.2 Segmentation Results on Individual Images

Since both the RG and the MG perform discontinuity preserving image smoothing, a reasonable way to compare their performance is to compare the edges detected by the networks with the results of human performance. In the remainder of this section, we describe the results of such a comparison.

We detect edge pixels in the image depending upon the condition of the memristors or the resistors connecting it to its right and bottom neighbors. For RG− and RG+, if the absolute value of the voltage difference across either of the resistors is greater than or equal to $V_t$, then we define the pixel to be an edge pixel. For the MG, if the absolute value of the charge in either of these memristors exceeds $Q_t$, then we define the pixel to be an edge pixel.

Each network results in a sequence of edge images. For the MG, there is an edge map for each point in time. For RG− and RG+ there is an edge image for every threshold value. For the MG, the edge map is initially empty since the memristors

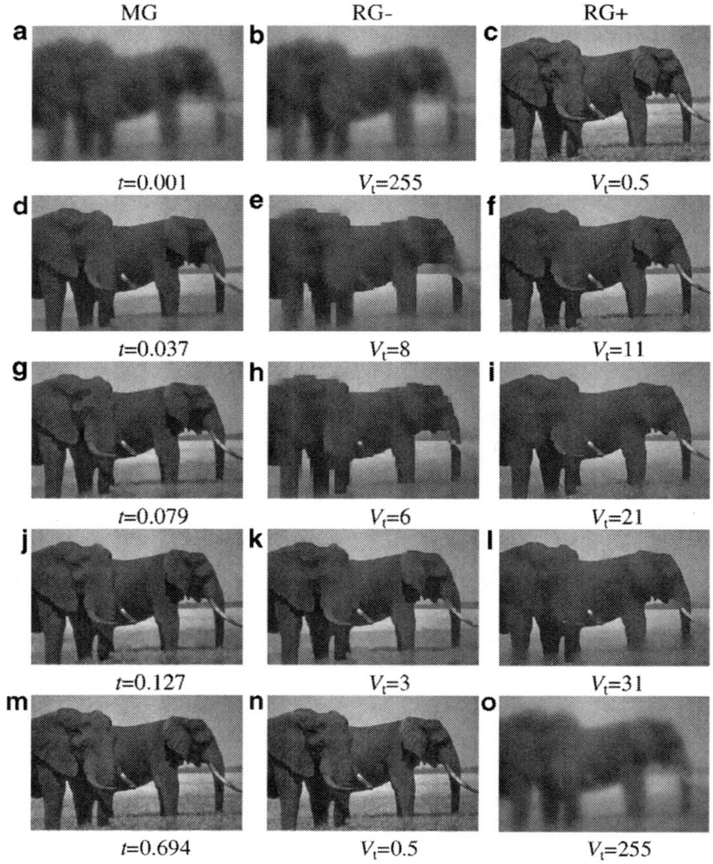


**Fig. 6** The output images from three grids for the input in Fig. 3**c**. The figure is organized similar to Fig. 4

are initialized with zero charge, and edges emerge over time. Similarly, on the one hand, for RG−, the threshold is large so that the edge map is initially empty. Edges emerge as the threshold is decreased. On the other hand, for RG+, the sequence starts with zero threshold, where most pixels are labeled as edges. Edges disappear as the threshold increases. Intuitively, we expect that for earlier points in time for the MG or larger thresholds for the RG, fewer edges are extracted but these are the more "important" or "reliable" edges in the scene.

To choose which of the edge maps to compare, we use the comparison methodology described by Martin et al. [14, 15]. For each image, we have ground truth edges obtained by human segmentation. We used the code provided with the dataset to benchmark the performance of the three networks. From each sequence of edge images completed by a grid, the code first performs edge thinning and then computes precision and recall values. Precision ($P$) is defined as the fraction of the detected

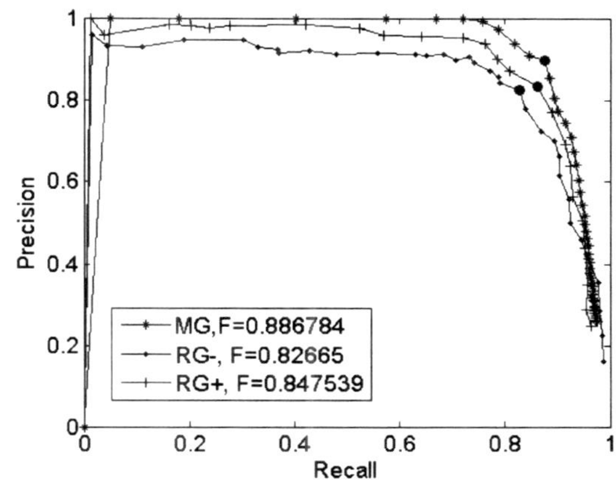


**Fig. 7** The Precision-Recall curves generated by the edge maps for the input image in Fig. 3**c**. Points corresponding to edge maps adjacent in time or in sequence are connected by *lines*. Points with the largest $F$-measure are shown with large filled *circles*

edge pixels that correspond to edges in the ground truth. Recall ($R$) is defined as the fraction of the edge pixels in ground truth that are detected.

Applying the reasoning above, for the MG, we expect recall to increase with time, but precision to decrease. Similarly, for the RG, we expect recall to decrease as threshold increases, but precision to increase. Thus, there is a trade-off between precision and recall.

Figure 7 shows the relationship between the precision and recall for the three networks operating on the input shown in Fig. 3c. This figure clearly illustrates the trade-off between precision and recall. For the MG, points corresponding to edge maps earlier in time lie in the upper left hand corner (low recall, but high precision). Points corresponding to edge maps later in time lie in the lower right hand corner (high recall, but low precision). For the RG, points in the upper left correspond to high threshold values. Points in the lower right correspond to low threshold values.

Points corresponding to desirable edge maps lie in the upper right hand corner of the plot (high precision and high recall). Following Martin et al., we use the $F$-measure, defined as $F = 2PR/(P+R)$ to select the best point along the curve. The $F$-measure increases the closer we get to the upper right hand corner. For each set of edge maps, we choose the one with the largest $F$-measure. In Fig. 7, points with the largest $F$-measure are shown with large filled circles. The $F$-measure also gives us a way to compare the quality of the "best" edge maps extracted by the networks quantitatively.

To illustrate the qualitative differences between the operations of the three networks, Fig. 8 shows the edge maps with the largest $F$-measure obtained from the input images shown in Fig. 3a, b. These edge maps were computed based upon the network outputs shown in Figs. 4g–i and 5g–i. The edges extracted by the MG are clearly the closest to the true edges, which are defined by the noise-free image. The edges extracted by the two RGs differ qualitatively from each other. For RG-, the output image is initially smooth. This effectively removes noises, but smoothes over the edge positions. Interference between edges causes the locations of maximum gradients in the output image to differ from the positions of the actual

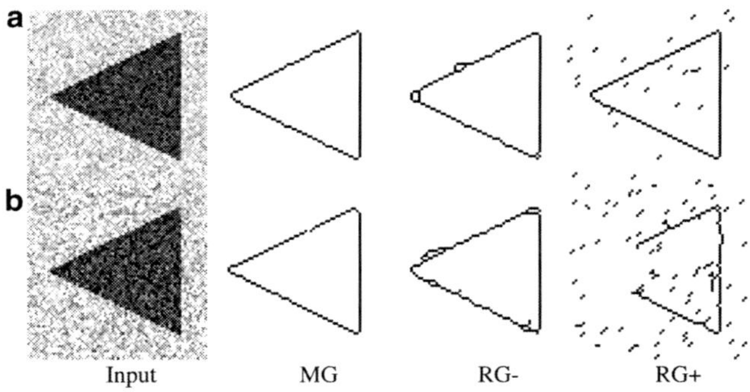


**Fig. 8** The input image and edge images extracted from the three grids

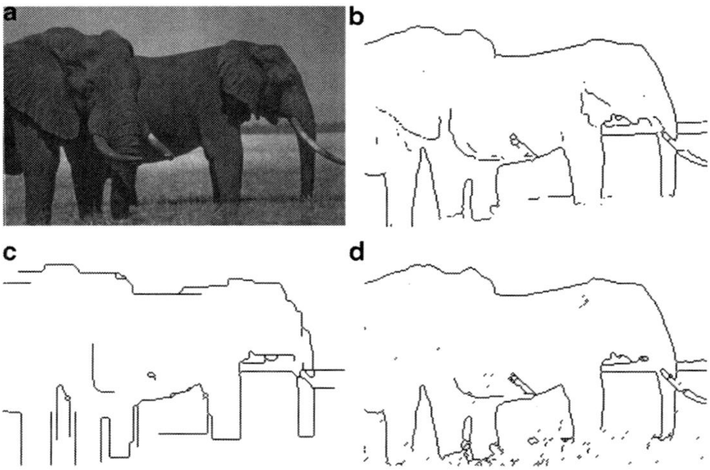


**Fig. 9** The input image (**a**) and the edge images extracted from three grids: MG(**b**), RG−(**c**), RG+(**d**)

edges in the input, leading to the extra or false edges detected. For RG+, the output image is initially close to the input image. This preserves the edge locations, but has the disadvantage that it also preserves the noise. In the second image, where the noise level is much higher, many of the true edges are lost while much of the noise remains.

Figure 9 shows the edge maps with largest $F$-measure obtained from the input image shown in Fig. 3c. These edge maps were computed based upon the network outputs shown in Fig. 6g–i. We observe similar phenomena as noted above. For example, for RG−, we observe false edges between the legs of the elephant on the left. For RG+, the fine texture in the grass at the bottom leads to small isolated points or regions in the edge image.

## 4.3 Benchmark Results from the Image Library

For a more quantitative measure of performance, we used three grids to produce edge maps for the entire image library of 100 natural images in the dataset of D. Martin et al. [14, 15]. These images were all $481 \times 321$ pixels. We sub-sampled the images and ground truth edge maps by a factor of two to reduce the amount of time to calculate the edge map sequences.

Figure 10 compares the average $F$-measure taken over the entire dataset for the three networks. The average $F$-measure is plotted as a function of the pixel tolerance used to determine whether an edge pixel detected by a network matches with an edge pixel in the ground truth. We can see that the MG is always the best of the three. As the pixel tolerance increases, the performance of the RG− (implementing the GNC algorithm) approaches that of the MG, since the incorrect edge localizations become less important. The performance of the RG+ is always lower than that of the other two, because of its inability to eliminate erroneous edge pixels due to fine texture.

As another comparison, Fig. 11 plots the percentage of images for which each network produces the "best" edge map as measured by the $F$-measure, again as a function of pixel tolerance. For each tolerance, the three percentages sum to 100. Consistent with our previous result, the MG produces the best result, especially for low-pixel tolerances. However, as the pixel tolerance increases, the MG and the RG− begin to perform almost equally well. The RG+ always has the worst performance.

## 4.4 Stability of the Edge Maps

As mentioned above, each network generates a sequence of edge maps, either as the threshold changes (for the RG) or over time (for the MG). We measure the stability

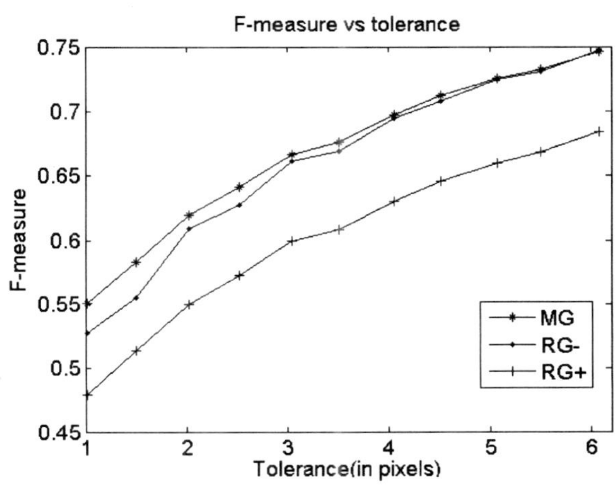


**Fig. 10** The average $F$-measure for the three grids plotted as a function of the tolerance used in matching detected and ground truth edges

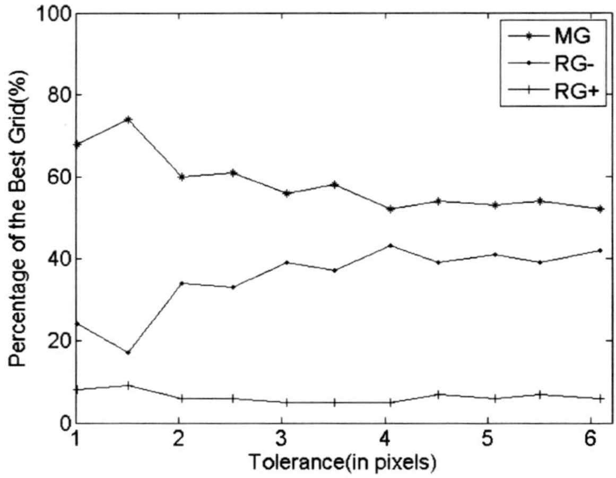


**Fig. 11** The percentage of images for which each grid produced the edge map with the largest $F$-measure as a function of the matching tolerance

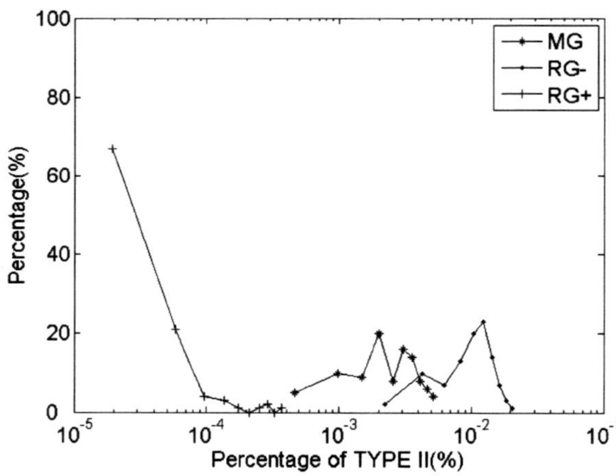


**Fig. 12** The histogram of the percentage of Type II pixels in the edge map sequences for each grid type. The histogram is computed over the entire database

or constancy of the edge map by measuring the stability of the classification of each pixel over the sequence. Each pixel in an edge map is binary, assuming either value 0 (if the network decides the pixel is not an edge) or value 1 (if the network decides that it is an edge). For each pixel, we obtain a sequence of edge decisions. We classify pixels into two types depending upon this sequence. Type I pixels are stable. They either maintain the same classification throughout the sequence, or make a transition from edge to nonedge or vice versa. Type II pixels are unstable, changing more than one time. For example, for the RG−, a pixel might start out as a nonedge, be classified as an edge, but later return to being a nonedge as the location of the edge shifts. Thus, we expect the percentage of Type II pixels in an edge map sequence to increase with the number of false or shifting edges.

Figure 12 plots the histogram of the percentage of Type II pixels taken over the entire database for each network. Consistent with our previous results, the histogram for the RG− exhibits the largest shift toward the right (larger percentages of Type II pixels, indicating that its edge maps are the less stable.)

## 5 Conclusion

This chapter has introduced an MG for edge preserving image smoothing. This network replaces the resistors in an RG with similar functionality by memristors. By designing the memristance appropriately, we can obtain a network that performs intensity discontinuity preserving image smoothing. Our experimental results indicate that this MG outperforms previously proposed RGs, in the sense that the edges or intensity discontinuities detected during the smoothing process more closely match with extracted by human observers. We conjecture that it may be possible to obtain significant performance gains in other parallel processing circuit architectures by appropriately replacing resistors by memristors.

**Acknowledgments** This work was supported in part by the Hong Kong Research Grants Council under Grant Number 619607.

## References

1. T. Poggio and C. Koch, Ill-posed problems in early vision: From computational theory to analogue networks, in *Proc. R. Soc. Lond. B*, vol. 226, pp. 303–323, 1985
2. D. Terzopoulos, Regularization of inverse visual problems involving discontinuities, *IEEE Trans. Pattern Anal. Mach. Intell.*, vol. PAMI-8, pp. 413–424, July 1986
3. A. Blake and A. Zisserman, *Visual Reconstruction*. Cambridge, MA: MIT, 1987
4. P. Perona and J. Malik, Scale Space and edge detection using anisotropic diffusion, *IEEE Trans. Pattern Ana. Mach. Intell.*, vol. 12, no. 7, pp. 629–639, July 1990
5. K. N. Nordström, Biased anisotropic diffusion: A unified regularization and diffusion approach to edge detection, *Image Vision Comput.*, vol. 8, no. 4, pp. 318–327, November 1990
6. J. G. Harris, C. Koch, and J. Luo, A two-dimensional analog VLSI circuit for detecting discontinuities in early vision, *Science*, vol. 248, pp. 1209–1211, June 1990
7. J. G. Harris, S. C. Liu, and B. Mathur, Discarding outliers using a nonlinear resistive network, in *International Joint Conference on Neural Network*, vol. 1, pp. 501–506, 1991
8. P. C. Yu, S. J. Decker, H. Lee, C. G. Sodini, and J. L. Wyatt, CMOS resistive fuses for image smoothing and segmentation, *IEEE J. Solid State Circ.*, vol. 27, pp. 545–553, 1992
9. W. Millar, Some general theorems for nonlinear systems possessing resistance, *Phil. Mag.*, ser. 7, vol. 42, no. 333, pp. 1150–1160, October 1951
10. J. L. Wyatt, Little-known properties of resistive grids that are useful in analog vision chip designs, in *Vision Chips: Implementing Vision Algorithms with Analog VLSI Circuits*, C. Koch and H. Li, Eds. Los Alamitos, CA: IEEE Computer Society Press, 1995, pp. 72–104
11. L. O. Chua, Memristor-The missing circuit element, *IEEE Trans. Circ. Theor.*, vol. CT-18, pp. 507–519, September 1971
12. D. B. Strukov, G. S. Snider, D. R. Stewart, and R.S. Williams, The missing memristor found, *Nature*, vol. 453, pp. 80–83, May 2008
13. B. E. Shi and L. O. Chua, Resistive grid image filtering: input/output analysis via the CNN framework, *IEEE Trans. Circ. Syst.-I*, vol. 39, pp. 531–548, July 1992
14. D. Martin, C. Fowlkes, D. Tal, and J. Malik, A database of human segmentated natural images and its application to evaluating segmentation algorithms and measuring ecological statistics, in *International Conference Computer Vision*, vol. 2, pp. 416–423, July 2001
15. D. Martin, C. Fowlkes, and J. Malik, Learning to detect natural image boundaries using brightness and texture, in *Advances in Neural Information Processing Systems*, vol. 14, 2002

# Bionic Eyeglass: Personal Navigation System for Visually Impaired People

**Kristóf Karacs, Róbert Wagner, and Tamás Roska**

**Abstract** The first self-contained experimental prototype of a Bionic Eyeglass is presented here, a device that helps blind and visually impaired people in basic tasks of their everyday life by converting visual information into audio signal. The indoor and outdoor situations and tasks have been selected by a technical committee consisting of blind and visually impaired persons, considering their most important needs and potential practical benefits that an audio guide can provide. The prototype system uses a cell phone as a front-end and an embedded cellular visual computer as a computing device. Typical events have been collected in the Blind Mobile Navigation Database to validate the algorithms developed.

## 1 Background

In spite of the impressive advances related to retinal prostheses [1], there is no imminent promise to make them soon available with a realistic performance to help blind or visually impaired persons in everyday needs. An electronic device that monitors the surrounding environment through its sensors and translates the information for the blind individual about it can be realized much sooner, and still provide real help [2]. The device is based on the Cellular Neural/nonlinear Network – Universal Machine (CNN-UM) and the underlying Cellular Wave Computing principle [3–5]. This paper presents the first prototype of the Bionic Eyeglass.

The project has three main distinctive features with respect to similar previous works [6–10]:

1. Frequent communication with blind and visually impaired people. This helps us to identify the main situations that a blind person faces in everyday life and which tasks are to be solved in these situations.

K. Karacs (✉)
Faculty of Information Technology, Pázmány Péter Catholic University, Budapest, Hungary
e-mail: Karacs.Kristof@itk.ppke.hu

Á. Zarándy (ed.), *Focal-Plane Sensor-Processor Chips*,
DOI 10.1007/978-1-4419-6475-5_10, 

2. Standardized clinical tests. Testing of the prototype system is carried out using standardized methods by doctors, taking into account the type of visual injury of the patients.
3. Neuromorphic solutions. The system is based on an intensive multichannel retina-like preprocessing of the input flow with semantic embedding. The CNN-UM proved to be a suitable tool for modeling the processing in the retina, where each of the channels extracts different spatiotemporal features of the flow. [4] The system also uses a neuromorphic attention model to focus on the information that is most important for humans.

The next section outlines the system requirements and architecture of the prototype system. The third and fourth sections present color processing and some details of the partially neuromorphic saliency and event recognition system. Section 5 presents the video database that we created to train and test the algorithms.

## 2 System Architecture

The Bionic Eyeglass provides a wearable TeraOps visual computing power to advise visually impaired people in their daily life: at home, at work, and on the way between them. The basic tasks are summarized in Table 1.

Two types of cellular wave computing algorithms are used: (1) standard templates and subroutines and (2) bioinspired neuromorphic spatial-temporal event detection. Examples for the former one are door handle detection, corridor sign extraction, and banknote and letter recognition [11, 12]. The second type of algorithm is a neuromorphic saliency system [13] using the recently developed

**Table 1** Typical tasks considered for the bionic eyeglass

| | Home | Street | Office |
|---|---|---|---|
| | Color and pattern recognition of clothes | Recognition of crosswalks | Recognition of control signs and displays in elevators |
| | Banknote recognition | Escalator direction recognition | Navigation in public offices and restrooms |
| *User initiated functions* | | Public transport sign recognition | Identification of restroom signs |
| | | Bus and tram stop identification | Recognition of signs on walkways |
| | | Recognition of fluorescent displays | |
| | Recognition of messages on LCD displays | | |
| *Autonomous warnings* | Light left on Gas stove left turned on | Obstacles at head and chest level (i.e., boughs, signs, devices attached to the wall, trucks being loaded) | |

multi-channel mammalian retinal model [14] followed by a classifier using the semantic embedding principle. The system architecture is shown in Fig. 1.

The prototype system has been built to show the feasibility of the system and to facilitate clinical and pilot tests. It consists of a cell phone with built-in camera and loudspeakers, the Bi-i visual computer [15] and a wireless adapter (Fig. 2.).

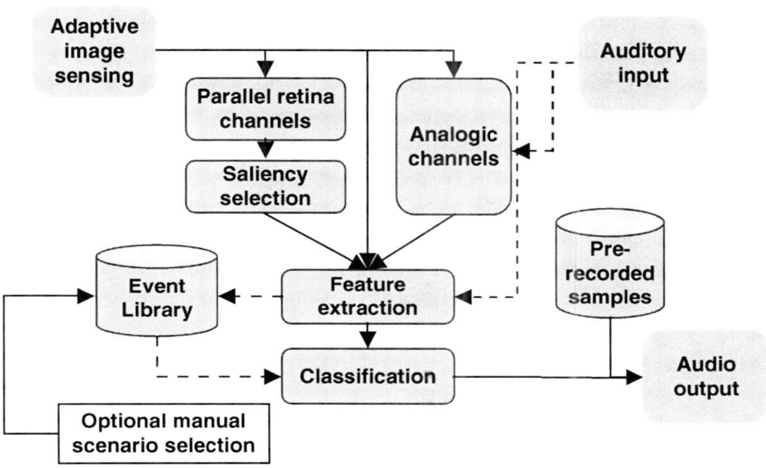


**Fig. 1** Overview of the system architecture

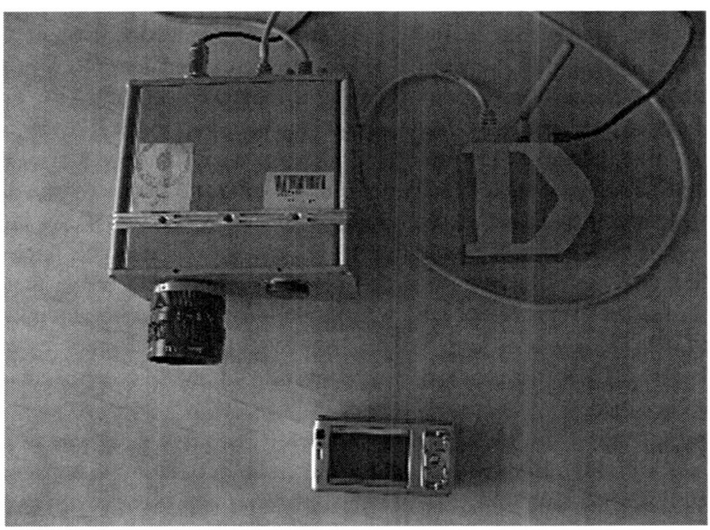

**Fig. 2** The prototype system consists of a cell phone (*bottom*), the Bi-i visual computer (*left*), and the wireless adapter (*right*)

Although the Bi-i has on-chip visual input, we did not use it in the configuration described, since the form factor of this version of the Bi-i version made this rather impractical.

We used a WiFi-enabled Nokia N95 cell phone in the prototype system. We implemented a communication framework for it that streams the image flow recorded by the cell phone camera wirelessly to the Bi-i. The Bi-i visual computer runs the algorithms on its parallel architecture and returns the result via the wireless connection to the cell phone. The cell phone plays a pre-recorded audio sample based on the result received from the Bi-i.

The prototype system is portable. The Bi-i and the wireless adapter are powered from 12 V and 5 V batteries respectively. These devices of the prototype system weigh less than 3 lbs and can be carried in a briefcase or a backpack.

The final Bionic Eyeglass will feature a tiny integrated eyeglass-mountable unit using neuromorphic hardware. The first step towards a wearable or handheld version is the integration of a many core processor into a cell phone. The first samples are already coming to the market in the form of handsets with a built in GPGPU (general-purpose graphics processing unit).

## 3 Parallel Image Sensing-Processing

### *3.1 Adaptive Image Sensing*

Adaptive image sensing is important if we deal with scenes that have large intra-scene dynamic range, like in real-world street image flows. Recent works on adaptive image sensing [17] using CNN-UM are developed using locally adaptable sensor array. A retina-like adaptation can be achieved by adjusting the integration time so, that the local average of an image region becomes the half of the maximum value. This eliminates the intra-frame DC differences. In outdoor scenes where the variations of illumination might be large – both in time and in space – the adaptation is a useful property that enables the operation of the recognition steps. Single chip vision systems with adaptive image sensors are going to be available soon [18].

The first and best-known part of the visual system is the retina that is a sophisticated feature preprocessor with a continuous input and several parallel output channels [19]. These interacting channels represent the visual scene by extracting several features. These features are filtered and considered as components of a vector that is classified.

Beyond reflecting the biological motivations, our main goal was to create an efficient algorithmic framework for real-life experiments, thus the enhanced image flow is analyzed via temporal, spatial and spatiotemporal processing channels that are derived from retinal processing and semantic understanding of the task. The outputs of these subchannels are then combined in a programmable configuration to form new channel responses.

## 3.2 Color Processing

Visually impaired people need color information in some cases, such as to be able to choose clothes of matching colors. Thus we included a function that informs the user about the color and texture of the objects seen.

The system we extracts the colors of the scene and retrieves their location, which task can be interpreted as a color segmentation problem. Color segmentation algorithms have two main aspects: the representation of the color space, and the method used to group the pixels [16].

Color space representations are linear or nonlinear transformations of the RGB color space. For the computation of the perceived color we use the nonlinear CIE Luv color space. A great advantage of it is that distances of the stimuli are similar to the human perceived chromatic distance. Hence the transformation of the RGB channels corresponds to the human perception.

The methods used for the grouping of pixels can be classified into three main classes:

1. Pixel-based methods. These algorithms try to find groups of pixels in the 3D color histogram. Common techniques for this are thresholding or clustering.
2. Region-based methods. These consist of growing regions from an initial seed until color boundaries are reached.
3. Edge-based methods. Similar to the region-based methods, but at first the boundaries of color objects are specified and a region-based method is applied afterward.

We applied $k$-means clustering (a pixel based method) with $k = 16$. Steps of the algorithm are as follows:

1. RGB $\rightarrow$ Luv color space conversion
2. **Luminance adaptation**
3. Clustering
4. Merging similar clusters
5. **White correction**
6. **Identification of color names for all regions.**

Steps on the CNN-UM are shown in bold, other operations such as clustering and color space conversion require less regional processing and they are performed on the accompanying digital processor.

The luminance adaptation reduces the distorting effects of the illumination by eliminating the differences in the spatial low-pass component of the luminance of the perceived image (see Fig. 2a, b). Since the "L" channel of the Luv space represents the luminance we performed our local luminance adaptation on it. The advantage of the CNN-UM architecture comes with local processing, which enables the easy computation of the local average.

A common problem of $k$-means clustering is oversegmentation, because we do not know the proper number of initial cluster centers. We overcame this problem with a postprocessing step, in which we merged similar clusters, whose centers are closer to each other then a given threshold. The result of clustering and merging can be seen in Fig. 3.

The postprocessing step "white correction" (see steps of the algorithm) exploits the fact that the image is clustered. If there is a cluster that lies close to saturated white color, we multiply its channels to become white. Channels of the other clusters are also scaled with the same multiplication factors. This correction reduces chromatic distortion (Fig. 3).

The last stage determines the location of the clusters and gives a verbal classification of their location using speech samples (Fig. 4).

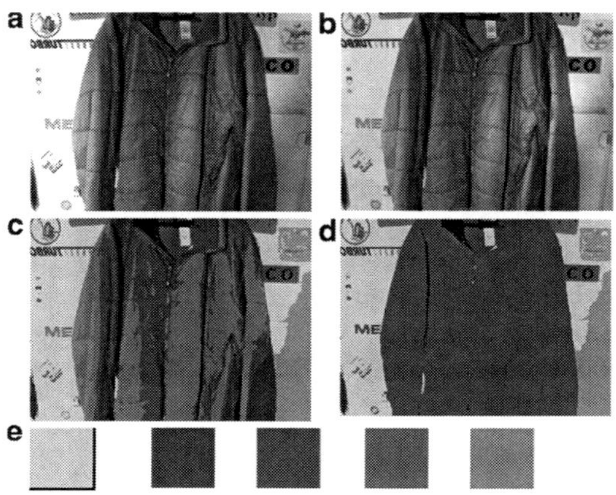

**Fig. 3** Clustering and merging of clusters. (**a**) shows the original picture taken by a standard digital camera. (**b**) shows the effect of the luminance adaptation. In (**c**), we can see the result of clustering. In this picture, the pixels have the color of the cluster they assigned to. Regions of a given color represent a cluster. (**d**) shows the clusters after merging of similar clusters. The main cluster colors can be seen in (**e**)

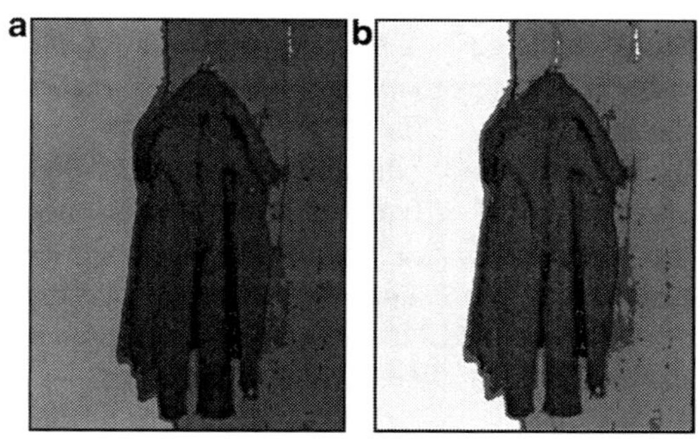

**Fig. 4** (**a**) Sample input image taken by a mobile phone, which has extra sensitivity to *blue*. (**b**) Corrected clusters. Note the white color of the wall

## 3.3 Detecting and Recognizing Signs

An example for this is the detection and recognition of signs posted on public transport vehicles. The processing uses subsequent frames of the input video flow to detect and recognize the sign. Eventually determining the location of the sign is much more difficult than locating it. Different algorithms are needed for black and white signs, and color displays [12, 20].

### 3.3.1 Sign Localization

Determining whether a route number sign is present on an image and localizing it is a very nontrivial task, especially on low-resolution, noisy images. The purpose of the sign localization step is to find the numbers to be recognized. But the intuitive way to tell the location of the sign is by looking for numbers on the image. In other words, the presence of a sign is only justified by the existence of numbers on it. This detection-recognition paradox is similar to the one defined by Sayre for the machine reading of human handwriting: the aim of locating the sign is to recognize the numbers included thereafter, but to find the location one needs to recognize the content first.

Different types of signs and displays can be detected in different ways. We have developed algorithms to detect signs with a white background and ones that have fluorescent numbers with a dark background. A black and white sign is defined as a rectangular shaped, almost white spot inside a big dark area (window) in the lower part of the image. This semantic definition is represented by algorithms modeling some channels. An example is shown in Fig. 5.

The particular difficulty in detecting a display is that in general it cannot be identified by some distinctive features as in the black and white case. To overcome this paradox we defined some simple but characteristic properties of displays:

- Bright and large figures are displayed on them;
- Typical colors are yellow, green, and red;
- They have high contrast for good legibility;
- Form of figures is not patch-like, rather stroke-type.

The first three properties refer to color and luminance, whereas the last one is a morphological property that, despite being ambiguous, enables basic differentiation from other formations with similar color and luminance patterns.

However, to better cope with the great variety of displays, we incorporated the possibility to give specific properties of displays. These may include the font(s) used, exact foreground and background color ranges, size of figures with respect to the display background, as well as other information on neighboring objects. This allows the system to perform more robustly in case of frequent display models and typical scenarios.

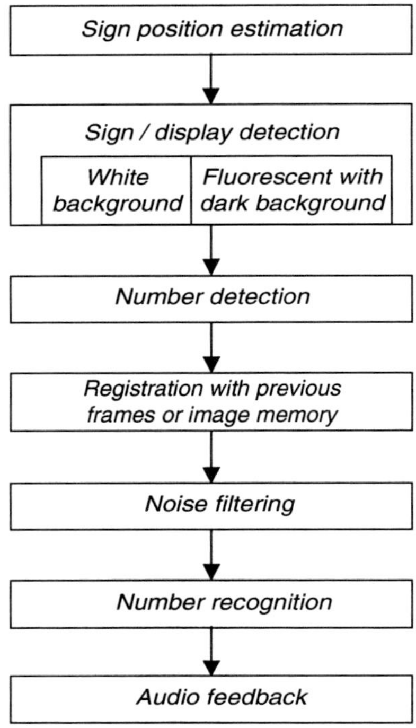


**Fig. 5** Block diagram of the sign detection and recognition framework

To overcome the aforementioned problems, we have developed a heuristic algorithm to locate white background signs. Basic steps are shown in Fig. 6., in the form of a universal machine on flows (UMF) diagram [21], the templates used can be found in [22] and [23]. The idea behind the algorithm is as follows. After thresholding the image, dark window areas are detected first and then the algorithm looks for almost white holes in the window areas with certain size constraints. The result of the algorithm for a sample frame is shown in Fig. 7.

The sign location is tracked through the frames with a simple linear kinematical model. The model gives the probable location of the sign on the next frame based on the location in the previous two frames. This makes the localization of the sign faster since the target area is much smaller. If the sign cannot be located in the estimated area, then the track is assumed to be lost and the algorithm is rerun for the whole frame.

### 3.3.2 Combining Colors and Morphology

Due to lack of availability of a locally adaptive sensor, automatic white correction is performed on the input as a first step, using the method described in [17] to deal with different light conditions and to retain color constancy. Then three different bright color range filters and a background (dark) color range filter are applied (Fig. 11). Bright color range detectors have both a wide and a tight detection range.

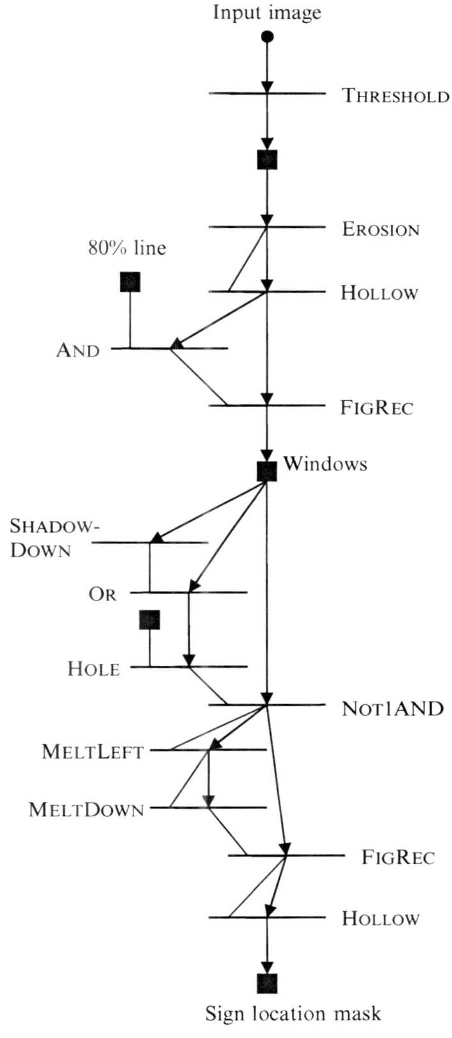


**Fig. 6** UMF diagram of the algorithm to locate signs with a *white* background

Wide-range filters detect the foreground pixels of the display of the given color robustly, but cover many other irrelevant objects too, whereas narrow-range filters are much less sensitive to noise, but may miss some foreground pixels (see Fig. 12b, c). To determine proper wide and tight color ranges for specific display types, we used sample training videos taken at different hours of the day and under different light conditions, and analyzed the colors manually. Ranges were determined for three typical display colors: yellow, green, and red.

Pixel noise is removed from the output of both the wide and the tight color filter by using the **MELTDOWN**, **SMALLKILLER** and **FIGREC** templates, respectively. We use the output of the denoised tight range filter to reconstruct the other denoised image and noise is removed from this one too with the same method (Fig. 12f, g).

**Fig. 7** Sign localization on a tram. (**a**) Original input frame (**b**) Binarized and eroded image (**c**) Smoothed window areas (**d**) Sign location

The use of the dark filter is motivated by trying to locate those bright formations where they do not form patches, that is every pixel has dark pixels in a close neighborhood, where the diameter of the neighborhood is the stroke width. A closing operation is performed on the dark filtered image to achieve this goal, and the result is used as a mask to find the right formations on the combined image from the denoised bright filters. This is carried out by an **AND** template. Closing is the key operation to distinguish strokes from patches and the length of its operation depends on the stroke width. All these operations are performed in parallel for all three colors.

Finally, the channels are verified if they contain patches with the correct width-height ratio and if so, their output is sent to the recognition module.

### 3.3.3 Number Localization

Once the sign area is located, it is converted into a binary image to enable number localization. We need to extract actual numbers by getting rid of other text and noise present in the sign area. It is not critical to remove all the noise here, because a final noise removal step takes place before the recognition.

Noise removal is realized in two parallel ways to make the extraction more robust. On the one hand, the frame is removed together with patches lower than a certain value due to the fact that the numbers are printed with the largest font size. The threshold was determined to be $1/40^{\text{th}}$ of the vertical resolution of the camera by taking into account possible distances from the vehicle and the typical size of figures of the route number signs. On the other hand, we also use the a priori information on the number location based on previous frames (they are usually in the center).

### 3.3.4 Registration with Previous Frames

Binary images of the numbers often become vague due to low resolution input and the high level of noise present on them. To overcome this problem, we make use of the a priori knowledge that signs normally do not change in time, which means we can superpose subsequent sign images to achieve better image quality. For this purpose, we use a fading memory calculated as a weighted average of a new frame and the previous memory image (Fig. 8).

As a first step of this process, the images need to be registered, because detected borders of the sign can differ due to noise and changes in light conditions. Since rotation in the plane of the image is negligible (the user is expected to and can generally easily avoid twisting wrist moves), registration can be done by shifting. Optimal shifting parameters are calculated via cross-correlation:

$$c_n(u,v) = \sum_{x,y} f_n(x-u, y-v) f_{n-1}(x,y), \tag{1}$$

where $f_i$ denotes the sign image of the $i^{\text{th}}$ frame, $u$ and $v$ are the horizontal and vertical shift distances and $c_i(u,v)$ is the cross correlation matrix of frame $i$ and $i-1$ ($i \geq 1$). The values of $u$ and $v$ are determined by maximizing the cross-correlation:

$$(u,v) = \arg\max_{u,v} c(u,v). \tag{2}$$

The size of the correlation window has been determined – based on experimental data – to be ± 10% (i.e., 20%) of the image size, both vertically and horizontally.

The shift operation accounts for translation in the plane, but it cannot handle the remaining three degrees of freedom (DOF):

- Translation toward the camera (zooming – 1 DOF),
- Rotation out of the plane (changing perspective – 2 DOF).

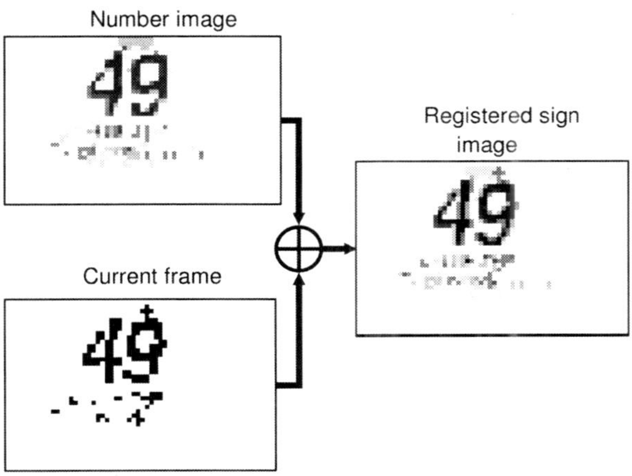


**Fig. 8** Enhancement of the number image by superposing the actual frame and the image memory

According to our experiments, these two changes can be neglected through 2–3 frames (at 15 frames $s^{-1}$), but not longer. Therefore, we maintain a gradually fading memory of the number image and we use the weighted sum of the memory and the actual frame to determine the new "aggregated" number image.

Let $a_i$ denote the aggregate memorized number image in step $i$. At the beginning of a new sign track, the number memory is initialized with $a_0 = f_1$. Registration works the same way as described above but changing the right factor $f_{n-1}(x,y)$ in Eq. (1) to $a_{n-1}(x,y)\ f_{n-1}$. The new number image is given by the following homotopy:

$$a_n = \alpha f_n + (1-\alpha)a_{n-1}, \tag{3}$$

where $\alpha$ is a function of normalized correlation. The optimal function depends on usage habits, assumptions made on the input image flow, and the threshold value used for binarization (recognition uses a binary number image as input). We used a threshold value of 0.1 and a piecewise linear function

$$\alpha = \left\{ \begin{array}{ll} 0.8, & \text{if } \hat{c}_{\max} < a \\ 0.8 + 0.3\frac{a-\hat{c}_{\max}}{b-a}, & \text{if } a \le \hat{c}_{\max} \le b \\ 0.5, & \text{if } \hat{c}_{\max} > b \end{array} \right\}, \tag{4}$$

where $\hat{c}_{\max}$ is the maximum normalized correlation, $a$ is the average normalized correlation of some random sign images, and $b$ is an adaptively corrected maximum value of the normalized correlation. These were determined based on the following considerations:

- Image memory should be preserved as long as the number features are detectable from the aggregated image to compensate for binarization noise.
- In case of fast camera moves or a relatively high speed vehicle, binary number images may change in size or orientation in under three frames, too much for the correctly super-positioned numbers to remain recognizable.

The parameters $a$ and $b$ were tuned on five sample video flows. If the sign track gets lost, then the image memory is cleared.

### Number Recognition

For number recognition, we use topographic shape features that can be extracted by cellular wave algorithms. In the first step, the number of figures in the number is determined by counting connected objects on the image that are bigger than a threshold. This threshold can be higher than the one used for localizing the number, because at this stage we assume the figures are already fully connected. In the second step, feature maps are generated (Fig. 5).

Features used include holes and lines. Holes are classified based on shape, size, and position, whereas lines are classified according to orientation (horizontal or vertical) and position. The method is based on algorithms developed for recognition of handwritten words described in detail in [23].

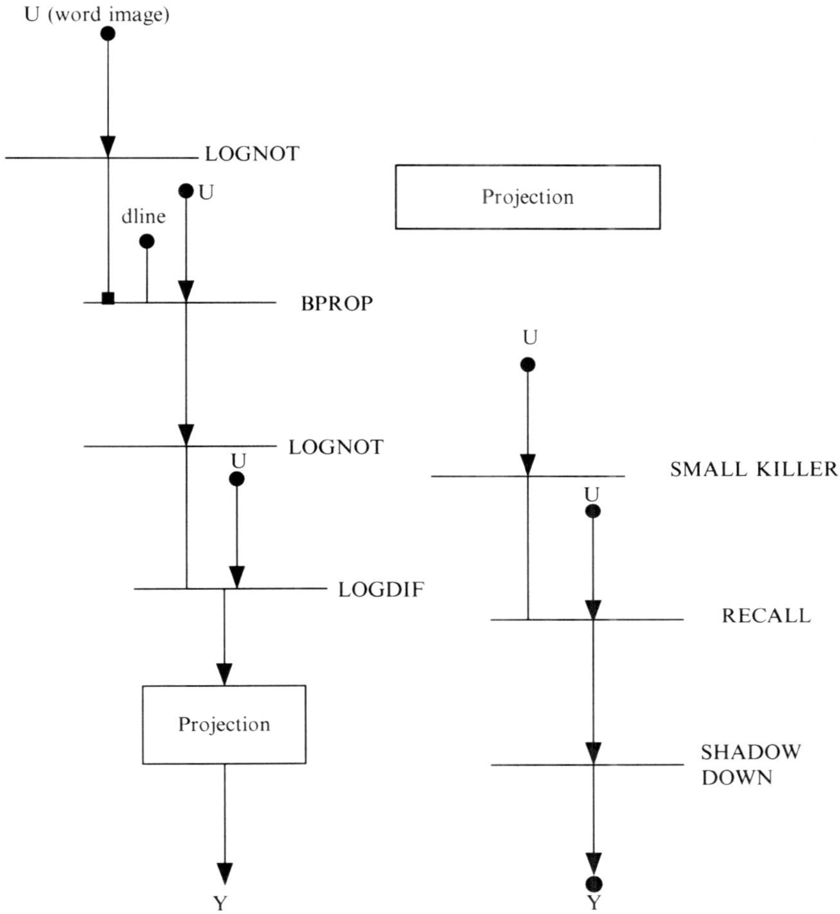


**Fig. 9** UMF diagram of the hole detection algorithm

Holes are defined to be areas with a size not smaller than a threshold that are not connected with the background. The UMF diagram of the hole detection process is shown in Fig. 9. Lines are detected as narrow but long rectangles, with threshold parameters scaled according to the height of figures, in two different combinations for vertical and horizontal lines (Fig. 10).

Size classification parameters are also derived from figure height. The figure the feature belongs to and its horizontal position is determined by vertical projection, whereas horizontal projection enables the calculation of the vertical position.

Shape classification is needed to differentiate between round and triangular holes. Classification is based on vertical and horizontal histograms. Histogram value increases downward and to the right in case of triangular holes ("4" is the only number having this feature).

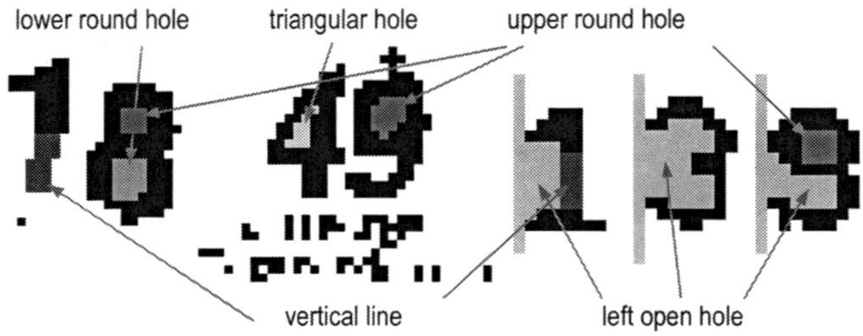


**Fig. 10** Feature maps of route numbers: *vertical line (blue), upper hole (red), lower hole (green), triangular hole (yellow)*

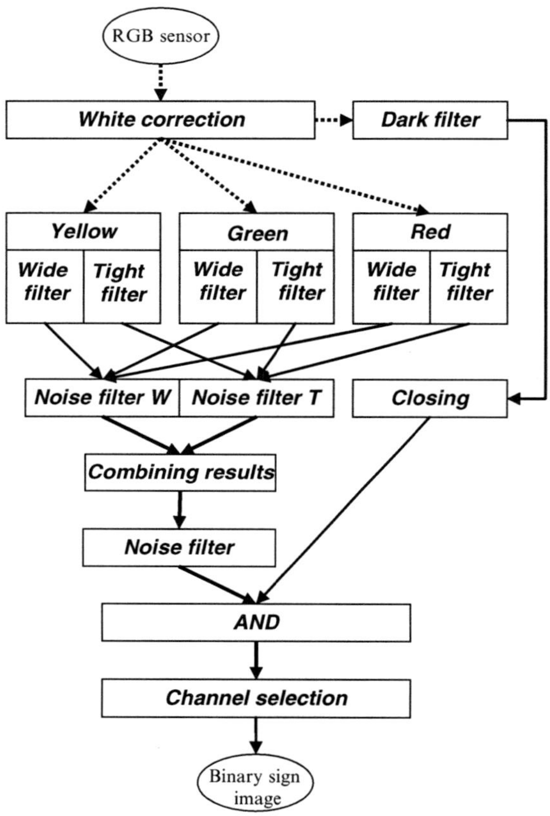


**Fig. 11** Flow diagram of display detection. *Dotted lines* refer RGB data flows, *bold lines* refer to multiple binary flows and *narrow lines* refer to single binary flows

Features are converted into numbers based on a feature allocation table (see Table 2). In a general case, this method is not restrictive enough to give robust results. But given an a priori knowledge on the lines serving the actual stop where the user is located, in pairwise comparisons the lack of features becomes much more relevant.

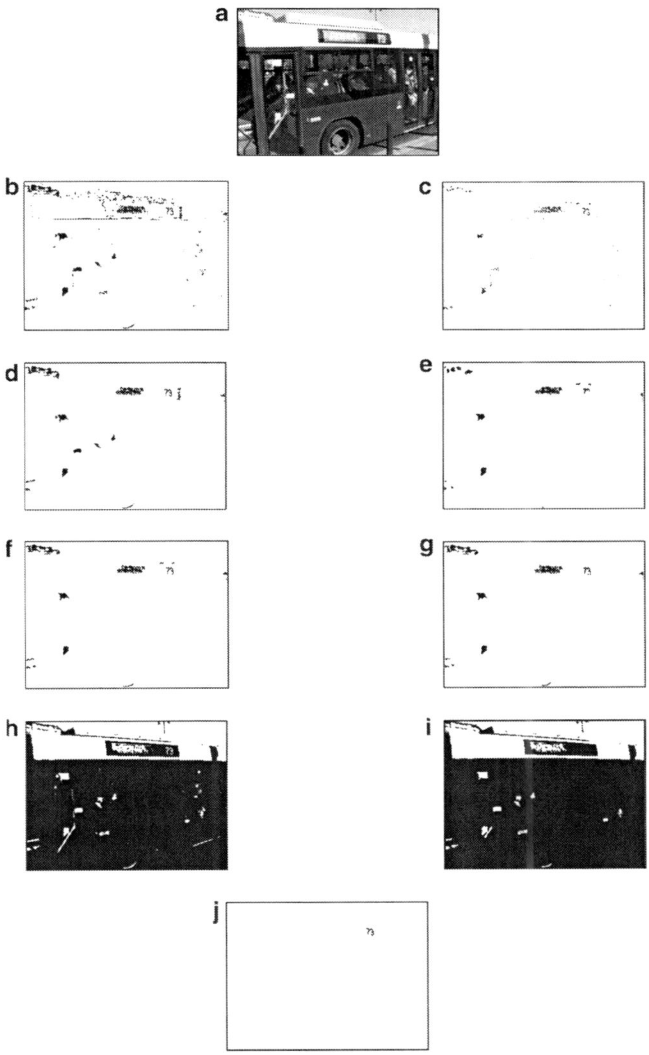

**Fig. 12** Image sequence showing the detection process. (**a**) Original image. (**b**) Wide-range filter for yellow channel. (**c**) Tight-range filter for yellow channel. (**d**) and (**e**) Noise removed from (**b**) and (**c**), respectively. (**f**) Figure reconstruction using (**d**) as input and (**e**) as initial state. (**g**) Noise removed from (**f**). (**h**) Result of dark filter on (**a**). (**i**) Closing performed on (**h**). (**j**) Number image

The feature detection method has been extended by checking for open holes in figures to make the recognition more robust and make it able to distinguish figures without holes (such as "2" and "3"). This is carried out by drawing side bars on the figure image, and checking for holes on these modified images. These holes are classified in the same manner as normal holes.

**Table 2** Feature conversion table

| | Hole | | | Line | |
|---|---|---|---|---|---|
| Fig. | Big | Round | Triangle | Horizontal | Vertical |
| 0 | + | | | | |
| 1 | LTO | | | D | M, D |
| 2 | LO, DRO | | DRO | D | |
| 3 | LO | | | | |
| 4 | | | U | M | |
| 5 | | URO, DLO | | U | |
| 6 | | URO, D | | | |
| 7 | LO | | | U | |
| 8 | | U, D | | | |
| 9 | DLO | U | | | |

+: Present, D: Down, U: Up, M: Middle, L: Left, R: Right, O: Open, T: Tight

### *3.4 Semantic Embedding through the Spatiotemporal Event Library*

The Event Library contains descriptions of events in the expected scenarios (Table 1). Parallel scenarios are activated by salient features extracted from the scene. If a scenario is active, it has an influence on the attention direction. The scenarios are weighted by a priori information and by the identified events, and the more weight a scenario is assigned the bigger the influence it will have on decisions and attention direction.

The classification task can be greatly enhanced by using semantic embedding, a formal method to systematically match the sensory context against the static and dynamic information in the spatiotemporal event library.

## 4 Blind Mobile Navigation Database

We established a database for training and testing the algorithms developed for the Bionic Eyeglass. The database is continuously being expanded more than 500 video flows of lengths between 10 s and 90 s have already been recorded by a blind person in different situations mentioned in Table 1.

We used commercial cellular phones and digital cameras to record videos. The resolution of the videos is either QCIF or QVGA. Presently, no visual microprocessors are available with a resolution higher than QCIF, these recordings were taken for performance comparison purposes. Phones appropriate for this task must have a camera capable of video recording with at least QCIF resolution, and there must be a hard-button by which recording can be started and stopped (soft buttons on a touch-screen are too vague for a visually impaired user).

**Acknowledgment** The support of the Bólyai János Research Scholarship, the Swiss Contribution, the Hungarian Academy of Sciences, the Pázmány Péter Catholic University, the Office of Naval Research and the Szentágothai Knowledge Center as well as the contributions of Róbert Wagner and Mihály Szuhaj are kindly acknowledged.

## References

1. D. Yanaia, J.D. Weilanda, M. Mahadevappaa, R.J. Greenbergd, I. Fine, M.S. Humayun, Visual performance using a retinal prosthesis in three subjects with retinitis pigmentosa, *American Journal of Ophthalmology*, Vol. 143, Issue 5, pp. 820–827, May 2007
2. T. Roska, D. Bálya, A. Lázár, K. Karacs, R. Wagner, System aspects of a bionic eyeglass, in *Proc. of the 2006 IEEE International Symposium on Circuits and Systems (ISCAS 2006)*, Island of Kos, Greece, May 2006, pp. 161–164
3. L.O. Chua, T. Roska, Cellular Neural Networks and Visual Computing, Cambridge University Press, Cambridge, UK, 2002
4. B. Roska, F.S. Werblin, Vertical interactions across ten parallel, stacked representations in the mammalian retina, *Nature*, Vol. 410, pp. 583–587, 2001
5. A. Zarándy, Cs. Rekeczky, P. Földesy, I. Szatmári, The new framework of applications – The Aladdin system, *Journal on Circuits Systems Computers*, Vol. 12, pp. 769–782, 2003
6. M. Mattar, A. Hanson, E. Learned-Miller. Sign classification for the visually impaired, *Technical Report*, 2005–014 University of Massachusetts Amherst, 2005
7. P.B.L. Meijer, An experimental system for auditory image representations, *IEEE Transactions on Biomedical Engineering*, Vol. 39, No. 2, pp. 112–121, Feb 1992. Reprinted in the 1993 IMIA Yearbook of Medical Informatics, pp. 291–300
8. A. Amedi, F. Bermpohl, J. Camprodon, L. Merabet, P. Meijer, A. Pascual-Leone, LO is a metamodal operator for shape: an fMRI study using auditory-to-visual sensory substitution, in *12th Annual Meeting of the Organization for Human Brain Mapping (HBM 2006)*, Florence, Italy, June 11–15, 2006
9. C.-H. Cheng, C.-Y. Wu, B. Sheu, L.-J. Lin, K.-H. Huang, H.-C. Jiang, W.-C. Yen, C.-W. Hsiao, In the blink of a silicon eye, circuits and devices magazine, *IEEE*, Vol. 17, No. 3, pp. 20–32, May 2001
10. A. Dollberg, H.G. Graf, B. Höfflinger, W. Nisch, J.D. Schulze Spuentrup, K. Schumacher, A fully testable retinal implant, *Proceedings of Biomedical Engineering*, 2003
11. Á. Zarándy, F. Werblin, T. Roska, L.O. Chua, Novel types of analogic CNN algorithms for recognizing bank-notes, *Proceedings of IEEE Int. Workshop on Cellular Neural Networks and their Applications*, pp. 273–278, 1994
12. K. Karacs, T. Roska, Route number recognition via the Bionic Eyeglass, in *Proc. of 10th IEEE Int. Workshop on Cellular Neural Networks and their Applications*, Istanbul, Turkey, pp. 79–84, Aug. 2006
13. L. Itti, Modeling primate visual attention, in J. Feng, *Computational Neuroscience: A Comprehensive Approach*, CRC Press, Boca Raton, pp. 635–655, 2003
14. D. Bálya, B. Roska, T. Roska, F.S. Werblin, A CNN framework for modeling parallel processing in a mammalian retina, *Int'l Journal on Circuit Theory and Applications*, Vol. 30, pp. 363–393, 2002
15. Á. Zarándy, Cs. Rekeczky: Bi-i: A standalone ultra high speed cellular vision system, *IEEE Circuits and Systems Magazine*, Vol. 5(2), pp. 36–45, 2005
16. H.D. Cheng, X.H. Jiang, Y. Sun, J. Wang Color image segmentation: advances and prospects, *Pattern Recognition*, Vol. 34, pp 2259–2281, 2001
17. R. Wagner, Á. Zarándy, T. Roska, Adaptive perception with locally-adaptable sensor array, *IEEE Transactions on Circuits and Systems I: Regular Papers*, Vol. 51, No.5, pp. 1014–1023, 2004

18. A. Rodríguez-Vázquez, R. Domínguez-Castro, F. Jiménez-Garrido, S. Morillas, A. García, C. Utrera, M. Dolores Pardo, J. Listan, R. Romay, A CMOS vision system on-chip with multi-core, cellular sensory-processing front-end, in C. Baatar, W. Porod, T. Roska, *Cellular Nanoscale Sensory Wave Computing*, ISBN: 978–1–4419–1010–3, 2009
19. J.E. Dowling, The Retina: An Approachable Part of the Brain, The Belknap Press of Harvard University Press, Cambridge, 1987
20. K. Karacs, T. Roska, Locating and reading color displays with the bionic eyeglass, in *Proc. of the 18th European Conference on Circuit Theory and Design (ECCTD 2007)*, Seville, Spain, Aug. 2007, pp. 515–518
21. T. Roska, Computational and computer complexity of analogic cellular wave computers, *Journal of Circuits, Systems, and Computers*, Vol. 12, pp. 539–562, 2003
22. K. Karacs, Gy, Cserey, Á. Zarándy, P. Szolgay, Cs. Rekeczky, L. Kék, V. Szabó, G. Pazienza, T. Roska, *Software Library for Cellular Wave Computing Engines V. 3.1*, MTA-SZTAKI and Pazmany University, Budapest, Hungary, 2010
23. K. Karacs, T. Roska, Holistic feature extraction from handwritten words on wave computers, in *Proc. of 8th IEEE International Workshop on Cellular Neural Networks and their Applications*, Budapest, Hungary, pp. 364–369, July 22–24, 2004

# Implementation and Validation of a Looming Object Detector Model Derived from Mammalian Retinal Circuit

Ákos Zarándy and Tamás Fülöp

**Abstract** The model of a recently identified mammalian retina circuit, responsible for identifying looming or approaching objects, is implemented on mixed-signal focal-plane sensor-processor array. The free parameters of the implementation are characterized; their effects to the model are analyzed. The implemented model is calibrated with real stimuli with known kinetic and geometrical properties. As the calibration shows, the identified retina channel is responsible for last minute detection of approaching objects.

## 1 Introduction

It is essential for a living creature to identify and avoid approaching objects, whether it is an attacking predator or an obstacle in the locomotion path. When an object is approaching, the patch caused by the projection of its silhouette on our retina is expanding. If the object is on a collision course, the expansion is symmetrical. Looming object detector neural circuit was identified in insect visual system earlier. *Locusta migratoria* is exceptionally good at detecting and reacting the visual motion of an object approaching on a collision course. As it turned out, some of the largest neurons in the Locust brain are dedicated to this task [1]. After successful identification, measurement, modeling, and characterization of this neural circuit of the locust, a technical team built and verified a visual sensor-processor chip for automotive application, which could detect collision threat [2] applying the same principles what the brain system of the Locust does.

The common understanding among neurobiologists was that in more developed animals (e.g., mammals) the cells responsible for detecting approaching objects are located in the higher stages of the visual pathway, most probably in the visual cortex. Therefore, it was a surprise when a looming object sensitive neuron type was

Á. Zarándy (✉)
Computer and Automation Research Institute of the Hungarian Academy of Sciences, (MTA-SZTAKI), Budapest, Hungary
e-mail: zarandy@sztaki.hu

Á. Zarándy (ed.), *Focal-Plane Sensor-Processor Chips*,
DOI 10.1007/978-1-4419-6475-5_11, 

identified, called the Pvalb-5 ganglion cell, in mouse retina. The identified retina circuitry, the electrophysiological measurement results, and a qualitative mathematical model are described in [3].

We have simulated the phenomenon in Matlab environment. It turned out that the Pvalb-5 ganglion cells calculate/measure a nonlinear spatial summation of the temporal brightness changes. The measured value (which depends on the approaching speed, the size, the distance, the contrast, and the color pattern of the moving object) can be interpreted as a collision threat indicator. To make quantitative analysis possible, we have generated an experimental framework using a 3D plotter, in which moving objects were recorded with known geometrical and kinematical parameters. Using these recordings, we have verified the qualitative model, calculated its parameters, and identified its operational gamut and sensitivity in different geometrical and kinematical situations.

Besides Matlab, we have made an optimized focal-plane array processor implementation of the mathematical model on the Eye-RIS [4] system as well. In this way, we made a visual approaching object detector what we can also call collision threat sensor. This device makes possible to perform real-time experiments, which is very important to characterize the model under different circumstances, because the response of the Pvalb-5 ganglion cells depends on many parameters of the approaching object. Other advantage of the looming sensor device is that it makes possible to predict the architecture of a higher level neural circuit, which evaluate the output of the modeled ganglion cells. In the future, the continuation of these studies may lead to a micro-sensor devices – similar to the mentioned Locust collision warning chip [2] – which can call the attention to approaching objects.

In this chapter, the neurobiological architecture, the key physiological experiments, and the original qualitative model are shown. Then, the experimental setup, and the characterization, verification, and the sensitivity calculation are described. Finally, the efficient Eye-RIS implementation is introduced.

## 2 The Retinal Circuit

Botond Roska's neurobiologist team has identified a ganglion cell called type, Pvalb-5, in the mouse retina, which responses to dark looming objects, while it does not respond to lateral or recess motion, or static stimuli [3]. To be able to measure these cells, they had isolated a transgenic mouse line, in which only the Pvalb-5 ganglion cells were fluorescently labeled. This means that the genetically modified (labeled) cells contained fluorescent materials, which were easily recognizable and accessible in the transparent retinal tissue under fluorescent microscope. This enabled them to morphologically identify the shape and the size of its dendritic tree. It turned out that the Pvalb-5 has huge dendritic tree ($\sim$350 μm diameter), which receives visual information from $\sim 10^\circ$ of the visual field. Under the fluorescent microscope, it was clearly seen that the density of the Pvalb-5 ganglion cells is very low (Fig. 1). They were distributed equally in a way that the dendritic tree of a cell was just reached the soma (body) of the neighboring cell.

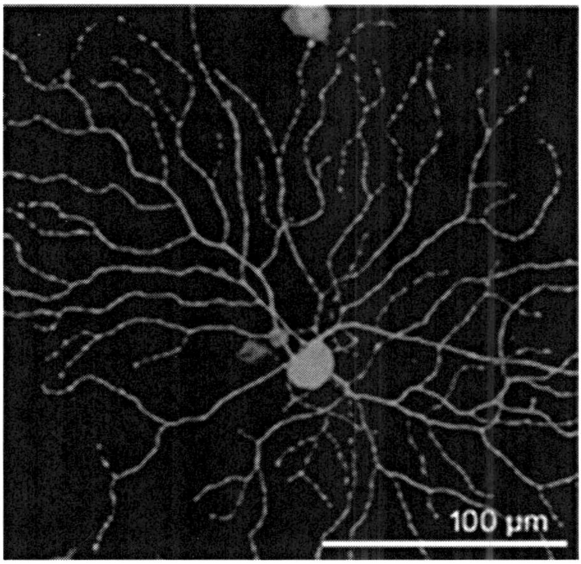


**Fig. 1** The dendritic tree of the Pvalb-5 cell. The large body on the *top* is the soma of the next Pvalb-5 cell. The distance between the somas is roughly half of the diameter of the dendritic tree

The behavior of the Pvalb-5 is depicted in Fig. 2. As it is shown, it selectively responses to dark looming objects, while it does not respond either to lateral or recess motions or to static stimuli. Since the expanding black body in the stimulus leads to an overall intensity falling (dimming) in the receptive field, the neurobiologists had to exclude that the cell provides simply a dynamic off response for the light intensity [3]. To exclude it, a specially generated pattern was projected to the retina, in which the shade of the expanding dark object was permanently lightened in a way that the overall average illumination level of the pattern did not change. Since the cell responded to this stimulus, the pure dimming detection explanation was excluded.

After extensive electrophysiology measurement series, the retinal circuitry was identified. According to the measurements, the ganglion cells average (sum) inputs coming from uniformly distributed inhibitory and excitatory channels from their entire visual fields.

## 3 The Qualitative Model

The qualitative model – proposed by the neurobiologist team – is constructed of a number of equally distributed, equal density inhibitory and excitatory channels (subunits). Each of the channels receives continuous input from one single sensor Fig. 3, hence no spatial interaction is performed at this level. The channels

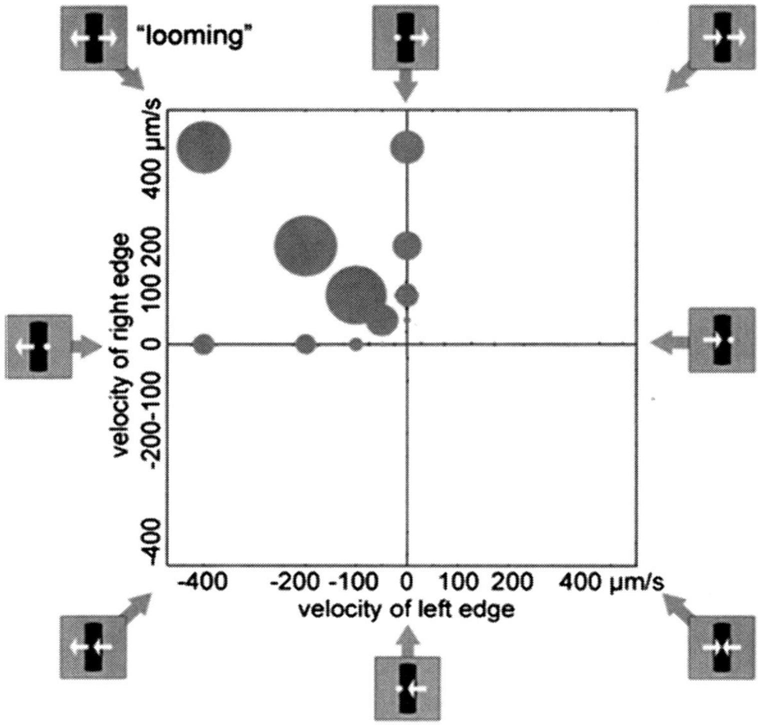


**Fig. 2** Response to different dynamic patterns. The cell fired in those cases only, when the *black* bar against *gray* background were expanding (looming). It did not fire to lateral movements or shrinking (recess motion)

apply linear temporal filtering with the curve showed in Fig. 3b. The two temporal linear filters are roughly each-others inverse. These linear filters generate inhibitory and excitatory (roughly inverted) signals inside the channels. Each channel has a rectification-like output characteristic. As it is shown, the inhibitory channel respond is large positive signal, when its input is changing from dark to light, and generates a small negative response to negative intensity changing. As a contrast, the excitatory channel responds with a large positive signal to negative light changes on the input, and with small negative signal to positive light changes. This shows that the characteristics of the two channels are roughly opposite. An engineer would ask why the retina needs two opposite channels. The most feasible answer is that the neurons are not bipolar devices; hence, the negative signals should be carried in inverted forms in off channels. Others would ask why the inverting channel is called excitatory, while the noninverted is the inhibitory. The reason is because the cell, which sums up the output of these channels, is an off ganglion cell, which reacts positively to the expanding dark objects. Hence, its positive input is the inverted excitatory channel and its negative input is the inhibitory channel. The outputs of

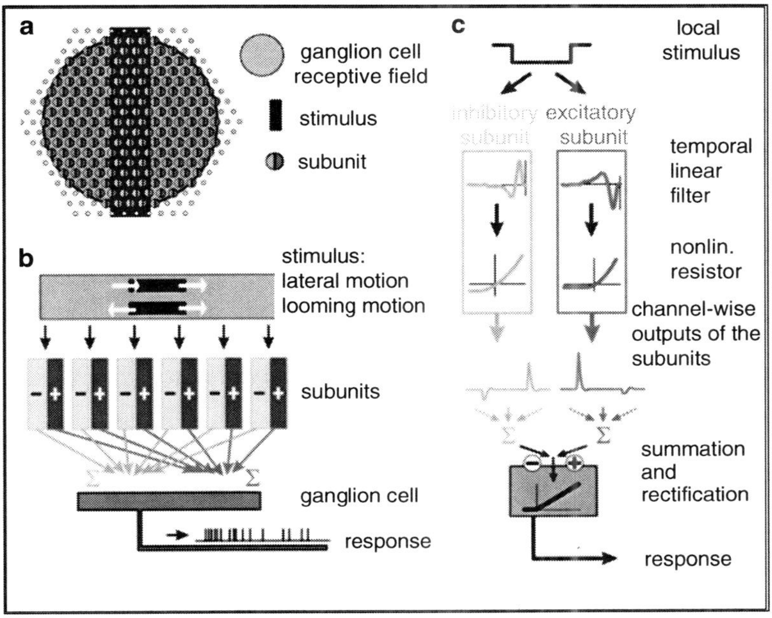


**Fig. 3** The qualitative model. (**a**) shows the receptive field with the excitatory and the inhibitory channels, and the stimulus. (**b**) shows the overall behavior of the model. (**c**) shows the details of the channel responses

the channels are summed up by the ganglion cell in a circle, which covers roughly $10^\circ$ of the visual field (Fig. 3a). In the large sum, the outputs of the inhibitory channels are taken with negative sign, while the outputs of the excitatory channels are taken with positive sign. The ganglion cell has a rectification type output characteristic also (Fig. 3c). Its output is coded in spike activity.

## 4 The Mathematical Model

The cell level signal processing in the retina can be described with mathematical equations which are continuous in time and discrete in space. Since our computers are discrete time machines we have to discretize equations in time also. In the following, we give a discrete time mathematical model which reflects the measurement results. The input of the model is the intensity of the sensed optical signal, while the output is the firing level of a Pvalb-5 ganglion cells.

The first steps of the mathematical model are the spatial filtering in both channels:

$$e_{i,j}(t) = \sum_{n=0}^{s-1} u_{i,j}(t-n)w_n^e \tag{1}$$

$$i_{i,j}(t) = \sum_{n=0}^{s-1} u_{i,j}(t-n)w_n^i, \tag{2}$$

where:

$u_{i,j}$ are the intensity values of the light reached the photoreceptors in position $i,j$ (input);
$s$ is the number of discrete snapshots involved in the temporal convolution;
$w_n^e$ are the weighting factors of the temporal convolutions in the excitatory channels;
$w_n^i$ are the weighting factors of the temporal convolutions in the inhibitory channels;
$e_{i,j}(t)$ is result of the temporal convolution in the excitatory channel in position $i,j$ (output);
$i_{i,j}(t)$ is result of the temporal convolution in the inhibitory channel in position $i,j$ (output).

The spatial convolution is followed by a nonlinear transfer functions. On the one hand, from neurobiological aspects, the rationale of this nonlinearity is twofold. First of all, it is a rectification, since the neural communication channels are unipolar. On the other hand, from signal processing point of view, it is important that it zeros the channels, which carry negative values, and those one also, which carry small positive values as well. Small positive values are generated by temporal noise, or by irrelevant slow temporal intensity changes, which should be cancelled before the spatial averaging to avoid or reduce false alarms. To simplify the mathematical model, we use a single breakpoint piece-wise linear functions mimic the measured nonlinear characteristic, because this reflects rationale of this functionality. The piece-wise linear function is as follows:

$$h_e(x) = \begin{pmatrix} (x+o_e), & \text{if}\,(x+o_e)>0 \\ 0 & \text{otherwise} \end{pmatrix} \tag{3}$$

$$h_i(x) = \begin{pmatrix} (x+o_i), & \text{if}\,(x+o_i)>0 \\ 0 & \text{otherwise} \end{pmatrix}, \tag{4}$$

where:

$o_e$ is the offset value in the excitatory channel;
$o_i$ is the offset value in the inhibitory channel;
$h_e$ is the transfer function of the excitatory channel;
$h_i$ is the transfer function of the inhibitory channel.

The outputs of the channels ($h_e(e_{i,j}(t))$ and $h_i(i_{i,j}(t))$) are spatially summed up by the Pvalb-5 ganglion cell, and rectification is applied on its output:

$$g_{k,l} = r\left(\sum_{i,j\in N_r(k,l)} (h_e(e_{i,j}) - h_i(i_{i,j}))\right), \tag{5}$$

where:

$N_r(k,l)$ is the receptive field of the ganglion cell in position $(k,l)$;
$o_i$ is the offset value in the inhibitory channel;
$r(x)$ is the rectification function: $r(x) = x(\text{sign}(x)+1)/2$.

The implementation and analysis of the mathematical model and its characteristic parameters will be shown in the next section.

# 5 Implementation on a Focal-Plane Sensor-Processor Device

The model was implemented on a standalone vision system, Eye-RIS [4]. It is a small embedded industrial vision system, based on a general purpose focal-plan sensor-processor (FPSP) chip, called Q-Eye. The section starts with a brief description of this system, before the implementation details are introduced.

## 5.1 The Eye-RIS System

The Eye-RIS system (Fig. 4), developed by AnaFocus Ltd, Seville, Spain [5] is constructed of an FPSP chip (Q-Eye) [4], a general purpose processor, which is used for driving the chip and for external communication.

The Q-Eye chip is constructed of a $176 \times 144$ sized locally interconnected mixed-signal processor array (Fig. 5). Each processor cell is corresponding to one pixel

**Fig. 4** The Eye-RIS system

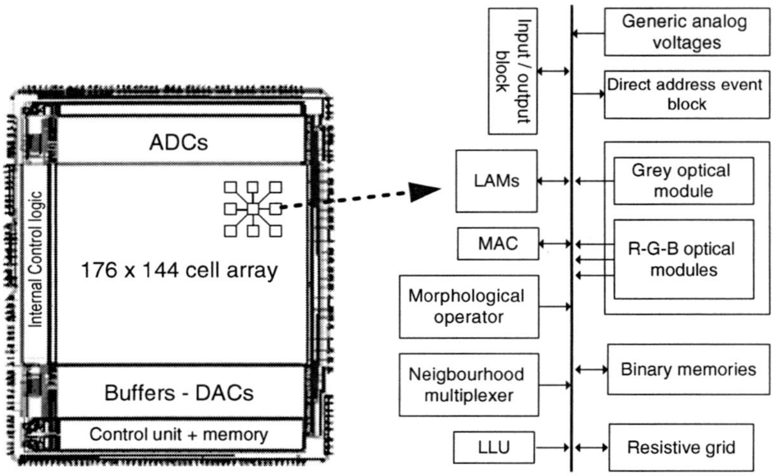


**Fig. 5** Architecture of the Q-Eye chip

(fine-grain); hence, the system can process 176 × 144 sized images. Each of the cells is equipped with photosensor, analog arithmetic and memory unit, and logic unit with logic memories. It can capture images, store them in its analog memories, and perform analog operation on them without analog to digital (AD) conversion. The execution of the operators takes a few microseconds only, thanks to their fully parallel execution. Therefore, the chip can perform above 1,000 fps image capturing and processing (real-time visual decision making). The power consumption of the chip is a few hundred mWs only, depending on its activity pattern. It was fabricated by 0.18 μm technology. The pixel pitch is roughly 25 μm.

The functionality of the chip is summarized by the following list:

- Grayscale
  - Diffusion (Gaussian, directional, masked);
  - Multiple-add (MAC);
  - Shift;
  - Threshold;
  - Mean;
  - Difference (positive, negative, absolute, signed);
- Image capture
  - Four photosensors/cell for color image sensing;
  - Nondestructive repetitive readout;
  - Masking → different integration time per pixel;
- Morphologic
  - Arbitrary 3 × 3 morphologic operations;

- Local logic
  - AND, OR, EQU, XOR, NOT, etc.
- Image I/O
  - Separate grayscale and binary readout
  - Readout of a few rows
  - Address event readout (coordinate of active pixels)

## *5.2 Implementation Details*

As we have seen, the first step of the mathematical model is the channel calculation. It starts with temporal convolution. We have tested various kernels and learned that the simplest convolution, which already leads to good results, is built up from the weighted summation of three snapshots only. We used $[-1/2, -1/2, 1]$ weights in the inhibitory channel and its opposite in the excitatory channel (Fig. 6c). It is important to use zero sum kernels, to cancel out the DC level of the intensity

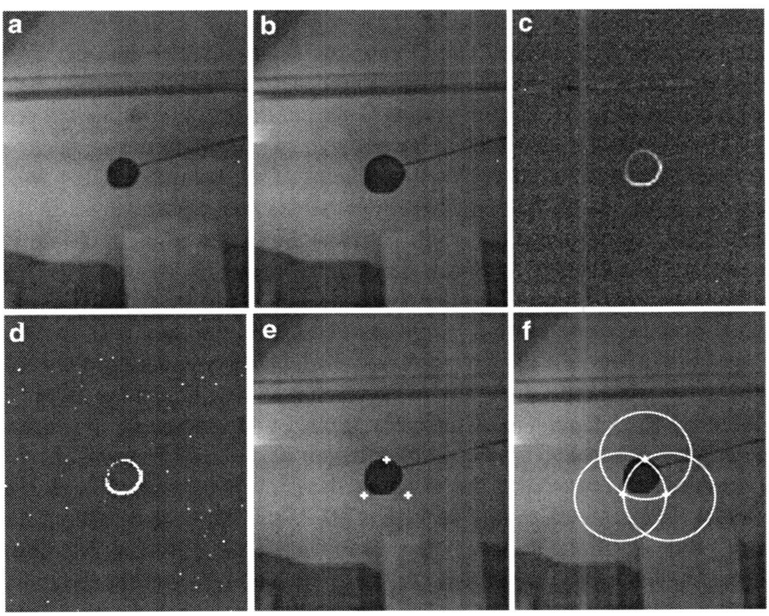


**Fig. 6** Snapshots of the calculation. (**a**) and (**b**) are two snapshots of the input with an approaching black object. (**c**) and (**d**) show the response of the excitatory channels before and after the rectification. The noise canceling role of the rectification is clearly seen. (**e**) shows the three responding Pvalb-5 ganglion cells (r = 25). (**f**) shows the boundaries of the receptive fields of the responding ganglion cells

of the image. Physically, this means three weighted pixel-by-pixel additions of the three consecutive snapshots. The temporal convolution takes 16 µs on the Eye-RIS system.

Larger temporal kernels naturally lead to more accurate approximation of the measurement results. However they increase the computational complexity, require more memory and data transfer, and modify the dynamic performance of the system, because the length of the temporal convolution increases. The length of the temporal convolution window is the first free characteristic parameters of the algorithm. The effects of the tuning of the characteristic parameters of the system will be examined later.

In the mathematical model, the second step is the application of the piecewise linear approximation of the nonlinear output characteristics. On the Eye-RIS, this is done by the addition of the offset and a thresholding, followed by a conditional overwriting of the pixels, which were below the threshold level. The operation takes 4 µs. The offset values ($o_e$ and $o_i$) are the second characteristic parameter of the system.

The third operation is the subtraction and the spatial summation of the output of the two channels. The spatial summation can be done in three ways.

- If the entire 176 × 144 image is considered as the input of a single Pvalb-5 ganglion cell we have to apply a mean instruction, which calculates the normalized sum of the whole array. This takes 12 µs.
- If the receptive field of the Pvalb-5 ganglion cell is smaller than the 176 × 144 image, we have to calculate the summation separately in each receptive field. This can be achieved by using constrained Gaussian diffusion within each receptive field. Technically it requires the usage of the fixed state mask during the diffusion. The fixed state mask contains the boundaries of the receptive fields. Inside the receptive fields, the diffusion fully smoothens the image part; hence its DC level is calculated. In this case, the result contains the output of the multiple Pvalb-5 ganglion cells before the rectifications (Fig. 6e).
- In one step, only nonoverlapping receptive fields can be calculated with this method. If we assume Pvalb-5 ganglion cells distribution as it is shown in Fig. 7, we need to use four sets of mask with nonoverlapping boundaries of the receptive fields, to calculate the summation in each cell position. This takes 50 µs.
- In the third case, Gaussian diffusion is applied to approximate the summation. The radius is controlled by the running length of the diffusion only. In this case the summation is not exact, but as an exchange, it is calculated in all the pixel positions. The calculation this way takes 10–15 µs. We have measured the error of this method. (Since the exact characteristics of the diffusion function of the Q-Eye chip is not known, we could not calculate the difference analytically.) The measurement was done in a way that we calculated the spatial summation using the second and third methods, and compared the results for different running lengths. We made the comparison for different radiuses. It turned out that the accuracy is within the LSB of the system (Fig. 8); hence this fast method can be also used.

**Fig. 7** The Pvalb-5 ganglion cells distribution in our model. *Small solid circles* are the cells, *large circles* are the receptive fields. Cells with nonoverlapping receptive field can be calculated parallel

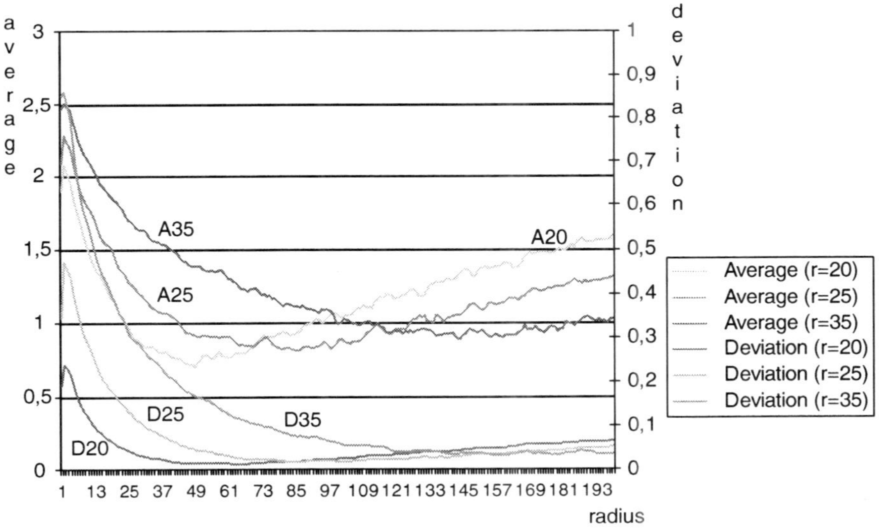


**Fig. 8** Evaluation of the measurement results. The average of the absolute errors for different radiuses (A20, A25, A35) and the deviance (D20, D25, and D35) are shown. The average absolute error and the deviance is measured in the LSB of the Q-Eye chip

The third characteristic parameter is the radius of the receptive field of the ganglion cell.

The last step of the model is the rectification. It is done in the same way as it was discussed previously. The threshold of this ganglion level rectification is the fourth characteristic parameter of the system.

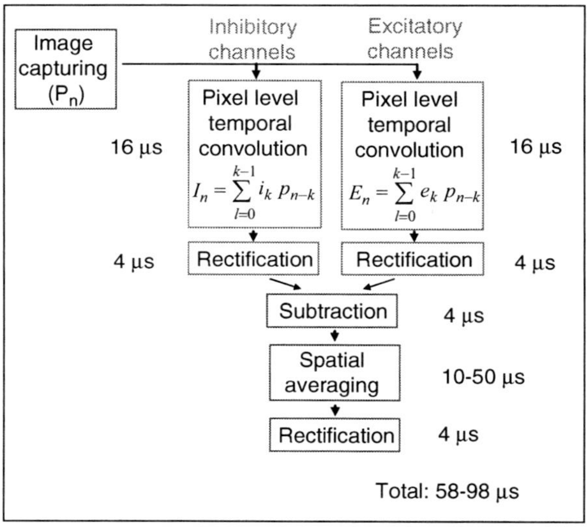


**Fig. 9** Flowchart of the implemented retinal circuit model

The flow-chart of the calculation is shown in Fig. 9. It starts with the parallel implementation of the inhibitory and the excitatory channels. Naturally, those are calculated one after the other on the Eye-RIS; hence their time is added up. The total processing time is 58–98 μs according to the selected calculation method. If the two channels use the same time convolution window, their calculation can be done in one step.

# 6 Calibration

We have built a setup, which contained the Eye-RIS system and a 3D plotter, holding a black circle. By using the 3D plotter, we could generate real spatial–temporal stimuli patterns with known registered geometrical and kinematical information. Image sequences were recorded with the Eye-RIS system. We made the recordings to be able to recalculate the different stimuli patterns with different parameter sets over and over again. We did the recalculations both on the Eye-RIS system and in Matlab. Figure 10 shows a snapshot of our calculation results. As it can be seen, there are seven ganglion cells are firing for the incoming black object. The radius of the receptive field is the distance of two neighboring ganglion cells. The ganglion cell in the middle generated the strongest respond, because all the increasing periphery of the approaching object is in its receptive field. The strength of the response is indicated with the sizes of the white "+" signs.

We have tested the model with different parameter sets, and identified the effect of the tuning of the characteristic parameters. The conclusions are as follows.

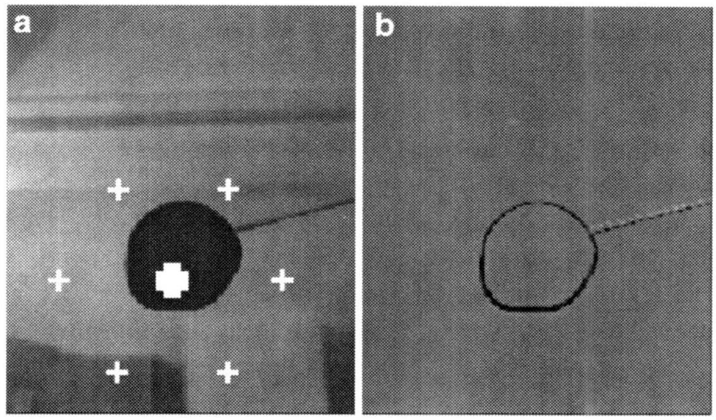


**Fig. 10** Firing ganglion cells (**a**) for approaching object stimulus. Combined outputs of the excitatory and inhibitory channels (**b**)

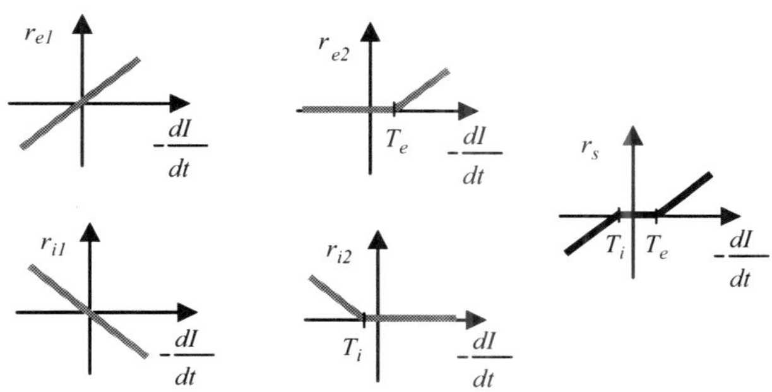


**Fig. 11** The excitatory (*upper*) and the inhibitory (*lower*) channel responses to the intensity changes. The rightmost chart shows the combined response. $T_i$ and $T_e$ are the channel thresholds

The first characteristic parameter of the system (the length of the temporal convolution window) is responsible for the sensitivity and the latency. The longer the window is the more sensitive the model. At the same time the latency is increased with the opening of the time window.

The second characteristic parameter of the system (the inhibitory and excitatory threshold levels) is responsible for the elimination of the small changes. This parameter also tunes the sensitivity, and on the other hand, it is an excellent way to reduce the sensor noise. The effect of this threshold parameters are shown in Fig. 11. The horizontal axis shows the intensity changes (darkening) in time, while the vertical axis shows the induced channel responses in the excitatory (upper) and in the inhibitory (lower) channels before and after the rectification. The rightmost diagram shows the combined (after subtraction) channel response. As it can be seen (Fig. 6c, d), the effects of the small changes are eliminated with the thresholds $(T_e, T_i)$. When there is a lateral movement, the same number of pixels become black at the head, as becomes white at the tail; hence, they cancel out each other in the spatial summation. However, in case of approaching object, the increasing number of black pixels generates positive response only.

The third characteristic parameter of the system (the receptive field of the Pvalb-5 ganglion cells) is responsible for the size of the looming object to be detected. If it is a narrow angle, it will notice a larger distant or a small close object earlier. However, it will not be able to cancel out the lateral movement of a larger object, because front and tale part of the object do not fit to the same receptive field at the same time.

The fourth characteristic parameter of the system (threshold of the ganglion cells) is responsible for general sensitivity. It sets the minimal ganglion cell signal level, which is needed for the cell to fire.

From the analysis of the responses, it turned out that the ganglion cells are not responding to lateral movement, as long as the moving object is entirely within the receptive fields. For approaching objects, we learned that the response is getting stronger as the object is approaching. Figure 12 shows the respond characteristics to a constant speed approaching object in the function of the distance. As we can see, the response is proportional with $1/x^2$. ($x$ is the distance from the sensor). It is not surprising, hence the response is proportional with the area increase of the projected image of the approaching object on the sensor surface, which is naturally proportional with $1/x^2$.

The strong decay of the response in the function of the distance indicates that this retina channel provides a last minute warning signal of an approaching object. By plugging in the parameters coming from the physiological measurements made in the mice retina, and the dynamics of an attacking hawk, it turns out that the ganglion cells starts responding less than 2 s before the predator arrives.

It is important to note that this model and the implemented approaching object detector device responds to an approaching dark object against lighter background by nature. It is very simple to make it sensitive to approaching light objects, just by skipping the last rectification step in the ganglion cell. In that case, the large positive response is a reaction to approaching dark objects, while the large negative response indicates the approaching light object.

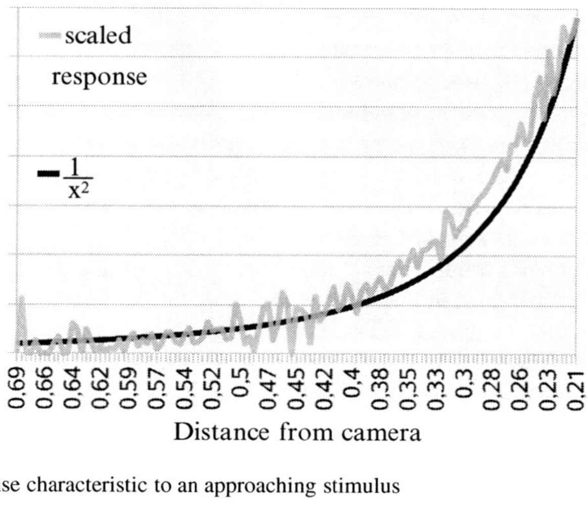


**Fig. 12** Response characteristic to an approaching stimulus

To understand the operation of the model, we have to discuss those situations, when the pattern on the surface of the object or the background is structured with different colors on it. In this case, the model is not behaving correctly. For example, if an object with checkerboard pattern is approaching, and the background is mid-gray, the changes in the receptive fields from gray to black and from gray to white will be in balance, hence the output will be silent as long as the individual dark areas of the checkerboard pattern dominate receptive fields. In these situations it helps, if we can somehow segment the dark or the light parts of the approaching object. In this case, we have to compute the ganglion cell response in each position (as we have seen using the third method in Sect. 5), and sum it app to the bright or the dark areas. However, we have to make sure that we do not include the background areas to the summation.

## 7 Conclusions

Recently identified mammalian retina circuit model, responding to looming object, was implemented on a mixed-signal FPSP array, called Q-Eye. The steps of the implementation were detailed. The characteristic parameters of the implementation are analyzed. The implemented circuit model was quantitatively characterized via stimulus with precisely known geometry and kinetics. It turned out that the retinal circuit is responsible for generating last minute warning signal attention call to approaching objects.

**Acknowledgement** The explanation of the neurobiological system level background of the model by Botond Roska is greatly admired.

## References

1. C. Rind, P.J. Simmons, Seeing what is coming: Building collision sensitive neurons. Trends in Neuroscience, 22, 215–220, 1999
2. G. Linan-Cembrano, L. Carranza, C. Rind, A. Zarandy, M. Soininen, A. Rodriguez-Vazquez, Insect-vision inspired colision warning vision processor for automotive, IEEE Circuits and Systems Magazine, 8(2), 6–24, 2008
3. T.A. Münch, R. Azeredo da Silveira, S. Siegert, T.J. Viney, G.B. Awatramani, B. Roska, Approach sensitivity in the retina processed by a multifunctional neural circuit, Nature Neuroscience, 12(10), 1308–1316, 2009
4. A. Rodríguez-Vázquez, R. Domínguez-Castro, F. Jiménez-Garrido, S. Morillas, A. García, C. Utrera, M. Dolores Pardo, J. Listan, R. Romay, A CMOS vision system on-chip with multi-core, cellular sensory-processing front-end, Cellular Nanoscale Sensory Wave Computing, edited by C. Baatar, W. Porod T. Roska, ISBN: 978–1–4419–1010–3, 2009
5. www.anafocus.com

# Real-Time Control of Laser Beam Welding Processes: Reality

**Leonardo Nicolosi, Andreas Blug, Felix Abt, Ronald Tetzlaff, Heinrich Höfler, and Daniel Carl**

**Abstract** Cellular neural networks (CNN) are more and more attractive for closed-loop control systems based on image processing because they allow for the combination of high computational power and short feedback times. This combination enables new applications, which are not feasible for conventional image processing systems. Laser beam welding (LBW), which has been largely adopted in the industrial scenario, is an example for such processes. Concerning the latter, monitoring systems using conventional cameras are quite common, but they do a statistical postprocess evaluation of certain image features for quality control purposes. Earlier attempts to build closed-loop control systems failed due to the lack of computational power. In order to increase controlling rates and decrease false detections by a more robust evaluation of the image feature, strategies based on CNN operations have been implemented in a cellular architecture called Q-Eye. They allow enabling the first robust closed-loop control system adapting the laser power by observing the full penetration hole (FPH) in the melt. In this paper, the algorithms adopted for the FPH detection in process images are described and compared. Furthermore, experimental results obtained in real-time applications are also discussed.

## 1 Introduction

The title of this paper is inspired by the publication "Process control in laser manufacturing – Dream or reality?" [1], in which, Schmidt et al. run through the state-of-the-art of monitoring and controlling techniques in laser manufacturing processes, including welding of metals. The authors conclude stating that "although nowdays process monitoring systems are suitable for various laser applications, a process control system to prevent weld seam defects on-line is still to come

L. Nicolosi (✉)
Technische Universität Dresden, Germany
e-mail: leonardo.nicolosi@tu-dresden.de

Á. Zarándy (ed.), *Focal-Plane Sensor-Processor Chips*,
DOI 10.1007/978-1-4419-6475-5_12, 

and a desire of likely all users." The scope of our work is to demonstrate that a real-time control of LBW processes can be made possible by adopting CNN-based architectures.

LBW is a widely used welding technique in manufacturing processes since it allows joining multiple pieces of metal with high process speed and minimal distortion. The main characteristic of the laser regarding to welding processes is the huge energy density that the beam can convey on the work piece ($10^6$–$10^7$ W cm$^{-2}$) allowing the production of deep and slender weld seams at high feeding rates (up to 50 m min$^{-1}$) [2].

In an early phase, a large percentage of the incident beam is reflected. However, the small amount of energy initially absorbed by the material surface allows creating local vaporization of the metal [3]. The production of metal vapor rapidly accelerates the absorption of the beam energy, which is focused on the work piece, creating a small cavity called *keyhole*. As the latter penetrates deeper into the joining partners, the laser light is scattered repeatedly within it, thus increasing the coupling of laser energy into the work piece. Furthermore, the keyhole is kept open by the vapor pressure, which also prevents the molten inner walls from collapsing. As the laser beam advances, the molten metal flows around the keyhole and solidifies after a certain time. If an opportune laser power is used, the weld seam extends over the whole cross section of the metal sheets forming a so-called *full penetration*. As shown in Fig. 1, the state of full penetration can be visible in the coaxial camera image as a dark area directly behind the laser interaction zone, i.e., the FPH.

In conventional LBW processes, a fixed laser power is used. It is measured experimentally and a safety surpass of about 10% is added to compensate for process drifts and other influences. As shown in Fig. 2, such uncontrolled process is characterized by significant imperfections such as smoke residue, spatters, and craters. Furthermore, this strategy does not allow changing welding conditions during the process, such as feeding rate or material thickness. Hereof comes the requirement for applications in practice of performing a permanent quality control during LBW processes to reduce the production rejects.

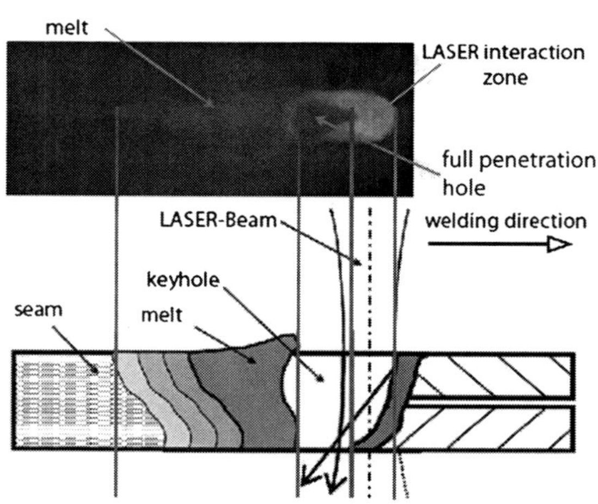


**Fig. 1** Schematics of an LBW process with two steel sheets in overlap-joint. The picture shows at the *bottom* the longitudinal section of the materials and at the *top* the resulting image of a coaxial process camera

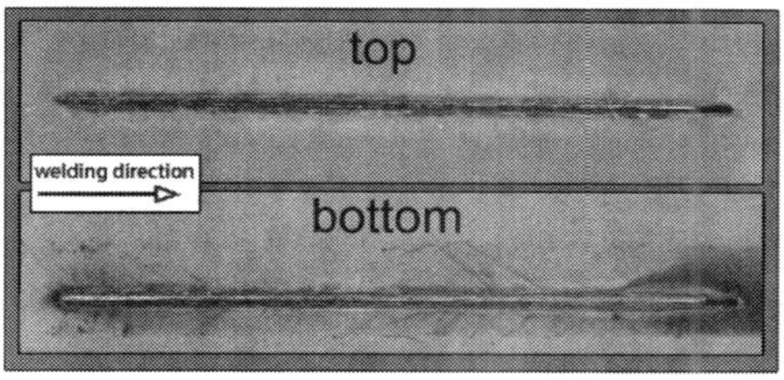


**Fig. 2** Uncontrolled full penetration weld of two zinc-coated steel sheets 0.7 mm thick with 0.1 mm gap in overlap joint. The process was performed at 9 m $min^{-1}$ by using a constant laser power of 5.5 kW with 10% power factor of safety

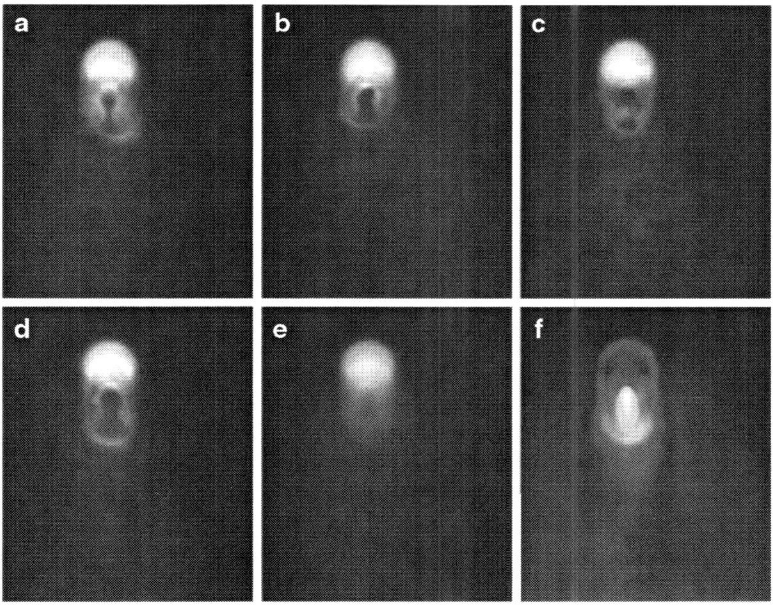


**Fig. 3** (**a–d**): Consecutive images of the FPH at a frame rate of 3 kHz. (**e**): Mean grey values per pixel over 100 images. (**f**): Corresponding standard deviations on a scale from 0 digits (*black*) to 40 digits (*white*)

To build a closed-loop control system, one needs controlled variables representing the state of the system and feedback parameters, which influence these variables on a suitable time scale. From the literature, a number of image-based quality features are known. In our work, the detection of the FPH has been used for an instant control of the laser power.

Figure 3 illustrates why a robust visual control system based on the FPH requires fast image processing. It shows consecutive images of the FPH in comparison with the mean grey values and standard deviations calculated for the grey value of every pixel over 100 images. All images were acquired from a coaxial camera position with an exposure time of 40 µs and a frame rate of approximately 3 kHz. The LBW process was uncontrolled, using a laser power of 6 kW for two 0.7 mm zinc-coated steel sheets at a speed of 9.5 m min$^{-1}$ in a constant full penetration state.

It is obvious that the contrast in the area of the FPH is much lower in the image (e) with the averaged grey values than in the single images. On the other hand, the standard deviation image shows that the area of the FPH contains the strongest fluctuations. A standard deviation of 40 digits in linear 8-bit grey images means that the intensity of the pixel varies over the whole scale from 0 to 255 digits. This means that due to the dynamics of the melt the position and the shape of the FPH fluctuate rapidly within a narrow area. Therefore, a fast contour detection in a large number of images is more suitable than an evaluation of the absolute intensities [4].

## 2 State of the Art

Error prevention and correction is a key to achieve high productivity and quality for fast and reliable LBW processes. In order to fulfill this requirement, several monitoring systems have been proposed and some of them have already been developed (such as Laser Welding Monitor LWM from Precitec, Welding monitor PD 2000 from Prometec, processobserver advanced from Plasmo Industrietechnik GmbH, Weldwatcher from 4D and Plasmo from ARC Seibersdorf research GmbH). On the contrary, no real-time closed-loop control systems have been integrated in commercially available architectures yet. Most of the proposed systems are mainly based on the analysis of the optical emissions, due to the interaction between the laser beam and the metal, by photodiodes, video cameras, or controllers.

Photodiodes are relatively inexpensive with a high temporal resolution while providing reduced spatial resolution. They allow for comparison of the detected optical emissions with previously recorded reference signals to estimate deviations of process parameters. Several sensors of this type have been developed for revealing defects: Thermal sensing of the welding pool [5], charge sensors detecting the electric potential between the welding nozzle and the work piece [6], photo sensors based on visible, infrared, ultraviolet, X-ray, acoustic and spectroscopic analyses of the radiation emitted by the plasma plume generated above the laser-metal interaction zone [7–10]. Despite also multi-features detecting systems have been proposed [11], photodiodes usually lead to problems in distinguishing between several possible defects (pore, full penetration losses, craters, spatters). Furthermore, the quality of imperfection recognition is very dependent on the quality of the reference data. Thus, such sensors require highly trained and skilled users and often destructive testing of random samples, which is both time and cost consuming.

The photodiode drawback regarding the reduced spatial information can be overcome by using spatially resolved detectors such as CCD, CMOS or thermal cameras. In this way, failures can be detected by extracting process image features [12, 13]. Nevertheless, as already mentioned, frame rates in the multi kHz range are necessary to improve the robustness of the welding process control against external influences. Conventional CCD cameras are typically limited by insufficient frame rates. An example of fast CCD camera is the JAI TM 6760, which allows reaching a frame rate of 1,250 fps at 228 × 160 pixels. Higher frame rates can be reached by using CMOS cameras and selecting an appropriate region of interest (ROI) of the image. For example, the MV-D752E-160-CL-8 from Photonfocus AG allows reaching frame rates around 5,400 fps at an exposure time of 10 μs and 2,300 fps at exposure times of about 250 μs at an ROI of 174 × 144 pixels. Furthermore, CMOS cameras offer a higher optical dynamic (up to 140 dB) compared to CCD cameras, allowing for the simultaneous observation of the keyhole area and the melt pool [14]. Nevertheless, it is necessary to consider additionally to the CCD/CMOS camera frame rates, the calculation time consumption depending on the image feature evaluation software for real-time control purposes. Therefore, most digital image processing systems cannot provide temporal resolutions in the multi kHz range, requested to follow rapid fluctuations. Furthermore, for simultaneous observation of keyhole opening and melt pool by means of CMOS cameras, the useful spectral range is rather limited. As shown in [14], at the upper boundary of the spectral range (at about 1,100 nm), the responsivity of silicon-based CMOS cameras is very low, leading to strong noise in the process image. Thermal cameras have the advantage of better visualizing the melt pool and the keyhole opening at the same time at long wavelengths. However, the use of thermal cameras in industrial applications is limited because of their overall size and high investment costs. Another technique is based on the simultaneous use of a camera and photodiodes, like in [15].

Standard PID controllers usually work for one material within a small range of laser power and welding velocity, which makes the applicability in varying industrial conditions difficult. Furthermore, the tuning of the control parameters is usually a very time-consuming job. Another solution is provided by the so-called "switching" controller, which tends to minimize the laser power to keep full penetration [16]. Nevertheless, such controllers often lack the necessary robustness.

In this paper, a new approach focused on the use of CNN-based architectures is described. The latter are cellular systems where each cell consists of a programmable processor merged with an optical sensor, and additional circuitry to be connected in several ways with its neighborhoods. Therefore, each cell can both sense the corresponding spatial sample of the image and process this data in close interaction and cooperation with other cells. This concept, which finds its basis in CNN theory [17], allows to benefit from the advantages of both video camera systems and photodiodes, providing high spatial and temporal resolutions. Concerning the latter, the Eye-RIS system v1.2 from AnaFocus lends itself as a good development platform. It is a visual system that includes a cellular smart image sensor (SIS) array called Q-Eye, which allows executing typical CNN operations.

In our recent investigations, different algorithms for the FPH detection have been implemented on such systems and tested in real-time experiments. In [18], the so-called *dilation algorithm* was introduced and in [19] some experimental results by the use of such an algorithm were described. It was implemented supposing a constant welding orientation during the process. Nevertheless, some real life applications are also executed by changing the welding orientation, e.g., weld of curved lines. Therefore, an orientation independent strategy called *omnidirectional algorithm* was also implemented and presented in [20]. Finally, a few experimental results obtained by the use of such algorithm were presented in [21]. In the following section, an extended description of the adopted visual closed-loop control system is given. Furthermore, some new experimental results will be discussed in order to provide a comparison between dilation and omnidirectional algorithms.

## 3 CNN

A CNN is a grid of interconnected cells, each containing linear and nonlinear circuit elements. The main feature that distinguishes CNN from classical neural networks is the local connectivity. In fact, every cell can directly interact with its neighboring cells and indirectly with other cells because of propagation effects of the continuous-time network dynamics. The original model of a CNN was proposed by Chua and Yang in 1988 [22]. In the following years, several kind of CNN models have been introduced. For the sake of brevity, we provide only a short description about the so-called *full range model*, which presents a reduction of the signal-range of the state variables, allowing the hardware realization of large-complexity CNN chips [23]. Let us consider, for example, a CNN having $M \times N$ cells arranged in $M$ rows and $N$ columns. The full range model can be described according to:

$$\dot{x}_{ij} = x_{ij} + \sum_{|k|\leq r, |l|\leq r} \hat{A}_{kl} y_{i+k,j+l} - g(x_{ij}) + \sum_{|k|\leq r, |l|\leq r} \hat{B}_{kl} u_{i+k,j+l} + \hat{I}, \tag{1}$$

where $x_{ij}$ and $u_{ij}$ represent the state and the input of the cell $(i, j)$, and $y_{ij}$ is the output defined by the following piecewise linear expression:

$$y_{ij} = \frac{1}{2}\left(\left|x_{ij}+1\right| - \left|x_{ij}-1\right|\right) \tag{2}$$

Furthermore, $r$ denotes the neighborhood of interaction of each cell, and $\hat{I}$, $\hat{A}$ and $\hat{B}$ are linear space invariant templates, respectively called *bias*, *feedback*, and *feedforward* operators. The function $g(x_{ij})$ is such that:

$$g(x_{ij}) = \begin{cases} m(x_{ij}+1) & x_{ij} < -1 \\ 0 & \left|x_{ij}\right| \leq 1 \\ m(x_{ij}-1) & x_{ij} > 1, \end{cases} \tag{3}$$

where $m$ must be large enough for approximating the nonlinear characteristic shown in [24].

CNNs are well suited for high-speed parallel signal processing. In fact, a gray scale image can be described as a grid of intensity pixels having values within 0 and 255. Thus, each pixel of the input image can be associated with the input or initial state of the spatially correspondent cell. Consequently, the input image evolves to the output image because of the CNN evolution, according to the state equation previously described. Early examples of CNN applications in the image processing field can be found in [25].

CNNs have also been employed to generate multiscroll chaos, to synchronize chaotic systems, and to exhibit multilevel hysterisis [26, 27]. Furthermore, since CNNs can be mathematically expressed as a series of differential equations, where each equation represents the state of an individual processing unit, they allow solving local, low-level, processor intensive problems expressed as a function of space and time. They can also be used either to approximate a wide range of partial differential equations (PDEs) such as heat dissipation and wave propagation or as reaction-diffusion (RD) processors [28, 29]. In the last few years, CNN theory has been used to develop several visual platforms, such as the family of the Eye-RIS systems produced by AnaFocus. The results shown in this paper have been obtained by the employment of the Eye-RIS system v1.2, which is described in the following section.

## 4 Experimental Setup

This work deals with the realization of a closed-loop visual control system for LBW processes. In the past, several authors have proposed feedback systems by using, e.g., the laser power, the focal-point position, or other parameters as an actuator [30]. In this paper, a strategy for the control of the laser power by the detection of the FPH in the process image is treated. The complete system consists essentially of a programmed logic device (PLC), the laser machine, the Eye-RIS system v1.2, and an interface board. The experiments discussed in the following section have been carried out with a 2D-laser scanner setup. The laser source is a 6 kW, 1,030 nm Trumpf TruDisk 6002 Yb:YAG thin disk with a 200 μm transport fiber. The laser scanner used, a Trumpf PFO-33, was equipped with a 450 mm focusing optic, which resulted in a focal diameter of 600 μm. The Eye-RIS system is connected to the scanner optic through a 90° beam splitter. Thus, the camera perspective is coaxial to the laser beam, allowing an invariant field of view regardless of the scanner position. Moreover, three achromatic lenses in combination with an optical band-pass filter were designed to achieve an optical magnification of about 4.6. The interface board was built to adapt the typical signals of PLC and laser machine into the range accepted by the Eye-RIS system. As shown in Fig. 4, the PLC sends a starting signal to initiate the control algorithm on the Eye-RIS system, and to activate the emission of the laser beam on the work piece and the machine movement. Concerning the latter, two possible strategies can be used, i.e., keeping fixed the position of the laser beam on the work piece and moving the material setup, or keeping fixed the material setup and using scanner optics to let the beam follow a specified path on the work piece. The experiments described in the following have been executed adopting the former strategy. As soon as the starting signal is received, the Eye-RIS system begins acquiring and evaluating process images and adjusting the laser power, e.g., if the FPH is found, the laser power is decreased or increased otherwise. The entire process is halted as the PLC sends the stopping signal.

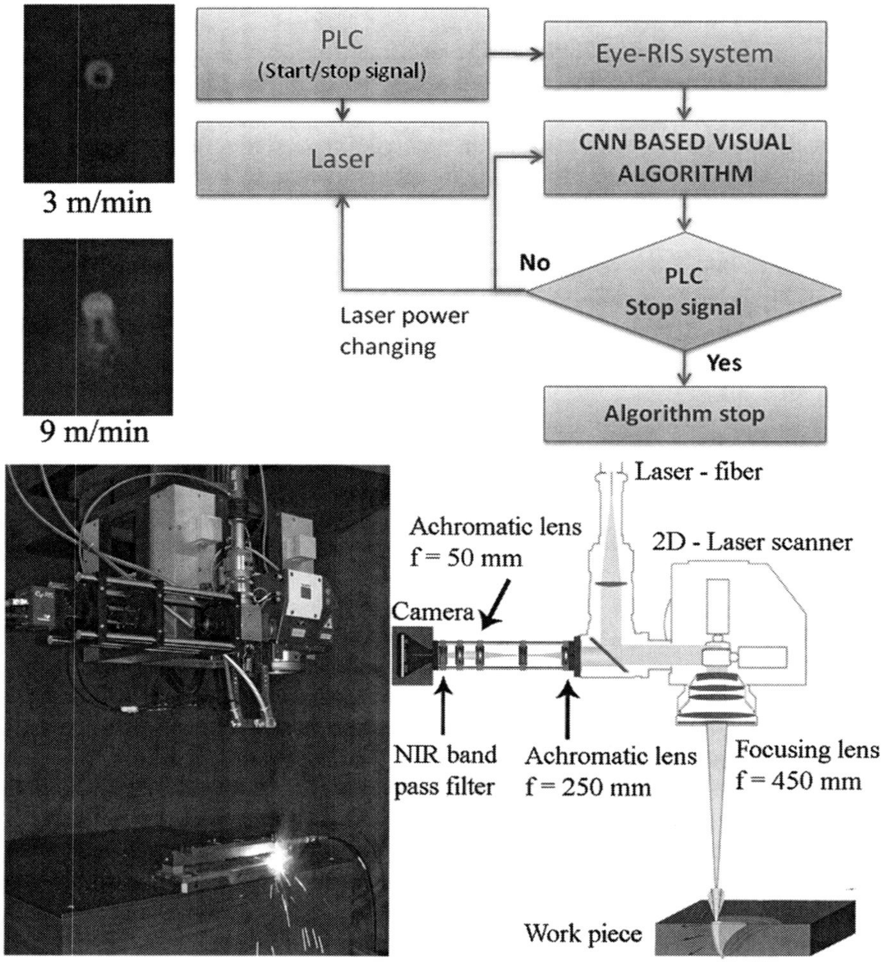


**Fig. 4** At the *top right*, the flow chart of the closed-loop control system is shown, while on the *left*, typical process images acquired by the Eye-RIS system point out the variation of the full penetration hole position with feeding rates. At the *bottom* the optical setup is described

## 4.1 Eye-RIS System v1.2

The Eye-RIS system v1.2 is a compact and modular vision system, which consists essentially of an Anafocus' Q-Eye SIS array, an Altera NIOS II processor, and I/O ports. The Q-Eye has a quarter common intermediate format (QCIF) resolution, i.e., 176 × 144 cells each containing a processor merged with an optical sensor, memory circuitry and interconnections to eight neighboring cells. Thus, each cell can both sense the corresponding spatial sample of the image and process this data in close interaction and cooperation with other cells. In particular, the Q-Eye hosts the whole image processing, which is obtained by the specification of typical CNN templates, allowing high-speed image evaluation tasks.

The Altera NIOS II processor is an FPGA-synthesizable digital microprocessor used to control the operations of the whole vision system and to analyze the information output of the SIS performing all the decision-making and actuation tasks. It manages the I/O port module, which includes several digital input and output ports, such as SPI, UART, PWM, GPIOs and USB 2.0, useful to interface the system with external devices. As regards the presented work, the NIOS II processor performs consecutive controlling steps between the starting and the stopping signals received from the PLC (through GPIOs). For each controlling step, the NIOS II processor recalls the Q-Eye chip in order to sense the image *i*; meanwhile, the image *i-1* is being evaluated. The result of the evaluation, i.e., the presence of an FPH, is returned to the NIOS II processor which adjusts opportunely the laser power by changing the duty cycle of a PWM signal. It is a duty of the external interface board to transform the PWM signal to the correspondent analog signal for the laser machine. As aforementioned, sensing of the image *i* and evaluation of the image *i-1* are simultaneously performed by the Q-Eye. Therefore, the exposure time adopted for image sensing strictly depends on the image evaluation time.

## 4.2 *Process Database*

The presented closed-loop control system has been tested in real-time applications to join two zinc-coated mild steel sheets under different welding conditions. As summarized in Table 1, our strategy allows the control of processes characterized by variable parameters such as either the material thickness or the feeding rate. Furthermore, processes with variable gap between the material sheets have been performed

**Table 1** Type of LBW processes controlled by the proposed closed-loop system

| | Thickness (mm) | Speed ($m\,min^{-1}$) | Gap (mm) | Welding shape | Joint |
|---|---|---|---|---|---|
| Constant parameters | 2 × 0.7–2.5 | 1–9 | 0.1–0.3 | Line/curve | Overlap/butt |
| Variable thickness | Bottom 0.7–1.5<br>Top 0.7–2.5<br>Step 0.3–1.3 | 3–6 | 0.1–0.3 | Line/curve | Overlap |
| Variable speed | 2 × 0.7 | 1 up to 9 | 0.1 | Line/curve | Overlap |
| | 2 × 1.0 | 1 up to 8 | 0.1–0.2 | Line/curve | Overlap |
| | 2 × 1.5 | 1 up to 5 | 0.1–0.2 | Line | Overlap |
| | 2 × 2.0 | 1 up to 3 | 0.2 | Line | Overlap |
| | 2 × 2.5 | 1 up to 2 | 0.3 | Line | Overlap |
| Variable gap | 2 × 0.7 | 3–9 | 0 up to 0.6 | Line | Overlap |
| | 2 × 1.5 | 2–4 | 0 up to 1.5 | Line | Overlap |

to confirm the high robustness of the control algorithm. Typical experimental results obtained by controlling certain processes are discussed in the following section, focusing more attention on those that allow a more detailed comparison between the control strategies.

## 5 Full Penetration Hole Detection

In this section, the evaluation algorithms for the FPH detection in LBW process images implemented on the Eye-RIS system v1.2 are described. As already mentioned, robust visual closed-loop control systems for LBW processes require fast real-time image processing, which would allow reaching controlling frequencies in the multi kilo Hertz range. Thus, the main goal was to minimize the visual algorithm complexity by the use of only a few CNN operations in order to reach such controlling rates.

The dilation algorithm has been developed supposing that the welding orientation does not change during the process. Such hypothesis simplifies the detection of the FPH, since its position in the image is not supposed to change remarkably. As shown in Fig. 4, only the distance of the FPH from the interaction zone can slightly change due to process speed variations. Nevertheless, this does not represent a serious problem for the evaluating algorithm.

The opposite consideration must be done, instead, in those processes characterized by variable welding orientation. In fact, in this case the FPH can be found everywhere around the interaction zone. The omnidirectional algorithm has been implemented as a general purpose strategy to be employed also in such situations. Nevertheless, since it deals with a more complex problem than the dilation algorithm, more operations have been necessary to detect the FPH, resulting in lower controlling rates. Both strategies are based on local neighborhood operations and, therefore, they are expected to run efficiently on any CNN architecture.

### *5.1 Dilation Algorithm*

For the sake of simplicity, the interaction zone is assumed, with its elongation toward the image bottom, as shown in Fig. 4. The source image is at first binarized. The threshold value depends on the image intensity and, therefore, on the exposure time. As previously mentioned, the exposure time is strictly correlated with the image processing time consumption, as image sensing and evaluating are simultaneously executed. Therefore, the best threshold can be found by an off-line analysis of process images and it can be used as far as the dilation algorithm is used. The second step consists in extracting the FPH from the binary image. Concerning the latter, morphological dilations [17] are executed along the image diagonal from the top to the bottom of the image. Afterward, a pixel-wise subtraction between the original

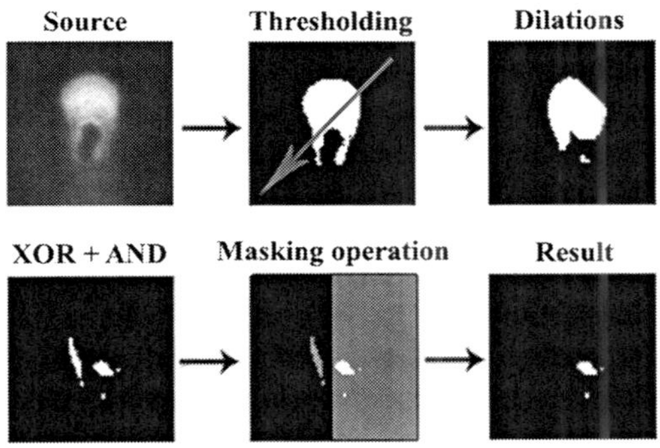


**Fig. 5** The dilation algorithm flow chart is shown. The source image is binarized and dilated in the direction of the *arrow*. Afterward, a logical XOR and a logical AND are applied in succession in order to extract the dilated area. The resulting image is masked to cut away the defects due to the dilation of the external edges of the interaction zone. The images have been zoomed to improve the visibility

binary image and the dilation result must be performed to extract only the dilated area. This is equivalent to apply first a logic XOR and consecutively a logic AND between the XOR resulting image and the dilated image. Finally, the application of a mask allows cutting away the defects due to the elaboration of the external edges of the interaction zone. Image analysis has revealed that the FPH is remarkably bigger than the possible defects. Thus, because the typical size of the FPH in the image is known, its presence is simply triggered by counting the number of white pixels in the final binary image. Figure 5 clarifies the dilation algorithm by an example. Using this strategy, controlling rates up to 14 kHz have been observed in real-time applications.

## 5.2 *Omnidirectional Algorithm*

Some real life processes are performed by changing the welding orientation, as in curved weld seams. In order to control such processes also, an orientation-independent strategy called omnidirectional algorithm can be adopted. As already explained in the Sect. 4, the experiments treated in this paper have been performed keeping the position of the laser beam fixed on the work piece and moving the material setup. This strategy does not allow reaching high speeds (around 1 m min$^{-1}$) by curved seams, leading to process images where the FPH can be found inside the interaction zone. Furthermore, because of the presence of vapor plume, interaction zone and FPH have small intensity differences. Therefore, the first step consists in enhancing the source image contrast to better distinguish the FPH. It was obtained by the application of sharpening filters on the source image. Another issue regards the threshold value. As revealed by recent image analysis [31], the omnidirectional algorithm evaluation result strictly depends on the quality of the binary image. In particular, the threshold value changes with the process speed. In fact, considering low speed processes, a high threshold value is necessary to distinguish the FPH inside the interaction zone. In processes characterized by higher speeds (greater than $3\,\mathrm{m\,min^{-1}}$), the FPH can be found outside the interaction zone, which presents an elongated shape. Thus, a low threshold value allows distinguishing these two areas. On the contrary, it has been experienced that in this case a high threshold value leads to defects into the interaction zone because of vapor plume intensity fluctuations. Concerning the latter, source image and filtering result are pixel-wise added. This operation provides the source image with the high frequency components (which are mostly located in the interaction zone area) further enhanced. It reduces the effect of the vapor plume fluctuation in the interaction zone and allows the application of a high threshold value also in high speed processes. Therefore, the subsequent binarization of the image can be performed by using the same global threshold value for a wide range of different processes and feeding rates. The second step consists in executing morphological closings in order to fill the FPH area. The application of logical XOR and AND comes after to extract the "closed" area only. The resulting image can be affected with defects due to the elaboration of the interaction zone edges. The last step, therefore, consists in reducing such defects by the use of morphological openings followed by the application of a mask. A detailed explanation of closings and openings can be found in [17]. Finally, as for the dilation algorithm, the presence of the FPH can be triggered by counting the number of white pixels in the resulting binary image. An example can be seen in Fig. 6. With the use of the omnidirectional algorithm, controlling rates up to 6 kHz have been observed in real-time applications.

## 5.3 Mask Builder

In this section, a technique to perform the aforementioned masking operations is discussed. As the initial position of the interaction zone in the image can vary with the specific optical setup, it is not assured that a unique mask can be adopted for every process. Concerning the latter, a mask builder was developed in order to create the mask automatically at the beginning of each process using the first available source image. It is based on the estimation of the position of the interaction zone center in order to align a default mask with it. For the dilation algorithm, the latter is vertically divided into two regions (black and white) of the same size as shown in Fig. 7d. For the omnidirectional algorithm, as shown in Fig. 7f, the default mask must have a roundish shape since the defects can be generated in all the directions.

The mask builder can be better understood considering the example in Fig. 7. The source image (a) is first binarized to obtain the picture (b). Afterward, by using in-built Q-Eye functions, the distance of the interaction zone center from the image center is estimated as in (c). In this way, the default mask can be shifted to overlap with the interaction zone center as in pictures (e) and (g). The mask builder time

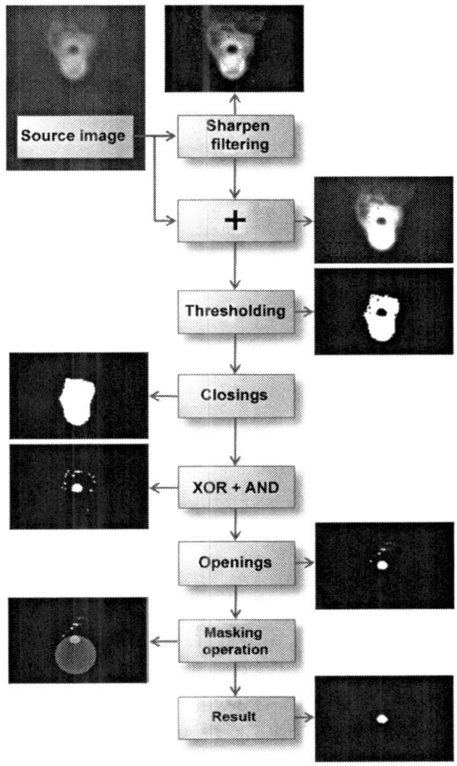


**Fig. 6** The omnidirectional algorithm flow chart is shown. Source image and filtering result are pixel-wise added. After a global thresholding, closings are executed and logical operations are applied in order to extract the full penetration hole. At the end, openings are performed to reduce the defects due to the elaboration of the interaction zone edges. The images have been zoomed to improve the visibility

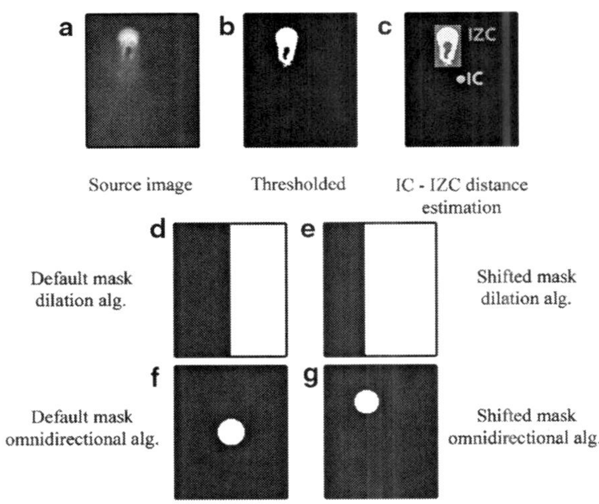


**Fig. 7** The mask builder is described. The source image (**a**) is binarized (**b**) and the distance of the interaction zone center (IZC) from the image center (IC) is estimated, as in (**c**). Afterward, the default mask – (**d**) for the dilation algorithm and (**f**) for the omnidirectional algorithm – is shifted to be overlapped with the IZC, as in (**e**) and (**g**)

consumption depends on the shifting operation. In the example of Fig. 7d, the mask builder executed a shift of 18 pixels in about 200 μs. Although it is more than the mean evaluation time, it must be executed only once at the beginning of the process and, therefore, it does not influence the process control.

# 6 Results

Dilation and omnidirectional algorithms have been first simulated off-line by the use of several image sequences acquired during LBW processes by the Eye-RIS system. The high evaluation precision revealed by simulation results has led to the employment of such strategies in real-time applications.

## 6.1 Simulation Results

The analysis of the simulation results by visual inspection is summed up in Tables 2 and 3. The former contains the results obtained with image sequences acquired during the weld of straight lines and simulated by the use of both the strategies. The latter, instead, lists the results obtained with images acquired during welds of curved seams and simulated by the use of the omnidirectional algorithm only.

The major number of false detections obtained with the omnidirectional algorithm can be a consequence of closing and opening operations. In fact, if

**Table 2** Omnidirectional vs. dilation algorithm: Simulation results

| | Image sequence | 1 | 2 | 3 |
|---|---|---|---|---|
| | Images with FPH | 390 | 207 | 129 |
| | Images without FPH | 46 | 165 | 240 |
| Dilation algorithm | False positives | 2 | 12 | 7 |
| | False negatives | 3 | 10 | 26 |
| | False detection (%) | ≈1 | ≈6 | ≈9 |
| Omnidirectional algorithm | False positives | 4 | 21 | 31 |
| | False negatives | 6 | 16 | 31 |
| | False detection (%) | ≈2 | ≈10 | ≈17 |

**Table 3** Omnidirectional algorithm: Simulation results

| Image sequence 05 | 004 | 021 | 022 | 023 | 026 |
|---|---|---|---|---|---|
| Images with FPH | 236 | 310 | 334 | 312 | 254 |
| Images without FPH | 147 | 77 | 35 | 74 | 137 |
| False positives | 20 | 11 | 12 | 7 | 13 |
| False negatives | 29 | 52 | 55 | 40 | 35 |
| False detection (%) | ≈13 | ≈16 | ≈18 | ≈12 | ≈12 |

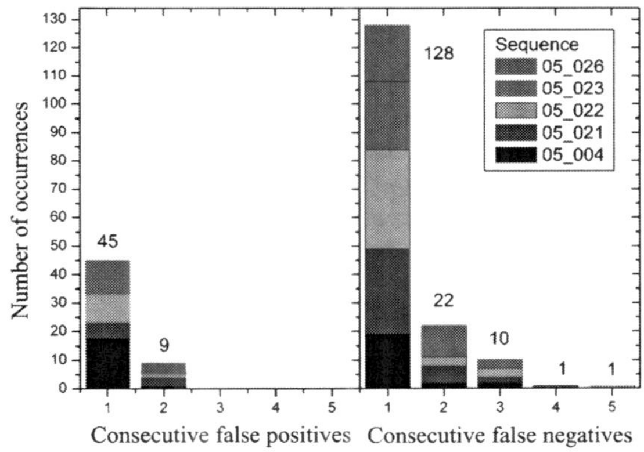


**Fig. 8** The picture shows the histogram of consecutive false detections during variable orientation welding processes. On the *x*-axis, there is the number of consecutive false detections (false positives and false negatives) for each image sequence. On the *y*-axis, the number of occurrences within each sequence is drawn

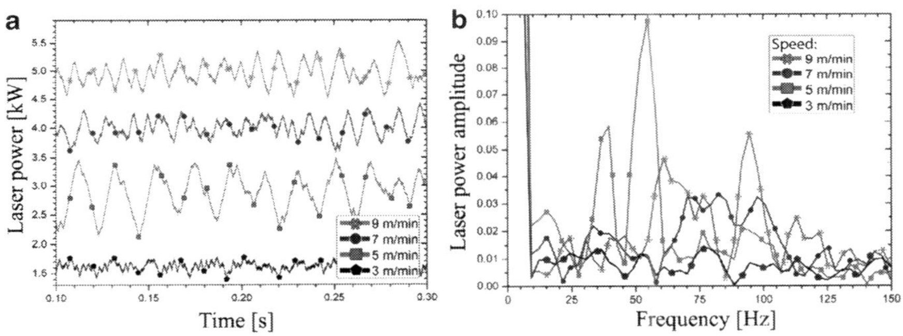


**Fig. 9** Picture (**a**) and (**b**) show, respectively, the laser power control signals used to join two 0.7 mm thick zinc-coated steel sheets at different feeding rates and the correspondent Fourier transform spectra

defects reach the typical size of the FPH, openings can be insufficient to delete it creating false positives. On the contrary, the application of openings can be too aggressive for small FPHs, increasing the number of false negatives. Nevertheless, the omnidirectional algorithm has provided false detections less than 18% and, as shown in Fig. 8, only few false detections of the same type occur consecutively (1–5). Figure 9a describes the Fourier transform of the laser power control signals presented in Fig. 9b. It allows to estimate the reacting time of the physical process within 4–20 ms. Assuming that in the examples of Fig. 9, the controlling rate was slowed down to about 2 kHz by the image storage (not needed in real-time control applications), the worst case of five consecutive false detections of the same type takes a total "false detection time" of about 2.5 ms. As it is considerably less than the reacting time of the physical process and made even worse by the image storing, such false detections do not significantly influence the real-time control of LBW processes.

## 6.2 Experimental Results

The first experiments to be discussed, have been executed to control the straight weld of two layers in overlap joint by both the dilation and omnidirectional algorithms, in order to compare the two strategies.

Figure 10 shows the controlled laser signal and the welding result obtained by joining materials 0.7 mm thick with gap of 0.1 mm at constant feeding rates (3, 5, 7, 9 m min$^{-1}$). The four experiments show that the laser power was initially brought to the opportune level to reach and keep the full penetration state and finally decreased due to the machine deceleration at the end of the process.

Figure 11 shows the controlled laser power used to join material layers 1.0 mm thick with 0.2 mm gap at constant feeding rates (4, 6, 8 m min$^{-1}$). In this case, the laser power signals controlled by the dilation and the omnidirectional algorithms have been overlapped to better explore the signal deviations.

Another experiment, shown in Fig. 12, regards the weld of material sheets 0.7 mm thick with a gap of 0.1 mm at variable feeding rates. The first one was executed with a speed variation from 3 to 9 m min$^{-1}$ through 5 and 7 m min$^{-1}$ and the second was performed reversing the speed profile. The goal is to prove that both the controlling strategies are able to change the laser power by the feeding rate variation in order to avoid full penetration losses in case of accelerations and cuts in case of decelerations.

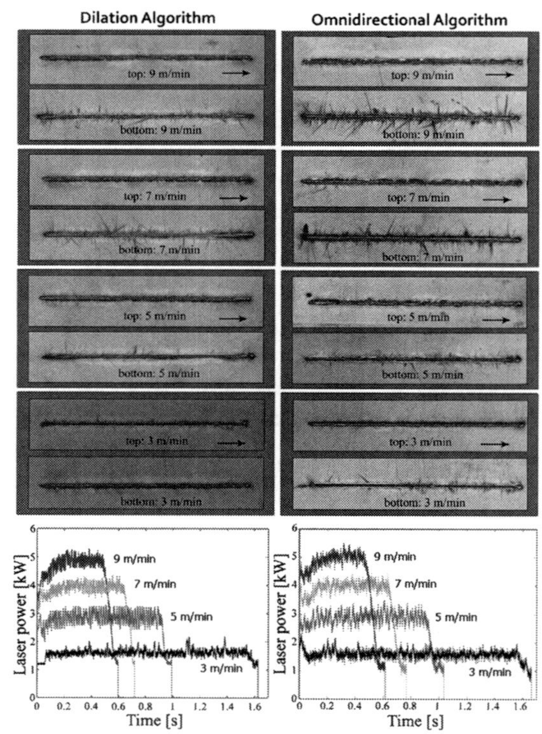


**Fig. 10** Controlled full penetration weld of two zinc-coated steel sheets 0.7 mm thick with 0.1 mm gap in overlap joint. For each algorithm, four experiments at different constant speed (9, 7, 5 and 3 m min$^{-1}$ from the *top*) are shown

**Fig. 11** Comparison between the laser power signals obtained by controlling through the dilation and the omnidirectional algorithm the weld of straight lines at different constant speeds (4, 6 and 8 m min$^{-1}$)

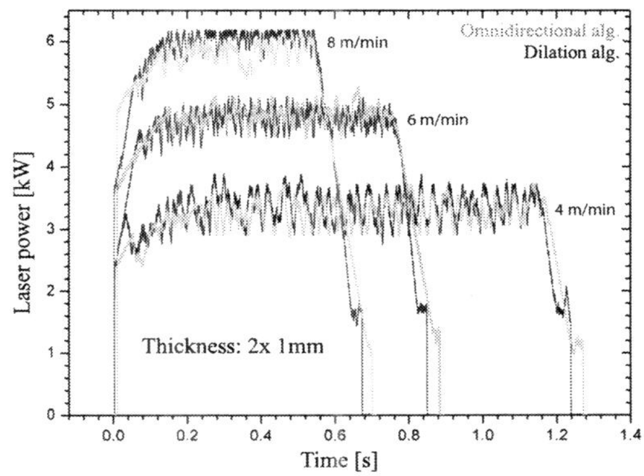


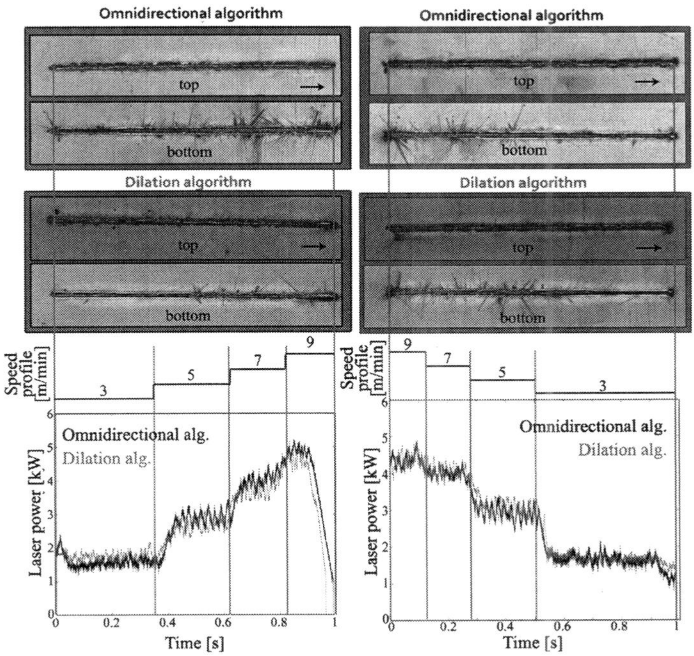


**Fig. 12** Controlled full penetration weld of two zinc-coated steel sheets 0.7 mm thick with 0.1 mm gap in overlap joint. For each algorithm, two experiments are presented: On the *left* varying the feeding rate from 3 to 9 m min$^{-1}$ through 5 and 7 m min$^{-1}$, while on the right reversing the speed. The pictures at the *top* show the welding results, while the pictures at the *bottom* present the laser power controlled by the dilation algorithm and the omnidirectional algorithm

The last experiment regards, instead, the control of weld of curved lines by the use of the omnidirectional algorithm. In the examples shown in Fig. 13, the results of welding two material sheets 0.7 mm thick with 0.1 mm gap and different initial speeds of 4 and 6 m min$^{-1}$ are presented. The seam shape is a semicircle with 5 mm

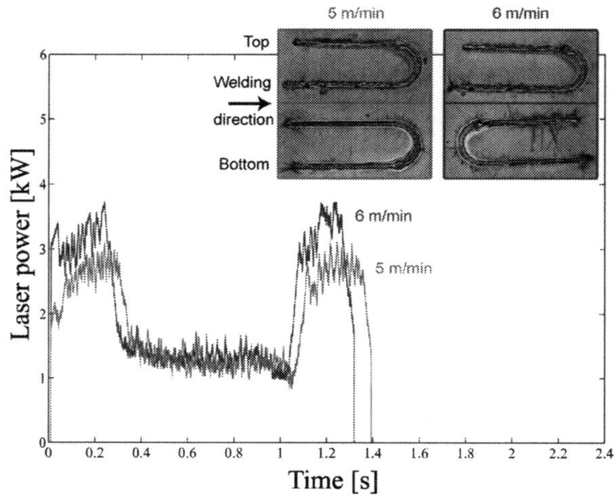


**Fig. 13** Controlled full penetration welds of two zinc-coated steel sheets 0.7 mm thick with 0.1 mm gap in overlap joint. Weld of *semicircles* with 5 mm radius, initial and final 25 mm *straight lines* and initial speed, respectively, of 5 and 6 m min$^{-1}$. At the *top* the welding results and at the *bottom* the controlled laser power are shown

radius and initial and final 25 mm straight lines. As explained before, curved seams are welded at lower feeding rates (1–2 m min$^{-1}$) and, therefore, decelerations and accelerations can be observed, respectively, by the beginning and the end of the semicircle. Also in this case the control system reacted by the occurrence of speed variations avoiding full penetration losses and craters in the extreme points of the curved lines.

More experimental results can be found in [32, 33].

# 7 Algorithm Comparison

The so-called dilation algorithm allows for reaching controlling rates up to 14 kHz but it has been developed to be used for high-speed LBW of straight lines. The omnidirectional algorithm is an alternative solution, which can also be used for controlling welding processes with variable orientation. Nevertheless, such strategy requires more complex operations reducing the controlling rate to 6 kHz.

Experimental results show that the quality of the controlled weld seam is remarkably better than the uncontrolled one. In fact, the visual control reveals a significant reduction of smoke residue and spatters on the bottom side of the material and avoids full penetration losses and craters. Besides, notably important is that the proposed closed-loop system can also handle particular kinds of LBW processes characterized by variable feeding rate or joining partners with variable thickness.

The direct comparison of the experimental results shows that similar laser power signals can be reached by using either the linear or the omnidirectional algorithm. The only difference is visible in the rise time, due to the lower controlling rate of the latter. Such lower controlling rate slightly limits the application of the omnidirectional algorithm in situations characterized by extremely big speed or thickness variations which can lead to a deterioration of the welding quality. Next studies, therefore, will be focused on reducing the omnidirectional algorithm complexity in order to speed up the real-time control. Furthermore, future developments will regard the system improvement to cover a wider variety of quality features and processes. Finally, one can state that the presented CNN-based control system has widely proved its ability to meet the requirements for the real-time high speed camera-based control in LBW.

# 8 Conclusion and Outlook

This paper proposes a closed-loop control system for keyhole welding processes, which adapts the laser power by evaluating the FPH. Two real-time control strategies have been presented. They have been implemented on the Eye-RIS system v1.2, which includes a cellular structure that can be programmed by typical CNN templates, making it possible to increase the image evaluation rate by an order of magnitude compared to conventional camera based control systems.

In contrast to monitoring systems, a feedback for the laser power is generated in order to maintain the full penetration state even when process conditions change. The FPH serves as a measure for the minimum energy density required by the welding process.

The major benefits of the closed-loop control system are a substantially augmented range of process parameters and an improved seam quality. The control system holds the process close to the energy density necessary for full penetration. Therefore, smoke residuals and spatters are reduced considerably, because no safety margin is necessary to compensate for process drifts or tolerances of components. As a result, the control system should increase process stability especially for complex welding systems or when samples with variable thickness are welded.

However, this special application benefits from an extremely strong laser as a light source. Therefore, it proves the computational properties of these camera systems, whereas other applications require different optical front ends and, in particular, light sources. In [34], such a light source allowing for exposure times of 20 μs at frame rates of 10 kHz is demonstrated for the inspection of metallic surfaces.

Furthermore, it was demonstrated that the CNN-based system in combination with external illumination is also applicable to laser ablation processes, e.g., to provide the trigger pulse for layer ablation by checking quality features [35].

**Acknowledgements** This work was financed by the Baden-Württemberg Stiftung gGmbH within the project "Analoge Bildverarbeitung mit zellularen neuronalen Netzen (CNN) zur Regelung laserbasierter Schweißprozesse (ACES)."

## References


1. Schmidt, M., Albert, F., Frick, T., Grimm, A., Kägeler, C., Rank, M., Tangermann-Gerk, K.: Process control in laser manufacturing – Dream or reality. Laser Materials Processing Conference. In Proc. of ICALEO 2007, pp. 1087–1096
2. Fortunato, A., Ascari, A.: Tecnologie di giunzione mediante saldatura, procedimenti non convenzionali e allo stato solido, vol. 2 (Societá editrice Esculapio S.R.L., 2008)
3. Dawes, C.: Laser welding, a practical guide (Abington Publishing, 1992), isbn: 1-85573-034-0
4. Blug, A., Abt, F., Nicolosi, L., Carl, D., Dausinger, F., Höfler, H., Tetzlaff, R., Weber, R.: Closed loop control of laser welding processes using cellular neural network cameras: Measurement technology. In Proc. of the 28th International Congress on Applications of Lasers and Electro Optics, ICALEO, November 2–5 2009, Orlando
5. Bicknell, A., Smith, J.S., Lucas, J.: Infrared sensor for top face monitoring of weld pools. Meas. Sci. Technol. 5 (1994), pp. 371–378
6. Li, L., Brookfield, D.J., Steen, W.M.: Plasma charge sensor for in-process, non-contact monitoring of the laser welding process. Meas. Sci. Technol. 7 (1996), pp. 615–626
7. Zhang, X., Chen, W., Jiang, P., Liu, C., Guo, J.: Double closed-loop control of the focal point position in laser beam welding. Meas. Sci. Technol. 14 (2003), pp. 1938–1943
8. Park, H., Rhee, S., Kim, D.: A fuzzy pattern recognition based system for monitoring laser weld quality. Meas. Sci. Technol. 12 (2001), pp. 1318–1324
9. Haran, F.M., Hand, D.P., Ebrahim, S.M., Peters, C., Jones, J.D.C.: Optical signal oscillations in laser keyhole welding and potential application to lap welding. Meas. Sci. Technol. 8 (1997), pp. 627–633
10. Sibillano, T., Ancona, A., Berardi, V., Lugará, P.M.: A real-time spectroscopic sensor for monitoring laser welding processes. Sensors 9 (2009), pp. 3376–3385, doi: 10.3390/s90503376
11. Otto, A., Hohenstein, R., Dietrich, S.: Diagnostik und Regelung beim Laserstrahlschweißen. Presented at the 5th Laser-Anwenderforum, 2006, Bremen
12. Beersiek, J.: New aspects of monitoring with a CMOS camera for laser materials processing. In Proc. of the 21st International Congress on Applications of Lasers and Electro-Optics, ICALEO, 2002
13. Chen, W., Zhang, X., Jia, L.: Penetration monitoring and control of CO2 laser welding with coaxial visual sensing system. Lasers in Material Processing and Manufacturing II. Edited by Deng, ShuShen; Matsunawa, Akira; Yao, Y. Lawrence; Zhong, Minlin. In Proc. of the SPIE, vol. 5629 (2005), pp. 129–140
14. Müller-Borhanian, J., Deininger, C., Dausinger, F.H., Hügel, H.: Spatially resolved on-line monitoring during laser beam welding of steel and aluminium. In Proc. of the 23rd International Congress on Applications of Lasers and Electro-Optics, ICALEO, 2004
15. Bardin, F., Cobo, A., Lopez-Higuera, J.M., Collin, O., Aubry, P., Dubois, T., Högström, M., Nylen, O., Jonsson, P., Jones, J.D.C., Hand, D.P.: Optical techniques for real-time penetration monitoring for laser welding. Appl. Optic. 44(19) (2005), pp. 3869–3876
16. De Graaf, M.W., Olde Benneker, J., Aarts, R.G.K.M., Meijer, J., Jonker, J.B.: Robust process-controller for Nd:Yag welding. In Proc. of the International Conference on Applications of Lasers and Electro-optics, ICALEO, October 31–November 3 2005, Miami, FL, USA
17. Chua, L.O., Roska, T.: Cellular neural networks and visual computing. Foundations and applications. Cambridge University Press, 2004, first published in printed format 2002
18. Nicolosi, L., Tetzlaff, R., Abt, F., Blug, A., Höfler, H., Carl, D.: New CNN based algorithms for the full penetration hole extraction in laser welding processes. In Proc. of the IEEE International Symposium on Circuits and Systems, ISCAS, pp. 2713–2716, May 24–27 2009, Taipei, Taiwan
19. Nicolosi, L., Tetzlaff, R., Abt, F., Blug, A., Carl, D., Höfler, H.: New CNN based algorithms for the full penetration hole extraction in laser welding processes: Experimental results. In Proc. of the IEEE International Joint Conference on Neural Networks, IJCNN, pp. 2256–2263, June 14–19 2009, Atlanta, GA, USA

20. Nicolosi, L., Tetzlaff, R., Abt, F., Blug, A., Höfler, H., Carl, D.: Omnidirectional algorithm for the full penetration hole extraction in laser welding processes. In Proc. of the European Conference on Circuit Theory and Design, ECCTD, pp. 177–180, August 23–27 2009, Antalya, Turkey
21. Nicolosi, L., Tetzlaff, R., Abt, F., Blug, A., Höfler, H.: Cellular neural network (CNN) based control algorithms for omnidirectional laser welding processes: Experimental results. In Proc. of the 12th International Workshop on Cellular Nanoscale Networks and Applications, CNNA, February 2–5 2010, Berkeley, CA, USA
22. Chua, L.O., Yang, L.: Cellular neural networks: Theory. IEEE Trans. Circ. Syst. 35(10) (1988), pp. 1257–1272
23. Corinto, F., Gilli, M., Civalleri, P.P.: On stability of full range and polynomial type CNNs. In Proc. of the 7th IEEE International Workshop on Cellular Neural Networks and their Applications, CNNA, pp. 33–40, July 22–24 2002
24. Linan, G., Dominguez-Castro, R., Espejo, S., Rodriguez-Vazquez, A.: Design of Large-Complexity Analog I/O CNNUC. Presented at the Design Automation Day on Cellular Visual Microprocessor, ECCTD, pp. 42–57, 1999, Stresa, Italy
25. Chua, L.O., Yang, L.: Cellular neural networks: Applications. IEEE Trans. Circ. Syst. pp. 1273–1290
26. Yalcin, M., Suykens, J., Vandewalle, J.: Cellular Neural Networks, Multi-Scroll Chaos And Synchronization. World scientific series on nonlinear science, series A, vol. 50, 2005, isbn: 978-981-256-161-9
27. Yokosawa, K., Tanji, Y., Tanaka, M.: CNN with multi-level hysteresis quantization output. In Proc. of the $6^{th}$ IEEE International Workshop on Cellular Neural Networks and Their Applications, CNNA, 2000
28. Gilli, M., Roska, T., Chua, L., Civalleri, P.P.: CNN Dynamics Represents a Broader Range Class than PDEs. Int. J. Bifurcat. Chaos 12(10) (2002), pp. 2051–2068
29. Adamatzky, A., Costello, B., Asai, T.: Reaction-Diffusion Computers. Elsevier Science and Technology, 2005, isbn: 978-0-444-52042-5
30. Bardin, F., Cobo, A., Lopez-Higuera, J.M., Collin, O., Aubry, P., Dubois, T., Högström, M., Nylen, O., Jonsson, P., Jones, J.D.C., Hand, D.P.: Closed-loop power and focus control of laser welding for full-penetration monitoring. Appl. Optic. 44(1) (2005)
31. Nicolosi, L., Tetzlaff, R.: Real Time Control of Curved Laser Welding Processes by Cellular Neural Networks (CNN): First Results. Presented at the Kleinheubacher Tagung, U.R.S.I. Landesausschuss in der Bundesrepublik Deutschland e.V., September 28–October 1 2009, Miltenberg, Germany. Adv. Radio Sci., 8, 117–122, 2010, www.adv-radio-sci.net/8/117/2010/, doi:10.5194/ars-8-117-2010
32. Abt, F., Blug, A., Nicolosi, L., Dausinger, F., Weber, R., Tetzlaff, R., Carl, D., Höfler, H.: Closed loop control of laser welding processes using cellular neural network cameras – experimental results. In Proc. of the 28th International Congress on Applications of Lasers and Electro Optics, ICALEO, November 2–5 2009, Orlando
33. Abt, F., Blug, A., Nicolosi, L., Dausinger, F., Höfler, H., Tetzlaff, R., Weber, R.: Real time closed loop control of full penetration keyhole welding with cellular neural network cameras. In Proc. of the 5th International Congress on Laser Advanced Materials Processing, LAMP, June 29–July 2 2009, Kobe
34. Blug, A., Jetter, V., Strohm, P., Carl, D., Höfler, H.: High power LED lighting for CNN based image processing at frame rates of 10 kHz. Presented at the 12th IEEE International Workshop on Cellular Nanoscale Networks and their Applications, CNNA 2010, February 3–5 2010, Berkeley, CA, USA
35. Strohm, P., Blug, A., Carl, D., Höfler, H.: Using Cellular neural network to control highly dynamic laser processes. Presented at Laser Optics Berlin, LOB, 2010

# Real-Time Multi-Finger Tracking in 3D for a Mouseless Desktop

**Norbert Bérci and Péter Szolgay**

**Abstract** In this chapter, we present a real-time 3D finger tracking system implemented on focal plane chip technology-enabled smart camera computers. The system utilizes visual input from two cameras, executes a model-based tracking and finally computes the 3D coordinate from the 2D projections. A reduced algorithm is also analysed, pros and cons emphasised. Measurements, robustness and auto calibration issues are also exposed.

## 1 Motivation

Nowadays, the stylish user interfaces are based on touch: instead of moving the mouse and clicking on one of the buttons, we have started to control the devices with touch, even with multiple finger gestures. The real problem – despite it is a huge step forward – is that it is only a 2D system. In contrast with the 3D world in which we live and the 3D way we control some equipments (cars, handheld machines), we cannot do this without computers. Yet.

Our goal was to implement some algorithm building blocks just to prove that under appropriate restrictions regarding the environment it is feasible to deploy 3D user interfaces. The possible application areas are almost unlimited: from GUI computer control through industrial automation, kiosk and advertisement to remote surgery and alternative actuation systems for disabled people the opportunities are extremely diverse.

N. Bérci (✉)
Faculty of Information Technology, Pázmány University, Práter u. 50/A, Budapest, Hungary
e-mail: berci.norbert@itk.ppke.hu

Á. Zarándy (ed.), *Focal-Plane Sensor-Processor Chips*,
DOI 10.1007/978-1-4419-6475-5_13, © Springer Science+Business Media, LLC 2011

## 2 Introduction

Our implementation is based on the visual tracking of the daily used hand gestures therefore, learning a new user interface is unnecessary: accommodation almost immediately occurs. The user need not wear gloves and his/her movements are not restricted by wires connecting the computer and the data-collecting devices. Our goal is to replace the mouse and the keyboard in equipment control.

### *2.1 Postures vs. Gestures*

We should emphasize the distinction between postures and gestures. A posture is a configuration of a 3D object (which can be thought of a vector of its describing parameters) while a gesture is the objects' temporal behavior, and can be thought of as a sequence of postures. Information is embodied in dynamics and succession.

Since gestures (and especially human movements) can be very diverse amongst people and even between the repeated gesture of the very same person, the main focus is shifted from precise parameter reconstruction to the relative and approximate computation of movements.

### *2.2 Tracking vs. Recognition*

The classical way of tracking is to compute features in the whole frame and try to assign them to tracked objects, thus updating their vector of parameters and searching for similar ones or trying to match via extrapolation.

The tracking becomes more efficient and robust if the frame rate is relatively high compared to the contained motion, as it results in less difference between the frames, which in turn means there is a high degree of overlapping of the same object in successive frames, so there is no need for high complexity parameter-based object recognition.

### *2.3 Model-Based Tracking*

General tracking does not have any information about the object, which means that any parameter can be set to any value imaginable. If we have some a priori data about the possible movements, we can restrict the parameter space, thus having the ability to elect one solution from the several possibilities.

Having a model is also a great support for error detection: if the values are certainly out of bounds, it must be some kind of feature detection error.

In our case, the human hand has well-known motion properties, the fingertips we track cannot move freely.

## *2.4 Initialization*

Due to the fact that we do not recognize the palm or the fingers in each frame, it is extremely important to have a high quality initial configuration to start with. It is something active contour techniques require to ensure their correct convergence.

Initialization also allows us to transform the recognition process to a real tracking task: we do not want to recognize the hand in each frame but to be able to follow it while moving. This requirement is manifested in a fixed, environmentally well-controlled starting position.

# 3 The Finger Tracking Algorithm

In the first version of the tracking algorithm, we have utilized all of the previously reviewed methods: an initialization aided, hand model-based tracking is performed. The algorithm consists of two main parts: an analogic image processing algorithm and a conventional algorithm for the hand model and error recovery. The flowchart of the image processing part can be seen in Fig. 1.

## *3.1 Preprocessing*

The input image should have some quality parameters met to be able to be used by the following processing tasks. The first algorithmic step is the removement of the irregularity artifacts from the input caused by the uneven parameters of the cells [1], which is achieved by subtracting a static image from every frame. After this, a diffusion is performed to aid the skeletonization process (see Sect. 5.1 for details).

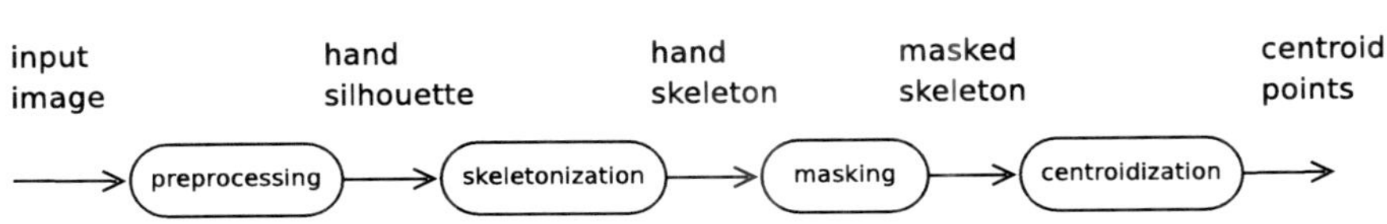


**Fig. 1** The flowchart of the image processing part of the finger tracking algorithm: all of these analogical operations run on the focal plane processor

### 3.2 Segmentation

At the end of the segmentation procedure, the hand silhouette should be produced, which is in our scenario (a human hand is the foremost object in the scene) equals to foreground–background separation. As we need high frame rate for tracking (as discussed in Sect. 2.2), high illumination is also required due to the low sensitivity of the optical image capturing, which will render the hand bright but more distant objects darker (more precisely radiance decreases squarely by distance). That said, the obvious solution is to use an adequate gray level value as the threshold for the segmentation process. This value is constantly updated by the hand model upon the detection of full scene illumination changes.

Should some other bright objects in the camera view appear, we get rid of them by the help of the hand model-assisted masking.

### 3.3 Skeletonization

The central part of the algorithm is the skeletonization. Its actual implementation has a number of effects regarding the robustness and the precision of the whole tracking algorithm and will be discussed in Sect. 5.1 where we also unfold the consequences.

The skeletonization is basically a thinning process resulting in a one pixel wide tree-like structure of the original object. It is a connection preserving transformation, so everything which was connected remains connected in the skeleton also. For some background information on the matter, see [2]. This way we lose the width parameter of the object but that is a useless parameter for the current tracking algorithm (we are doing tracking without recognition). An example skeletonization can be seen in Fig. 2.

As we are of no interest in the exact track of the skeleton (just the position of the fingertips), this procedure is not only advantageous in the sense of data reduction, but it also hides some image acquisition and segmentation errors.

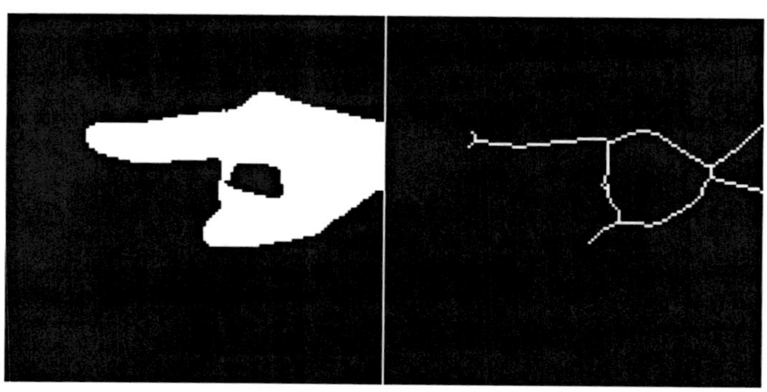

**Fig. 2** Skeletonization example: the input silhouette is on the *left* and the resultant skeleton is on the *right*

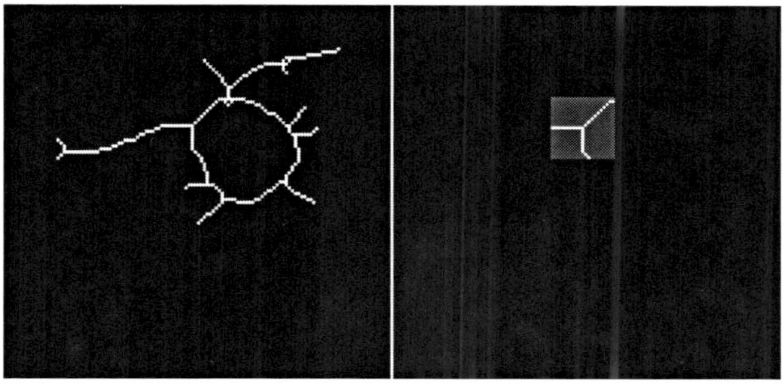

**Fig. 3** Multiple lines touching the masked skeleton section: skeleton image on the *left*, the masked section on the *right*

## 3.4 Masking

Having the skeleton of the whole image ready, the next task is to extract those parts, which carry valuable information. Since the current hand model only stores the fingertips we extrapolate the position the fingertip should be in from the parameters stored in the hand model and at that position we apply a rectangular mask. This masking is key to get rid of the remaining bright parts beyond the tracked fingertip.

At this step, some error detection could also be performed: if the masked region contains multiple lines such that they are touching the border of the mask (Fig. 3), the mask is reevaluated by using another prediction from the hand model or if it is not available, the algorithm increases the mask size. If neither leads to an acceptable solution, the algorithm stops, concluding the lost of the fingertip tracking.

## 3.5 Centroidization

A centroid function extracts the central point of the connected components from the previously masked skeleton section (Fig. 4), but the result differs significantly depending on the branching at the end of the masked skeleton (input skeletons on Fig. 5 and Fig. 3 with the corresponding centroid results on Fig. 6). This difference lowers the precision, since the centroid position is different based on the branch type.

Another possibility if there are multiple components in the masked skeleton section. The algorithm tries to recover by looking for the centroid pixel starting from the extrapolated position and progressing along a width increasing spiral.

If no errors occurred, at the end of this processing step we have obtained the fingertip position.

**Fig. 4** Centroidization example: the skeleton section input on the *left*, the computed centroid on the *right*

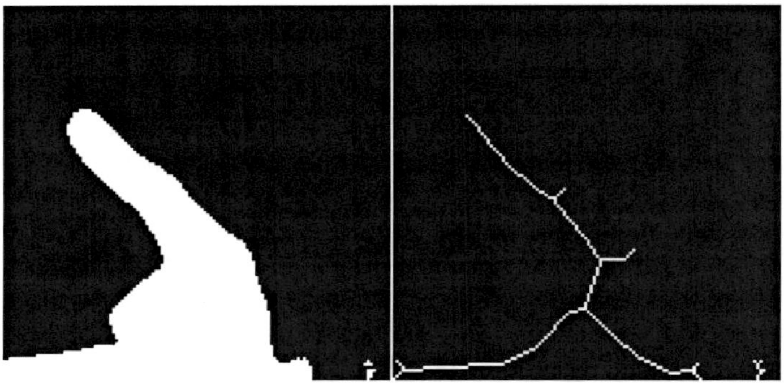

**Fig. 5** Depending on the finger orientation on the input picture (*left*), the end of the skeleton can be branchless (*right*). Compare with the skeleton on the *left* subfigure in Fig. 3

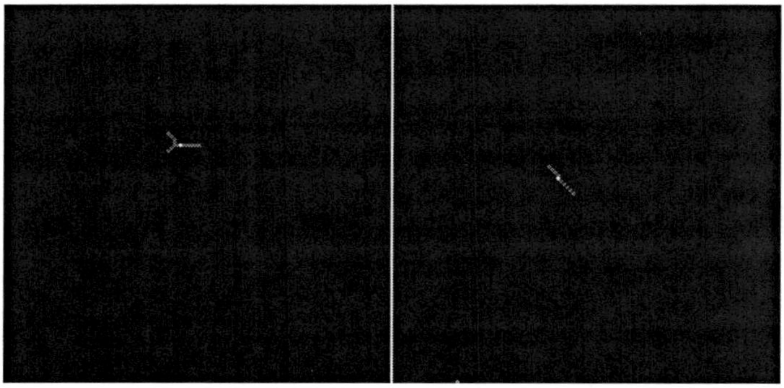

**Fig. 6** Difference of the centroid point on the skeleton section based on the finger direction

## *3.6 Hand Model*

Proper tracking cannot be done without an object model, and an appropriate model has just the information it needs to compute the next possible configurations of the object based on the previously acquired data. It has to balance between robustness and performance. The more accurate the model is the more processing power it needs to compute or decide between the possible values.

The implemented hand model is extremely simple: it is the wireframe representation of the palm storing the positions of the fingertips. This has the advantage of easy implementation and easy updating, but also serves our requirements.

With the masked skeleton and the centroid at hand, we are able to calculate the direction of the fingers, which will be also stored in the hand model. The computation is done in the following way: the masked centroid should have a skeleton touching the edge of the mask, since the skeleton should drive into the palm. Otherwise, we have found something which is not the intended fingertip in which case an error recovery method should be applied. From the centroid and the point where the skeleton touches the edge of the mask, the finger direction can be easily derived, which is also stored in the hand model.

### 3.6.1 Initialization

We have to emphasize again the key role of the initialization step: without it, the hand model would not be operable.

Recognition seems more difficult for dynamic image flows, but if we initialize the parameters of the hand model in the start position, we are able to collect lots of valuable information which can be utilized later. What is more important, the initialization makes it possible to transform the recognition process to a tracking task: we do not want to find the hand in every frame independently of each other, but be able to follow it while moving. This makes it also possible to have a convenient starting position to find the hand, and later only partial or defective information is enough to follow it.

### 3.6.2 Extrapolation

When the masks need to be defined in each frame, the underlying model stores the position and speed of the corresponding fingertip. The next position of the mask is solely defined by extrapolating the previous position and velocity of the fingertip. The characteristics of the fingers gathered at the initialization step is also consumed to create an appropriately sized and positioned mask window.

# 4 Algorithm Extensions

## 4.1 Multiple Finger Tracking

After some preliminary testing, it became clear that the algorithm can be used not just for one finger tracking but for all the five. The only problem occurs when the fingers are tightly crossed, in which case the algorithm cannot distinguish between the fingers, detecting the whole palm as a single (albeit very thick) finger. Example results are in Fig. 7.

The algorithm itself only needs obvious modifications to be able to handle the multiple masks, centroids and final positions.

## 4.2 3D Tracking

If we extend the system with another smart camera computer, tracking in 3D becomes possible. The system architecture in Fig. 8 is amended also with a host computer, which processes the two data streams coming from the camera computers and calculates the 3D coordinates. The OS driver interface provides a 3D mouse for the operating system.

The communication channel is over TCP/IP easing the connection of the system to any machine needing 3D input and also allows remote placement.

### 4.2.1 3D Reconstruction

The cameras have been placed in such a way that they are perpendicular to each other so the two cameras look at two perpendicular (virtual) projection planes. For 3D reconstruction, we produce two camera sight lines each one starting from the

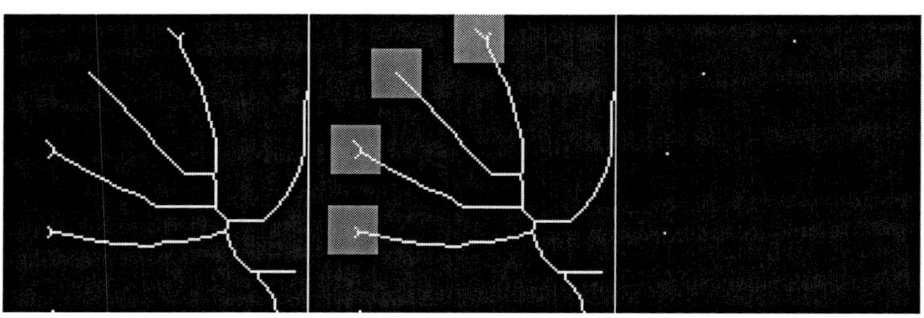

**Fig. 7** Example results of the different algorithmic parts when tracking multiple fingers: the skeleton on the *left*, the masks overlayed on the skeleton on the *middle*, and the resultant fingertip positions on the *right*

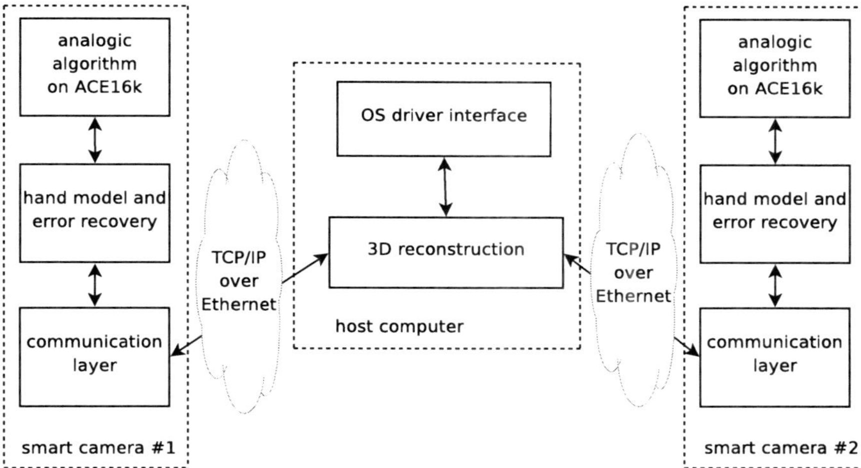


**Fig. 8** The architecture of the 3D tracking system

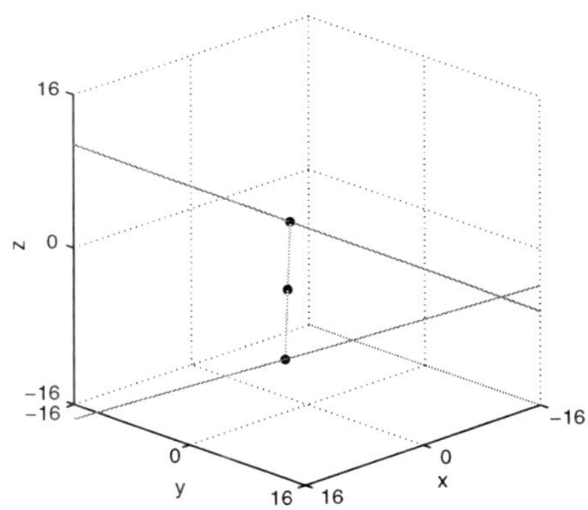


**Fig. 9** The coordinates represent the pixel resolution of the cameras with (0,0) at the center and they capture the $x-z$ and $y-z$ projected planes. The *lines* are generated from the detected planar points and the camera positions. The coordinate supposed to be detected (i.e., the computed one) is the *dot* in the *centre*

detected coordinate on the projection plane and ending on the known fixed camera positions. Theoretically, the two sight lines should intersect, which is the point at which the requested 3D coordinate. Practically, this case almost never occurs due to measurement and algorithmic errors resulting in skew lines. We have chosen to define the approximate 3D coordinate as the middle point between the shortest line segment connecting the skewing sight lines (Fig. 9).

The 3D reconstruction software running on the host computer receives the two 2D coordinates on their corresponding planes from the two cameras and executes the calculations described in this section.

Formally: The direction vectors $(\mathbf{s_1}, \mathbf{s_2})$ of the two sight lines are the difference of the camera positions $(C_1, C_2)$ and the detected 2D points $(D_1, D_2)$ on the sight planes:

$$\mathbf{s_1} = D_1 - C_1 \qquad \mathbf{s_2} = D_2 - C_2.$$

For the calculation of the middle point, we have to determine the points which are the closest to the other sight line. They can be computed with the help of any **c** vector connecting the two lines projected onto the normed direction vectors:

$$\mathbf{c}_1 = \left\langle \mathbf{c}\,,\, \frac{\mathbf{s}_1}{|\mathbf{s}_1|} \right\rangle \frac{\mathbf{s}_1}{|\mathbf{s}_1|} \qquad \mathbf{c}_2 = \left\langle \mathbf{c}\,,\, \frac{\mathbf{s}_2}{|\mathbf{s}_2|} \right\rangle \frac{\mathbf{s}_2}{|\mathbf{s}_2|},$$

where the **c** vector could be, for example, the vector connecting the camera positions: $\mathbf{c} = C_1 - C_2$. Having these results at hand, the detected 3D coordinate can be computed by:

$$D = \frac{1}{2}((C_1 - \mathbf{c}_1) - (C_2 + \mathbf{c}_2)) + (C_2 + \mathbf{c}_2).$$

If we define the coordinate system in such a way that two of its axes are equal to the axes of one of the cameras, and the third is the perpendicular axis of the other cam (since our camera placement is perpendicular, one axis from each of the projection planes are the same, so there is only one axis left on the other cam) the 3D reconstruction becomes extremely easy: we just have to augment the two coordinates from one cam with the remaining coordinate of the other cam. This way there is no need for any transformation for the 3D reconstruction.

## 5 Precision and Robustness

### *5.1 Skeletonization Consequences*

We have used the native skeletonization implementations of the smart camera computers, which have some effects we should be aware of.

It is worth to note that being connected does not necessarily mean a minimal connected graph. Sometimes only one pixel difference is enough to return a totally different skeleton. This circumstance fundamentally influences the precision: the centroid can oscillate between two more stable states.

Another problem is what we have mentioned briefly in the section about the centroidization (Sect. 3.5): the finger direction influences the position of the computed centroid point.

### *5.2 Automatic Calibration*

The objective checking of the 3D tracking requires a measure to be defined. We have chosen the distance of the two (skew) sight lines (the length of the transversal) as the error measure.

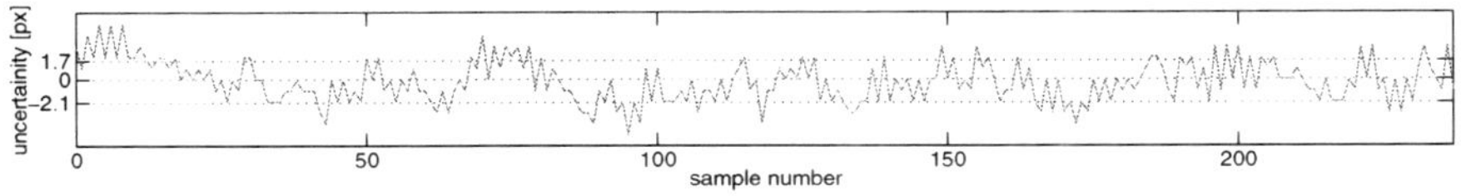


**Fig. 10** The computed length of the transversal (the uncertainty measure) in pixels (the sign means direction relative to one of the cameras) of almost 250 samples. The biggest is a 5-pixel difference, the data is centered approximately around 0 (the mean is $-0.2$) and the standard deviation is 1.9. The *dotted lines* sign the one sigma interval around the mean

Since the transversal is perpendicular to both lines, its normal vector $\mathbf{n_t}$ can be computed by the cross-product of the two direction vectors:

$$\mathbf{n_t} = \mathbf{s_1} \times \mathbf{s_2}.$$

If the $\mathbf{c}$ vector obtained by connecting any two points on the lines is projected on the normed $\mathbf{n_t}$, we obtain the transversal vector, which has got length equal to their inner product:

$$d = \left| \left\langle \mathbf{c}\, , \frac{\mathbf{n_t}}{|\mathbf{n_t}|} \right\rangle \right| .$$

This formula also verifies that this error measure is zero if and only if $\mathbf{c}$ and $\mathbf{n_t}$ are perpendicular (the inner product vanishes), which only happens when the lines intersect.

This error measure is just an approximation, since we do not know the exact coordinates of the hand, but it served well the purpose of representing the error. The actual measurements can be seen in Fig. 10.

We are also investigating the possibilities how we can increase the precision by utilizing more cameras and defining a higher dimensional error measure.

## 5.3 *Camera Placement and Its Effect on Precision*

The described setup needs precise camera placement, since the approximation of the 3D coordinate (if the sight lines do not intersect) is the middle point. The longer this line is, the lower the precision. Fortunately, misplacement makes a difference only in the error mean (Fig. 11). Following the mean, it is possible to adjust the position of the cameras and also possible to attract the attention of the user to this condition automatically.

## 5.4 *Failure Handling*

Apart form the failure handling in the masking step (see Sect. 3.4), using more than one camera not only enables us to step from 2D to 3D but also enables us

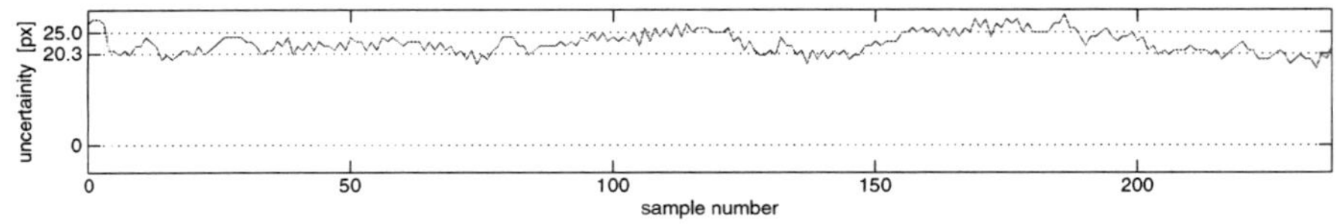


**Fig. 11** The computed length of the transversal (the uncertainty measure) in pixels of almost 250 samples with rough camera placement. Compared to Fig. 10, the difference is just an offset in the mean (it is 22.6), while the standard deviation remains almost the same (it is 2.4). The *dotted lines* bound the one sigma interval around the mean

to recognize object lost errors. If the error measure grows beyond a limit, we can consider the tracked object lost, and refuse to send the coordinates to the OS driver. The hand model-assisted failure recovery is activated, and in most cases it enables the system to return to a normal working state.

With the masked skeleton and the centroid at hand, we are able to calculate the current direction of the fingers, which will be also stored in the hand model. The computation is done in the following way: the masked centroid should have a skeleton segment touching the edge of the mask, since the skeleton should drive into the palm. Otherwise, we have found something which is not the intended fingertip, so an error recovery method like the one described in the previous section should be applied. From the centroid and the point where the skeleton touches the edge of the mask, the finger direction can be easily derived.

# 6 Implementation on the Bi-i

The algorithm has been implemented on the Bi-i smart camera computer [3, 4], which embeds a CNN-based ACE16k focal plane processor [1] and a Texas Instruments DSP. The full flowchart can be seen in Fig. 12.

The achieved frame rates are summarized in Table 1 and the time consumption of the algorithmic parts is given in Table 2. The data state that the system is able to track at about 51–52 FPS, which proved to be enough for normal speed human hand motions. The differences in the runtimes are mostly due to the error recovery.

We can conclude that in a proper environment the system can track the human fingers with reasonable resolution and frame rate.

# 7 The Reduced Finger Tracking Algorithm

As we noted earlier, we had believed that the reason of the fluctuation of processing time is the error recovery. We did an extreme reduction of the algorithm and wondered what will be the effect of it on these parameters. The new algorithm is given in Fig. 13.

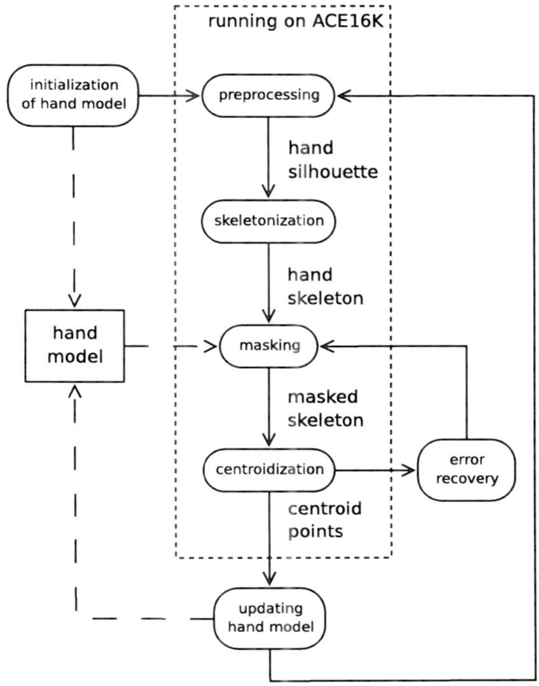


**Fig. 12** The flowchart of the first version of the tracking algorithm: the analogical fature extraction (drawn in the *dashed line rectangle*) runs on the ACE16k focal plane processor, whereas the hand model with auxiliary functions (drawn outside of the *dashed rectangle*) runs on the accompanying DSP

**Table 1** Framerate of tracking one finger

| Time from start (s) | Framerate (FPS) |
|---|---|
| 1 | 54 |
| 2 | 49 |
| 3 | 48 |
| 4 | 53 |
| 5 | 51 |
| 6 | 51 |
| 7 | 54 |
| 8 | 49 |
| 9 | 55 |
| 10 | 49 |
| 11 | 55 |
| 12 | 48 |
| 13 | 54 |
| 14 | 53 |
| 15 | 52 |

**Table 2** Delay caused by different algorithmic parts (ms)

| Capture | Diffusion | Skeleton | Centroidization | Search |
|---|---|---|---|---|
| 1.010193 | 0.054933 | 3.710227 | 5.620107 | 5.017680 |
| 1.010193 | 0.054933 | 3.710423 | 6.379907 | 4.954080 |
| 1.010190 | 0.054933 | 3.710400 | 7.169920 | 4.995547 |
| 1.010193 | 0.054933 | 3.994133 | 7.199973 | 5.035400 |
| 1.010193 | 0.054933 | 4.277653 | 7.266973 | 5.089867 |
| 1.010193 | 0.054933 | 4.564573 | 6.319160 | 4.840587 |
| 1.010193 | 0.054933 | 4.845933 | 4.956533 | 5.376413 |
| 1.010193 | 0.054933 | 4.847960 | 4.957080 | 5.405493 |

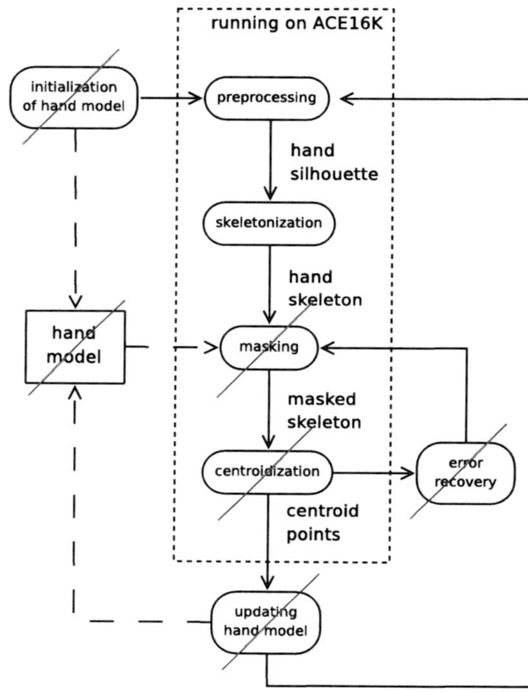


**Fig. 13** The flowchart of the reduced version of the algorithm: now all parts run on the Q-Eye visual processor

# 8 Implementation on the Eye-RIS

We have implemented the new algorithm on the Eye-RIS smart camera system. The difference is huge: we have left out all the conventional algorithmic parts and this way we were able to run the whole algorithm on the Q-Eye focal plane processor! The masking and the centroidization were fused into an endpoint computation, which was made possible by the native support of the hardware.

The real achievement was the improvement in the processing time, as it can be seen in Fig. 15. For reference purposes, the data gathered from the previous algorithm can be seen in Fig. 14, and a zoomed in version is in Fig. 16.

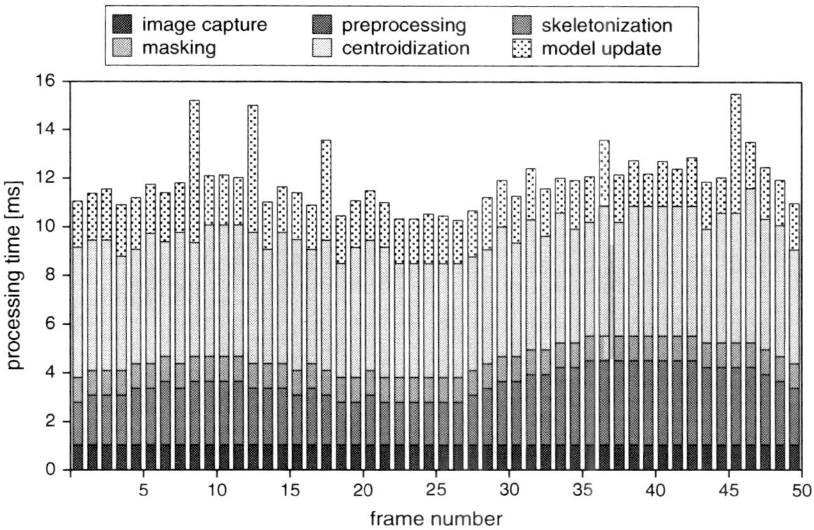


**Fig. 14** The runtime chart of the original version of the algorithm

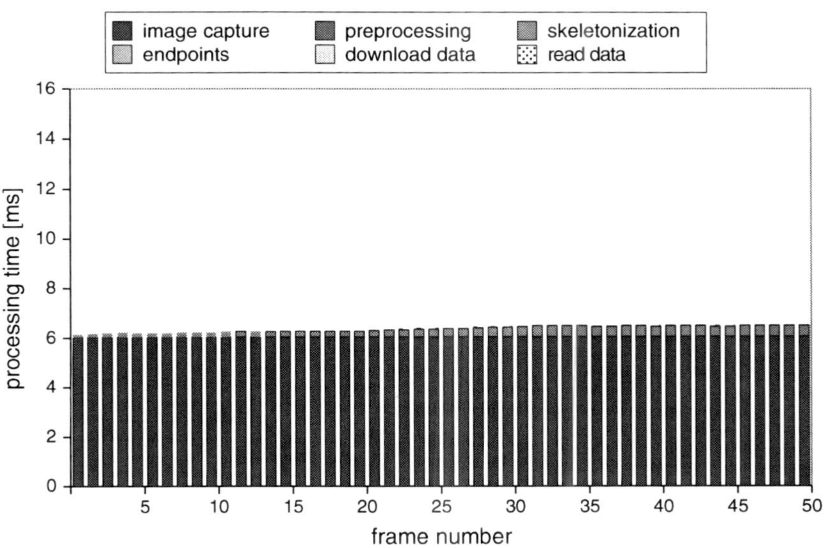


**Fig. 15** The runtime chart of the reduced version of the algorithm: now all parts run on the Q-Eye visual processor

On the one hand, the processing time of the old algorithm is about 12 ms, and it is unstable, depends highly on the properties of the visual input. On the other hand, the processing time of the new algorithm has lowered remarkably to about 6 ms.

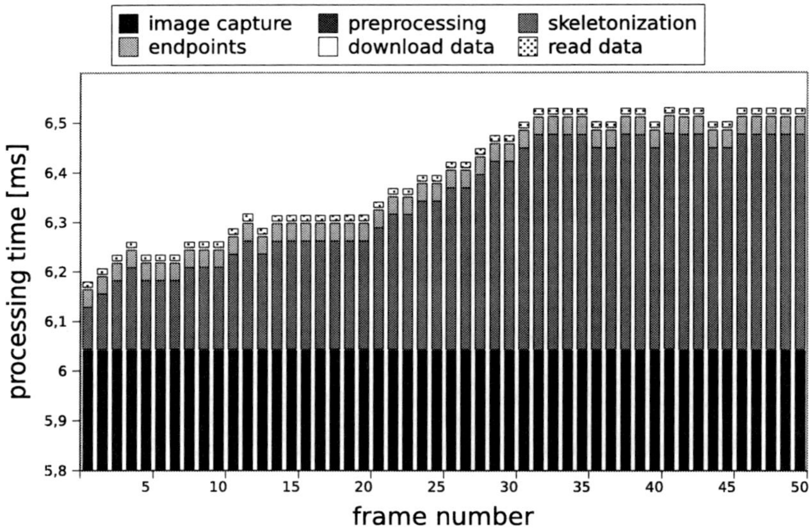


**Fig. 16** The runtime chart of the reduced version of the algorithm: now all parts run on the Q-Eye visual processor (zoomed in version)

**Table 3** Ratio between the different computation subtasks when computing one frame with the reduced algorithm

| Subtask | Avg. ratio of frame time | Avg. time spent |
|---|---|---|
| Image capture | 94.42% | 6.0399 |
| Preprocessing | 0.05% | 0.0034 |
| Skeletonization | 4.72% | 0.3038 |
| Endpoint detection | 0.55% | 0.0353 |
| Points download | 0.08% | 0.0054 |
| Read data | 0.17% | 0.0111 |

What is really interesting is that this new algorithm also solves the tracking task. Obviously, this new version is much more sensitive to the environment parameters as it does not have any error recovery procedure.

In Table 3, we have detailed the runtime of the different algorithmic parts. The key thing to note is that on average almost 95% of all processing time is wasted in the image capture. To put it the other way around, the algorithm could run much faster if the direct optical input was more sensitive. Nevertheless, the results show that the system is able to achieve about 150 FPS in ordinary office lighting conditions.

According to our tests, delay and jitter is not measurable and is negligible to the achieved precision.

## 9 Summary

The focal plane processing architecture enabled us to implement multiple finger tracking in 3D and also showed that the possible speed exceeds the normal frame rates giving chance to high speed tracking with proper illumination.

The two versions of the algorithm may be considered as a trade between speed and robustness. The hand model backed version is more robust and is capable to recover from situations where the latter one cannot. The price payed is performance.

The focal plane processing architecture extremely well suited to high-speed image processing tasks such as the ones described in this chapter.

## References

1. Liñán, G., Domínguez-Castro, R., Espejo, S., Rodrguez-Vzquez, Á.: ACE16k: A programmable focal plane vision processor with 128 × 128 resolution. In: Proc. Eur. Conf. on Circ. Theory and Design (ECCTD), pp. 345–348. Espoo, Finland (2001)
2. Golland, P., Grimson, W.E.L.: Fixed topology skeletons. Tech. rep., Artificial Intelligence Laboratory, Massachusetts Institute of Technology, Cambridge, MA, USA (2000)
3. Zarándy, Á., Rekeczky, C.: Bi-i: A standalone cellular vision system, Part I. Architecture and ultra high frame rate processing examples. In: Proc. 8th Int. Workshop on CNNs and their Appl. (CNNA), pp. 4–9. Budapest, Hungary (2004)
4. Zarándy, Á., Rekeczky, C.: Bi-i: A standalone cellular vision system, Part II. Topographic and non-topographic algorithms and related applications. In: Proc. Int. Workshop on CNNs and their Appl. (CNNA), pp. 10–15. Budapest, Hungary (2004)

# Index



Á. Zarándy (ed.), *Focal-Plane Sensor-Processor Chips*,
DOI 10.1007/978-1-4419-6475-5, 

CPSIA information can be obtained at www.ICGtesting.com
Printed in the USA
LVOW032220281111

256888LV00007B/15/P

9 781441 964748